Broken Movement

Broken Movement

The Neurobiology of Motor Recovery after Stroke

John W. Krakauer and S. Thomas Carmichael

The MIT Press
Cambridge, Massachusetts
London, England

This book was set in Times New Roman by Westchester Publishing Services.

Library of Congress Cataloging-in-Publication Data

Names: Krakauer, John W., author. | Carmichael, S. Thomas.
Title: Broken movement : the neurobiology of motor recovery after stroke / John W. Krakauer
    and S. Thomas Carmichael.
Description: Cambridge, MA : The MIT Press, [2017] | Includes bibliographical references and index.
Identifiers: LCCN 2017021092 | ISBN 9780262037228 (hardcover : alk. paper), 9780262545839 (pb)
Subjects: LCSH: Cerebrovascular disease—Patients—Rehabilitation. | Motor learning.
Classification: LCC RC388.5 .K67 2017 | DDC 616.8/1—dc23
    LC record available at https://lccn.loc.gov/2017021092

I dedicate this book to my husband, Omar, who patiently weathered my nocturnal writing binges, mountains of papers and books, and demands to be fed.

—JWK

I dedicate this book to my wife, Diana, whose unfailing calm and understanding balances impatience, impertinence, and indelicacy.

—TC

Contents

Preface

This book represents an attempt to provide a concise survey of the conceptual landscape of the neurobiology of motor recovery of the arm and hand after stroke. We take the position that empirical and theoretical rigor, along with definitional clarity, are essential in a field with diverse and confusing belief systems about motor recovery after stroke. We have not attempted to be either encyclopedic or systematic but instead have focused on ideas, open questions, and areas of disagreement, but of course doing our best to ground our positions on empirical observation. With the term *neurobiology*, we refer to a broad compass ranging from behavior to neurophysiology to molecular processes. The core idea of the book is that we will not make real progress in neurorehabilitation unless we get the biology right. As a corollary, we maintain that technology will only be useful for neurorehabilitation if it remains subordinate to the biology.

We wrote the book because we saw the need for a single readable volume that brought together the large number of domains and disciplines that are of direct relevance to the science of neurorehabilitation. The knowledge base required in this field is daunting. For example, it demands familiarity with motor learning concepts from experimental psychology, with brain and spinal cord anatomy and physiology in both humans and animal models, with the cellular and molecular neuroscience of neuronal plasticity, and with the principles underlying emerging technologies such as robotics and virtual reality.

As important as facts are, however, the ideas that we interpret them with are arguably even more important. It is here that we hope the book will make a unique contribution because we readily admit that our goal was to force readers to think and rethink. This emphasis on critically revisiting the concepts underlying the science of neurorehabilitation is because, in our view, the field is unfortunately replete with "neuromythology": neuroscience-flavored stories that float free of both logic and data but nevertheless stubbornly persist. For example, that spasticity impairs voluntary movement of the upper limb, that motor recovery is synonymous with motor learning, that noninvasive brain stimulation and robotics have already been shown to be efficacious, that task-oriented training can treat impairment and leads to brain reorganization, and that stem cells work by integrating into neural circuits to mediate recovery. The list goes on. Of course, the book is meant to stimulate dialogue and debate and be a spur to new discovery. Without these things, we would simply be replacing old myths with new ones.

Acknowledgments

Finding effective rehabilitative treatments for motor recovery after stroke is a daunting challenge that will require a huge collaborative effort across disciplines and scientific domains. Writing a book reflects this truth in microcosm: a long process that is sufficiently arduous to require pulling in as many other people as possible to share in the ordeal. Here we thank the many people who edited drafts of the chapters, provided figures from their own work, and engaged in protracted dialogues in person, through email, or videoconference about parts of chapters we had them read: Stuart Baker, Julie Bernhardt, John Buford, David Burke, Winston Byblow, Jules Dewald, Joern Diedrichsen, David DiTullio, Bruce Dobkin, Naveed Ejaz, Mike Ellis, Simon Gandevia, Amy Gleichman, Jeff Goldsmith, Jim Gordon, Robert Hardwick, Michelle Harran, Noriyuki Higo, Theresa Jones, Gert Kwakkel, Catherine Lang, Esther Nie, Richard O'Dwyer, Monica Perez, David Reinkensmeyer, Zev Rymer, Bob Sainburg, Bob Scheidt, Catherine Schweppe, Reza Shadmehr, Cathy Stinear, Theresa Sukal Moulton, Nick Ward, Carolee Winstein, Jing Xu, and Steve Zeiler.

John W. Krakauer

Thanks to all the members of BLAM lab; a port in the storm. Special thanks to lab codirector Adrian Haith for incomparable kindness and support, for book section rewrites and edits, and for all-round scientific and critical brilliance. Thanks to Jarreau Wimberly for a figure and for the wonderfully drawn cover illustration. Thanks to the members of the KATA project: Omar Ahmad, Kevin Olds, Promit Roy, Kat McNally and Jarreau Wimberly, for inspiration and designing the treatments of the future. Many thanks to my collaborators in stroke recovery: Amy Bastian, Meret Branscheidt, Pablo Celnik, Joern Diedrichsen, Argye Hillis, Tomoko Kitago, Andy Luft, Heidi Schambra, and Steve Zeiler. Thanks to the Champalimaud Center for the Unknown, in particular Rui Costa, Zach Mainen, and Alexandra Piedade, for giving me the yearly visiting scholarship and the beautiful Lisbon setting that allowed the first chapter drafts to be written. Thanks to my personal assistant Field Blauvelt who protected my time, and me, so that the book could be finished. Thanks to my department chair, Justin McArthur, for creating the conditions at Johns Hopkins that allowed me to think and write.

Very special thanks to Juan Camilo Cortés, who is really all but a third author on the book. He made sure the book came together both intellectually and practically—references, formatting, editing or creating figures and legends, and general conceptual coherence would have been impossible without him.

Special thanks also to Michelle Harran for scouring every line of the manuscript and finding errors missed by everyone else.

Thanks to my mother and stepfather for long-term love and short-term forbearance: putting up with me during the writing of the book and providing much tea and sympathy. Last but not least, much love and thanks to my husband Omar Semidey, without whom I would not have attempted this at all.

S. Thomas Carmichael

For the mentorship and training of critical consideration and open-mindedness, I wish to thank important mentors over the years: Joel Price, Tom Thach, Tom Woolsey, Bruce Dobkin, Bob Collins, Marie-Francoise Chesselet, and Bruce Dobkin.

1

The Current Landscape of Upper Limb Neurorehabilitation

1.1 Introduction

The emphasis of this book will be on neuroscience as it pertains to the mechanisms of and recovery from arm and hand paresis and to the development of new therapeutic interventions to potentiate motor recovery. To best see how findings in neuroscience could lead to new approaches to motor recovery, it is important to first provide some context by describing the current state of neurorehabilitation as it relates to the upper limb. For anyone outside of neurorehabilitation and probably even for those inside it, the landscape can appear bewilderingly large and hard to navigate, and the methods and language baroque. Different types of outcome can be emphasized as the goal of therapy: for example, at the levels of function/impairment, or quality of life (QOL). The choice of outcome will directly influence the kind of treatment the patient will receive. Various measurement scales are available to assess each type of outcome. Several different clinical specialties operate in the neurorehabilitation space, including physiatrists, occupational therapists, physical therapists, speech and language pathologists, neurologists, social workers, and clinical neuropsychologists. These specialists differ in their choice of outcome and outcome measure, their understanding of scientific principles, and their treatment approaches. Finally, adding to the confusion is the fact that the goals and outcome measures in current practice are incongruent with the goals and outcome measures in research studies and clinical trials. Needless to say, the survey of the landscape presented in this chapter is necessarily selective and critical. The central tenet of this book is that the landscape needs a makeover and neuroscience can help to achieve this.

1.2 The Many Faces of Motor Recovery

As this book is about the neurobiology of motor recovery after stroke, it is best to start by understanding the various ways that recovery is defined from the standpoints of public health, clinical practice, and research trials. This is important because we will argue that one of the reasons why findings in neuroscience have not made their way into practice as much as could be hoped for is because of the current treatment goals favored by clinicians and payers alike. The World Health Organization (WHO), in collaboration with the International Society of Physical and Rehabilitation Medicine, developed the International Classification of Functioning, Disability, and Health (ICF) in 2001 (World Health Organization 2002). At the top of this classification system are the two terms *functioning* and *disability*. These are homologous terms, but one emphasizes gains and the other losses at different hierarchical levels: impairment, activity limitation, or participation restriction. *Impairment* refers to problems in body function and structure. In the case of hemiparesis, this is loss of strength and motor control. *Activity limitation* refers to the decreased ability to complete an action or a task. For example, a patient with hemiparesis has motor impairment but could use a fork with his or her unaffected hand, avoiding limitation in that activity. *Participation restriction*, the term that has replaced *handicap*, is more vague and refers to decreased involvement in social situations. For example, a patient could still participate in a family outing to the park even if he or she had to use a wheelchair—as long as the premises are designed to facilitate its use. The main point here is that participation restriction can uncouple from levels of impairment and activity limitation. The participation level in the ICF overlaps considerably with QOL, which emphasizes physical, mental, and social well-being at any given level of disease. We will not be discussing either participation or QOL for the rest of this book, not because they are unimportant but because they are not the level at which repair and reorganization mechanisms operate, even though these mechanisms can have knock-on effects at other ICF levels. In support of this view are data showing that one year after stroke, arm motor impairment is associated with poor QOL (Wyller et al. 1997; Franceschini et al. 2010).

Although *disability* is now the umbrella term for impairment, activity limitation, and participation restriction, some confusion can arise because the term is still used in current neurorehabilitation to refer to just the activity level of the ICF, in particular *activities of daily living* (ADLs). Examples of ADLs include washing oneself, eating, and dressing. Thus, in most neurorehabilitation today, the emphasis is on training on real-life tasks to reduce disability by achieving the highest degree of independence possible in ADLs. This emphasis is reflected in the Functional Independence Measure (FIM) (Keith et al. 1987).

The terms *function* and *functional* are commonly used in relation to achievement on tasks that incorporate movements that are assumed to relate to ADLs. The favored outcome measures for the large multicenter neurorehabilitation trials to date have been such specially devised functional or activity-based tasks rather than global ADL or QOL measures. The narrowing of the meaning of disability and functioning to an almost exclusive focus on activity in the ICF classification has meant a turning away from attempts to reduce impairment. This is unfortunate because the neuroscience of plasticity and brain repair maps most logically onto reduction of impairment. In contrast, motor learning concepts are most readily applicable at the activity level, although, as we shall see, they can also be applied at the impairment level. We should clarify our position here a bit further—although the topic will be covered in much more detail in subsequent chapters. Brain repair and reorganization are processes that occur after stroke and are dependent on altered physiological and cellular conditions that are time limited and triggered by the ischemic lesion. These unique conditions early after stroke, which are likely modulated by training, can bring about true, if often only partial, restoration of normal movements. In contrast, we will argue that in chronic stroke, plasticity conditions are normal and so learning can only be used to teach compensatory strategies—references to reorganization and plasticity are misleading because restoration is not, in most cases, occurring at the chronic stage. Thus, two areas of neuroscience, repair and reorganization of neural tissue, on one hand, and normal motor learning, on the other, operate predominantly at two different levels of disability. As we shall argue, this distinction is important, but it can be hard to grasp and therefore has led to considerable conceptual confusion in the field.

1.3   True Recovery versus Compensation

Another way to think about neurorehabilitation is to distinguish between *compensation* versus true behavioral recovery—*restitution* (Rothi and Horner 1983; Levin et al. 2009; Krakauer et al. 2012). Restitution means a return to or toward premorbid levels of motor control and strength. In contrast, compensation refers to a patient's use of his or her residual effectors, muscles, or joints to accomplish the task. The most obvious example would be a patient with arm paresis compensating by using his or her unaffected arm instead. Another slightly less obvious example would be the difference between the patient who manages to increase the distance he or she can reach by gaining back some elbow extension (restitution) and the patient who flexes the whole trunk forward to cover the reach distance (compensation) (Cirstea and Levin 2000; Roby-Brami et al. 2003).

1.4   (Mis)measures of Arm Paresis

Here we will describe the impairment and activity scales that are most frequently used in studies of recovery of the upper extremity after stroke. For the rest of this book, we will call tasks that measure motor performance considered relevant to ADLs *functional* tasks, as they measure *motor function*. As we have already noted, although the term *function* covers all three levels of the ICF, it has come to be used synonymously with the activity level. In the book, we shall use *function* and *activity* interchangeably. Functional tasks will be distinguished from assessments of impairment and motor control (speed, accuracy, and coordination of movement). Global ADL scales such as the Barthel Index (BI) and the FIM will only be mentioned because of their ubiquity on rehabilitation units and, to some degree, clinical trials. The use of motion capture, force transducers, robots, and electromyography (EMG) to assess poststroke movement kinematics and dynamics will be covered in the context of relevant studies throughout the book (Figure 1.1).

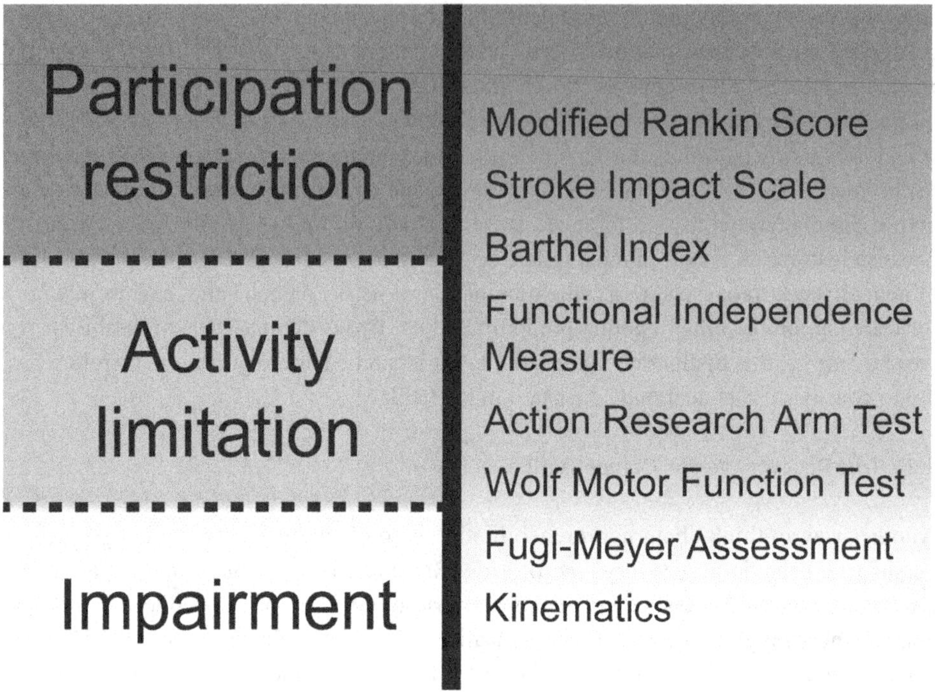

Figure 1.1
ICF levels and measures of arm paresis. Stroke measures mapped onto the International Classification
of Functioning, Disability, and Health (ICF) framework. Different measures capture stroke-related motor
disability at different levels; for example, quantifiable measures of movement kinematics are more specific
at the level of motor impairment while self-reported measures assess the impact of stroke at higher levels
of disability. Clinical measures are not specific and span disability levels with varying degrees of
overlap.

1.4.1   The Fugl-Meyer Scale and the Downfall of Impairment

The Fugl-Meyer Assessment (FMA) of Physical Performance was devised in 1975 to assess impairment after stroke (Fugl-Meyer et al. 1975) due to a perceived lack of a quantifiable measure to track recovery from hemiparesis after stroke. To quote from the original
paper: "With few exceptions, authors who have investigated the functional capacities of
hemiplegics have abstained from describing the neuromuscular capacity per se and apply
especially designed ADL-performance testing." It is apparent that the divide between impairment and activity was already present forty years ago. The authors of the original
paper based their design of the motor section of the FMA on the work of Thomas E.
Twitchell and then Signe Brunnstrom, who both emphasized that a patient with hemiparesis has not just lost muscle strength but also the ability to control joints in isolation because
of the intrusion of synergies (the concept of synergies will be discussed in detail in

chapter 2, but for now, the term can be understood to imply stereotypical activation of muscles or coarticulation of multiple joints during attempted behavior). Historically, two principal synergies have been defined after stroke—the *flexor* synergy, so called because obligatory flexion occurs at the shoulder, elbow, and wrist when attempting to flex any of these joints in isolation. The *extensor* synergy is the opposite (Twitchell 1951; Brunnstrom 1966). The main focus of the motor section of the FMA is therefore to test a patient's ability to isolate and control individual joints outside of flexor and extensor synergies; the movements tested are not related to functional tasks. Examples of what a patient is required to do include elbow extension and shoulder abduction. Importantly, the emphasis is on joint coordination rather than strength, so a certain force level exerted at the bicep within a synergy would get a lower score than if the same force was generated in isolation. The upper extremity score has a maximum of 66 points, and patients are often stratified using this score—for example, severe is approximately 25 or below, moderate is 26 to 45, and mild is 46 to 66 (Duncan et al. 2000; Gladstone et al. 2002). The FMA is subject to ceiling effects for patients with very mild impairment, and finger individuation and precision grip are minimally tested. The FMA has excellent interrater and intrarater reliability and validity, as well as a low minimal detectable change (Shelton et al. 2001; Gladstone et al. 2002; Platz et al. 2005; Page et al. 2012; See et al. 2013).

Due to its reliable characteristics, the FMA remains the main clinical instrument for the assessment of motor impairment in studies and trials. In contrast to its use in research, the FMA is rarely used today in clinical practice; one reason is that it is time-consuming, but the main reason is that it no longer reflects clinical priorities. The FMA has a therapeutic complement, previously dominant in neurorehabilitation, which has been referred to as either the facilitation or neurophysiological approach because it emphasizes the need to relearn normal patterns of movement and overcome abnormal reflexes to reduce spasticity. To achieve these two ends, the patient is moved passively or with some assistance by the therapist (Bobath and Bobath 1984; Bobath 1990; Kollen et al. 2009). Beginning in the 1980s, however, there was increasing awareness that teaching normal movement patterns out of context did not necessarily generalize to ADLs. This realization led to a shift away from the treatment of impairment toward training on ADLs (ICF activity level) with functional tasks, conceptualized within a motor learning rather than a physiological framework. Herein lies an irony: impairment's eclipse by function, although quite understandable historically, has made it more difficult to bring it back as an outcome target in the light of new discoveries and concepts in neuroscience.

1.4.2 The Ascendance of Activity

In a thoughtful essay on the recent history of the changes in the theoretical perspective underlying neurorehabilitation, James Gordon describes how treatment centered on facilitation; the encouragement of normal movement patterns through assistive guidance that reproduces normal sensory feedback, was subject to growing dissatisfaction due to the problem of what he calls functional carryover. He states, "It is easy enough to 'facilitate' a certain pattern of movement. What is difficult is to get patients to use that pattern when they are actually carrying out some functional activity. This is the fundamental challenge facing rehabilitation therapists" (Gordon 1987). In Gordon's view, this realization marked a shift from a neurophysiological toward a motor learning–based perspective in neurorehabilitation. We shall have much more to say about the applicability of motor learning to neurorehabilitation throughout the book, specifically in chapter 6. Here it will suffice to say that the conceptual shift toward motor learning had a practical consequence: patient care moved to a focus on practicing everyday tasks and learning coping strategies. The efficacy of this new rehabilitative approach was deemed to be best captured by global ADL scales. For example, in the United States, the FIM is the standard tool to measure responsiveness during inpatient stays. It was designed in 1984 with the aim of capturing global disability at the level of motor and cognitive activities by measuring degrees of independence in ADLs (Keith et al. 1987; Ottenbacher et al. 1996).

As outlined above, the change from a physiological perspective to one focused on motor learning and task practice led to a move toward global disability scales as the outcome measure in clinical practice. The change in focus had a different consequence when it came to research studies, namely, a switch to task-specific functional measures. For example, in the paper that introduced the Wolf Motor Function Test (WMFT), the authors state, "Yet the FMA is difficult to use and examines synergy patterns that no longer form the basis for many functionally oriented treatments" (Wolf et al. 2001). It is hard to imagine a statement that better encapsulates the conceptual shift that has occurred in neurorehabilitation research over the past 30 years.

The two most commonly used activity-level tests are the Action Research Arm Test (ARAT) and the WMFT (Lyle 1981; Wolf et al. 2001; Yozbatiran et al. 2008). The ARAT is based on the assumption that upper extremity movements used in ADLs depend on four movement components: grasp, grip, pinch, and extension and flexion at the elbow and shoulder. Examples of activities tested in the ARAT include grasping a wooden block, pouring a glass of water, pinching a marble, and placing the hand on top of the head. Scoring takes into account quality of coordination, time taken, and task completion. The ARAT has been shown to have high reliability and validity, and it correlates well with the FMA in the acute and chronic stages of stroke (Rabadi and Rabadi 2006; Nijland et al. 2010). Thus, a functional assessment scale correlates well with an impairment measure that does not have a functional component. If this is the case, however, why so much fuss over a change in perspective from impairment to function? The answer to this question is very important and is often unappreciated. A high correlation at baseline between two measures does not imply that a change in one of these measures in response to treatment will lead to a change in the other. This logical fallacy has been pointed out in the context of cognitive training: "Pretraining correlation does not imply training covariation" (Moreau and Conway 2014).

The reason for this uncoupling is that training may operate on those mechanisms that lie in the unexplained variance in the initial correlation. More concretely, a correlation between the upper extremity FMA and the ARAT can be expected initially, because being less impaired will necessarily be reflected in better functional performance across a broad range of impairment. This does not mean, however, that when training improves performance on the ARAT that this will lead to better control of joints out of synergy (i.e., a change in the FMA). Conversely, a small reduction in impairment may not translate into improvement in function. Why should this be? The reason is that the unexplained variance in the initial correlation between two clinical measures may reflect differences in the roles of compensation and restitution in the two measures. So in the case of an improvement in the ARAT, this may occur through compensation without any restitution. In the case where a small amount of restitution occurs with an increase in the FMA, this may not be sufficient to reduce the need to compensate.

It should be apparent from the discussion thus far that the restitution-compensation dichotomy maps to a large degree onto the impairment-activity dichotomy. The mapping is not complete, however, because improvement on functional tasks can be accomplished by either restitution or compensation. Neither the ARAT nor the WMFT has sufficient fine-grained behavioral resolution to disambiguate between restitution and compensation, although it has recently been argued that the WMFT has the capacity for such disambiguation (Levin et al. 2009). This argument is not entirely convincing; the WMFT is essentially a more elaborate ARAT, and its components do not correlate any better with the FMA than the ARAT does (Hsieh et al. 2009). In both cases, the amount of variance explained is under 50 percent. That the ARAT and WMFT pick up on different kinds of recovery compared to the FMA is apparent in the dissociation in these measures in recent clinical trials. For example, consider a comparison between robotic therapy and constraint-induced movement therapy (CIMT). In the Veterans Affairs (VA) Robotic-Assisted Upper-Limb Neurorehabilitation in Stroke Patients study, which investigated the effect of upper limb robotic therapy over twelve weeks in patients with chronic stroke, there was a small but significant effect on the FMA but no effect on the WMFT at week 12. The near identical result was obtained in a more recent robotic therapy trial (Klamroth-Marganska et al. 2014). The origin of these very small changes in the FMA is unknown—it could be a small amount of restitution in the chronic state but more likely reflects increases in strength from the robotic training. In contrast, in the Extremity Constraint Induced Therapy Evaluation (EXCITE) trial of CIMT, there was a significant (moderate to large) effect on the WMFT. The FMA was not an outcome measure in EXCITE, again reflecting the eclipse of impairment in neurorehabilitation trials (Wolf et al. 2010). Other smaller studies report an activity-level effect of CIMT but no effect on impairment (Boake et al. 2007; Dromerick et al. 2009). Two recently published meta-analyses of neurorehabilitation interventions also indicate that there is impairment versus function dissociation for robotics versus CIMT. Recent meta-analyses also provide several other examples of dissociations between impairment and functional outcome measures (Pollock et al. 2014; Veerbeek et al. 2014). The take-home point is that such dissociations are to be expected when such measures differentially detect improvements due to either restitution or compensation. It is to be hoped that as future designers of neurorehabilitation trials become more mechanistic in their *a priori* assumptions, the choice of outcome measure will be more focused and judicious. Our position is that brain repair is best assayed with impairment measures and quantitative movement analysis. This is because improvements on functional outcome measures are contaminated by compensation.

1.5   Current Upper Limb Neurorehabilitation: What Is Done, and Does It Work?

This will not be an attempt to systematically review the evidence for efficacy of interventions after stroke but will instead present a summary of the overall state of rehabilitative treatments for the arm and hand at the levels of both impairment and function. Particular forms of intervention such as robotic therapy and CIMT will also be covered in more detail later in the book as examples of how motor learning principles have been used to develop new therapeutic approaches.

Most studies and randomized controlled trials (RCTs) for neurorehabilitation have been conducted in patients with chronic stroke (more than six months out). Very few equivalently sized studies have been conducted in the first weeks to three months after stroke, which is curious given that this is the period when most conventional neurorehabilitation is given. Thus, RCTs have not been conducted to prove efficacy for the current standard of inpatient and outpatient care in the first three months after stroke or to test new interventions in this same time period (Stinear et al. 2013). This is an odd state of affairs to say the least and certainly counterproductive for progress in the field.

There are many explanations for the almost exclusive focus of research on chronic stroke. One of these is perhaps a tacit admission that the inpatient stay is so inadequate that new treatments are needed to make up for it once patients have been discharged. Is this a fair conclusion? In a study conducted in Australia titled "Inactive and Alone: Physical Activity within the First 14 Days of Acute Stroke Unit Care," patients in five acute stroke units (average length of stay ranged between six and fourteen days) were observed very closely over two consecutive days from 8 a.m. to 5 p.m. (Bernhardt et al. 2004). The investigators found that patients were alone in their room 60 percent of the time, in bed 50 percent of the time, and active only 13 percent of the time. Admittedly, patients had not yet been transferred to acute rehabilitation units, but as we shall see later, the first few weeks after stroke are of critical importance as this is when most spontaneous biological recovery occurs (Dromerick et al. 2015).

In an important study of conventional inpatient and outpatient rehabilitation, Catherine Lang and colleagues quantified the amount of upper limb task-specific practice that patients actually received by recording movement repetitions during 312 physical therapy (PT) and occupational therapy (OT) sessions at seven sites in the United States and Canada (Lang et al. 2009a). Functional movements were observed in only half of the 162 sessions that were ostensibly devoted to the upper limb. The sessions were dominated instead by active and passive exercises, defined as movement of the upper limb in a specific pattern outside the context of a functional task. Thus, even in the 2000s, one can see that therapy remains to some degree a mix of the earlier neurophysiological approach and the later motor learning approach. The average number of functional repetitions per session was only thirty-two, where a single repetition was defined as one completion of a functional task, for example, putting on a shoe. It should also be noted that the total number of thirty-two repetitions was usually divided across two to four functional tasks, which means that the amount of practice per task was very low indeed. Studies of motor learning in healthy nonhuman animal models suggest that hundreds of goal-directed movement repetitions are needed to induce changes in cortical synaptic density, which means that if one is going to use rehabilitative training to augment the effects of spontaneous biological recovery mechanisms early after stroke, much higher numbers of repetitions will likely be required (Nudo and Milliken 1996; Kleim et al. 1998; Plautz et al. 2000). In a study of squirrel monkeys, full recovery from a stroke in the primary motor cortex was accomplished by having them practice 500 to 600 skilled pellet retrievals per day for one to two weeks (Friel et al. 2007). Clearly, there is a glaring mismatch between the low repetition dose patients receive and the high dose suggested by animal models. Some caution is required, however, when comparing the repetition doses given in nonhuman animal studies with those seen in the study by Lang and colleagues. This is because poststroke training is nearly always given very early in animal models, whereas the average time since stroke in the study by Lang and colleagues was 118 days (Lang et al. 2009). Studies in animal models that have delayed poststroke training onset have reported diminished responsiveness (Biernaskie et al. 2004; Ng et al. 2015). Thus, two things must always be kept distinct and not conflated: dose and time of treatment onset. As we shall see later, these are both important and likely interact. It is nevertheless encouraging that in two recent studies, it has been shown that patients can accomplish and tolerate around 300 task repetitions in a one-hour inpatient session (i.e., doses comparable to those in animal models) (Birkenmeier et al. 2010; Waddell et al. 2014).

Despite the very low dose of rehabilitation provided for the upper limb during acute rehabilitation stays, conducting trials in this same time period remains logistically challenging as they might be deemed to interfere with the patient's mandated treatment schedule and cause excessive fatigue. It would be difficult to get an ethics board to countenance withholding regular rehabilitation to free up more time for a trial. More to the point is the fact that regular treatment, despite its inadequacy, must be given to pay for the hospital stay, at least in the United States. Finally, it has been reasoned (or rationalized) that the effect of a new intervention is more easily detected when spontaneous biological recovery has plateaued in the chronic state. As shall become apparent in subsequent chapters, this position has set the field back.

1.6 Conclusions

It is hard to be anything but underwhelmed when it comes to rehabilitation of the upper extremity in current practice. For chronic stroke, physical therapy (the term is used to cover all tested behavioral interventions compared to standard of care) has low to moderate impact at the levels of function and self-report and clinically negligible effects at the level of impairment.

Chronic stroke is what is studied in most RCTs. The control, standard treatment, is usually given during acute rehabilitation and has never itself been proven to be effective in this time period, and it never will be, as withholding therapy is not an option for obvious ethical reasons. To the degree that current rehabilitation achieves some independence in ADLs at three months, this is lost in one in six people by twelve months (MacLeod and Turner 2015), which is consistent with learned compensatory strategies that will be forgotten without continuous practice. Usual care in chronic stroke is currently sparse to say the least. To the degree that it has been looked at, it has no impact on impairment, function, or QOL (Lo et al. 2010).

In our view, only one treatment, CIMT, has shown a convincing and moderate effect size at the level of activity in a large trial—EXCITE (Wolf et al. 2010). It should be said, however, that there was no control group in this trial, which was not the decision of the investigators. In a recent Cochrane review of upper limb neurorehabilitation, it is stated, "Currently, no high-quality evidence can be found for any interventions that are currently used as part of routine practice, and evidence is insufficient to enable comparison of the relative effectiveness of interventions." (Pollock et al. 2014). This Cochrane review and another recent meta-analysis also addressed the critical issue of dose, with both concluding that benefits probably increase with dose but that nothing definitive can yet be said with existing data (Pollock et al. 2014; Veerbeek et al. 2014). Similarly, it is still hard to say anything definitive about earlier intervention—that is, that the first three months are better than later as so few RCTs have been conducted in this time period or comparisons made between early and late rehabilitation.

The authors of the Cochrane review go on to state, "Effective collaboration is urgently needed to support large, robust RCTs of interventions currently used routinely within clinical practice. Evidence related to dose of interventions is particularly needed, as this information has widespread clinical and research implications." (Pollock et al. 2014). Although undoubtedly well meaning, this recommendation is, in our view, misguided as it is putting the cart before the horse. The current crisis in neurorehabilitation is not because existing therapies have not been properly tested in RCTs but because existing and new experimental therapies themselves are often ill conceived from a biological and mechanistic standpoint. RCTs should not be conducted until more mechanistically informed therapies based on scientific principles have been devised. The premature rush to large phase III RCTs has, in our view unsurprisingly, led to a string of failures that will only make it more difficult in the future to convince funding bodies that this area is very much in need of support. We hope that the rest of the book will provide the mechanistic groundwork for new and more effective therapies to rehabilitate the upper limb after stroke.

2

Upper Limb Paresis
Phenotype, Anatomy, and Physiology

2.1 Introduction

In this chapter, we will discuss the phenotype of arm and hand paresis in humans and non-human primates mainly in cross section. The time course of recovery from hemiparesis will be addressed in chapter 3. We feel that this separation is justified because the physiology of any degree of initial paresis is not likely to differ in a fundamental way from paresis that evolves during recovery to an equivalent level. For example, a mild initial paresis is probably not different from a moderate paresis that then recovers to mild levels. Paresis comes from classical Greek and means "letting go." In modern medical usage, it has come to mean weakness and/or loss of voluntary movement in any neurological condition. Plegia refers to complete loss of voluntary movement. Thomas E. Twitchell in his classic paper titled "The Restoration of Motor Function following Hemiplegia in Man," published in the journal *Brain* in 1951, considered hemiplegia a disorder of movement (Twitchell 1951). It is in our view correct to consider arm paresis a movement disorder because it consists of changes in posture, in phasic and tonic reflexes, in strength, and in voluntary motor control. The challenge is to understand how these components combine to create the overall paresis phenotype and to determine the relative contribution of these components to the loss of motor function experienced by the patient. This chapter takes a decomposition approach both behaviorally and physiologically. On occasion, the hand and the arm will be treated separately, primarily because of how some studies tend to get done and reported. Careful consideration of the studies does not suggest fundamental differences between how the healthy human brain controls the hand and the arm, but this continues to be debated and investigated.

2.2 The Modern History of Hemiparesis (Part 1): Sherrington, Tower, and Walshe

The first half of the twentieth century saw a fruitful scientific dialogue between scientists performing physiological experiments in nonhuman animals and neurologists observing and examining patients with hemiparesis after stroke. The renowned British neurologist,

F. M. R. Walshe, made a convincing case that the study of the patients after stroke by clinicians was of crucial importance to elucidation of the role of the pyramidal tract in "willed movements." He argued that with patients, unlike with animal models, finer and more varied gradations of paresis were available for examination, and these could be studied over longer periods of time with a cooperative subject (Walshe 1947). The intellectual richness of the back and forth between physicians and physiologists is apparent in a series of classic publications that will be discussed here as historical prolegomena to more recent investigations of hemiparesis. The insights and the way of thinking about hemiparesis that these early publications offer remain invaluable. In addition, we also hope that they underscore the continued need for careful study of humans in the clinical setting by scientifically informed physician-scientists.

Thinking about hemiparesis in the first half of the twentieth century was heavily influenced by the physiologist Charles Sherrington's seminal studies in nonhuman animals on spinal reflexes. Sir Francis M. R. Walshe, in a classic 1919 paper, states, in the context of emerging studies on hemiplegia and spasticity in man, "For the first time a serious attempt was made to identify in man those principles underlying the physiological activity of the nervous system which the researches of Sherrington have revealed in the lower animals" (Walshe 1919). Of particular relevance were Sherrington's studies of *decerebrate rigidity* in the cat because it was largely assumed that this condition served as a model for spasticity seen after stroke—Sherrington himself, however, does not seem to have given much consideration to spasticity in humans. Decerebrate rigidity can be induced in cats by intercollicular transection and leads to immediate velocity-dependent increases in the gain of the stretch reflex, preferentially affecting antigravity muscles—the extensors of the hind limb (Sherrington 1898). Sherrington's influence meant that neurologists studying hemiparesis were just as interested in what they called "positive" or "release" manifestations of hemiparesis, such as hyperreflexia and hypertonus, as in the negative phenomenon of loss of voluntary movement. This Sherringtonian perspective has also projected forward to the present day with the continued emphasis placed by many physical therapists on the treatment of spasticity after stroke, the so-called physiological approach to neurorehabilitation discussed in chapter 1.

In 1919, F. M. R. Walshe published a paper in the journal *Brain* whose title is worth quoting in full: "On the Genesis and Physiological Significance of Spasticity and Other Disorders of Motor Innervation: With a Consideration of the Functional Relationships of the Pyramidal System" (Walshe 1919). What is critical about this paper is that Walshe already was aware that the Sherrington-inspired focus on the similarity between the heightened stretch reflexes observed in decerebrate cats and those in patients with hemiparesis was not sufficient to explain the full phenotype. The reflex view had to be fused somehow with Hughlings Jackson's hierarchical framework for the motor system, in which he posited that the motor cortex and the pyramidal tract control lower level bulbar and spinal reflex circuits through inhibition (Jackson 1884). Thus, Walshe intuited that the hemiparesis phenotype with its positive and negative signs (nomenclature also derived from Hughlings Jackson) resulted from the loss of the normal interplay between the pyramidal tract and segmental reflex circuits. This thinking was fully fleshed out in a follow-up paper in 1947, also published in *Brain*, titled "On the Role of the Pyramidal System in Willed Movements" (Walshe 1947).

The term *pyramidal tract* (PT) refers to axons in the medullary pyramids made up of corticospinal projections and those remaining corticobulbar projections that have not already synapsed on brainstem nuclei more rostrally. In his 1947 paper, Walshe made several important points that are surprisingly prescient for current-day findings and ongoing debates and nicely summarize the paresis phenotype for the upper limb. First, while recognizing that "willed movements" were dependent on the PT and to some degree did not depend on sensory input, Walshe made a very strong claim for the PT as an "internuncial" pathway that is directed by multimodal sensory input and thus should not be considered the output pathway from a "keyboard" in the motor cortex that initiates movements ab initio, uncoupled from sensory input (Walshe 1947). In this conception, the Sherringtonian reflex arc is maintained as the central unit of action and is simply extended up through cortex. Recent work spearheaded by Stephen Scott and his colleagues supports this neo-Sherringtonian sensory feedback-based view of the motor cortex, with long-latency transcortical reflex pathways posited as the substrate for voluntary movement (Kurtzer et al. 2008; Pruszynski and Scott 2012; Scott 2012).

Walshe then goes on to emphasize the "dual character of hemiplegia," the combination of loss of voluntary movement and the intrusion of positive phenomena, and asks if both these negative and positive phenomena can be attributed to a lesion of the PT or if, instead, another second descending system has also been destroyed that releases subcortical circuits. To attempt to answer this question, Walshe turned to the seminal work of Sarah Tower, in which she examined the behavioral consequences of sectioning the medullary pyramids in monkeys and chimpanzees. In human stroke, the lesion is uncontrolled and so there is always the possibility of interrupting multiple descending pathways. In contrast, pyramidotomy offers, as Tower states, "a unique opportunity to interrupt corticospinal action, and very nearly that alone; and thus to establish both the symptomatology of pyramidal lesions, and, by reinterpretation, the functions of the pyramidal tract" (Tower 1940). As Walshe himself recognized, the 1940 paper is full of rich and careful observations; reading it to this day is informative and rewarding. As a side note, Tower described almost complete elimination of use of the affected upper limb after unilateral pyramidotomy, what she called a "raised threshold," which could be overcome by "tying up the normal hand." Thus, Tower characterized limb nonuse and invented a form of constraint-induced therapy almost a half a century before the work of Edward Taub and colleagues in rhizotomized monkeys (Taub 1980).

Weakness is prominent after unilateral pyramidotomy—the monkeys could no longer support their body weight when hanging from a bar, although they could grasp the bar and hold on if given some partial body support. In terms of motor control, Tower focused primarily on the hands, a focus that continued in studies of hemiparesis throughout the twentieth century and into the twenty-first century. Such an emphasis on the hand is understandable from a functional perspective; moving the arm around with an uncontrollable hand at the end of it would be of limited usefulness, excepting a supportive role in bimanual tasks. Unfortunately, the focus on the hand has led to somewhat hand-centric views of the motor cortex and the PT. As shall be seen later, careful analysis of proximal arm movements reveals deficits that are arguably equivalent to loss of digit individuation. It is probably more accurate to consider a spectrum of control loss both distally and proximally, a point already apparent in Tower's description of grooming performance with the paretic upper limb. She first makes the distinction between behaviors that require use of the all the digits together (scratching, brushing, and grasping) versus those in which the digits must be used independently, for example, opposition of thumb and index finger to pick up a small object. She then states that PT lesions have a much more pronounced effect on the latter compared to the former. In the act of grooming, Tower contrasts brushing performance, although slow, clumsy, and poorly aimed, with the complete absence of picking behavior. Poor aiming, whereby the monkey would sometimes brush the air above the hair, appears to be considered by Tower as qualitatively different from the loss of digit control. This difference is captured by her distinction between more and less stereotyped behaviors, with the implication that the former are controlled by extrapyramidal systems. The alternative possibility, which she does not consider, is that aiming and speed for the proximal arm might require the same kind of PT contribution as digit individuation (Tower 1940).

Tone and phasic reflexes were reduced throughout the arm. That said, Tower observed that hypotonia was not equal throughout the arm and hand—the extensors were more hypotonic than the flexors both proximally and distally. The clasp knife phenomenon, clonus, and the Babinski were never observed. To summarize what Tower found after unilateral pyramidotomy in the monkey in her own words, there was "diminished muscle tone, . . . defective initiation and execution of all performance by skeletal musculature with elimination of non-stereotyped movements and elimination of all discrete usage of the digits" (Tower 1940). In a complementary fashion, Walshe summarizes his view of the PT, which he acknowledges derives in large measure from Sarah Tower's work. He states that "the cortex by way of the pyramidal system fractionates and combines the functional elements of reflex segmental mechanisms" (Tower 1940).

Thus, the pyramidal phenotype in monkeys is very similar to the negative signs seen in human paresis, but the prominent positive signs seen in humans were absent, which led Tower to write, "There is no evidence for an inhibitory function" for the PT; indeed, she used the term *hypotonic paresis* for the monkey and contrasted it with *spastic paralysis* in man. She concludes by stating "the ability to execute isolated movements, whether of the digits or of more proximal parts, is indubitable evidence of function of the pyramidal tracts."

In her view, positive signs and stereotypical multijoint movements in humans were the result of compensation by an extrapyramidal system. Walshe was not convinced that the PT itself did not have an inhibitory role. Nor did he believe that there is a true dichotomous separation between discrete movements controlled by the PT and stereotypical movements controlled by the extrapyramidal system. He did concede that at least some positive signs might be due to release of extrapyramidal systems from non-PT-mediated cortical control. Both Tower and Walshe agreed, however, that in the normal case, movements were a seamless blend of the pyramidal and the extrapyramidal, with both under cortical control (Tower 1940; Walshe 1947).

It is remarkable how many themes already present and speculated about in these classic papers by Tower and Walshe on arm paresis and spasticity remain debated today. As we shall see, the questions they raised and answers they proposed are recapitulated in both the studies that immediately followed theirs and in studies that continue up to the present day.

2.3   The Modern History of Hemiparesis (Part 2): The Lawrence and Kuypers Studies in the Macaque

As stated above, the study of the effects of pyramidal lesions in nonhuman primates was primarily undertaken to provide insight into the hemiparesis phenotype after stroke in humans. In what is widely considered the definitive study, following up on that by Tower, who only studied seven animals, six of whom died within five days, Donald Lawrence and Henry Kuypers reported the effects of bilateral pyramidotomy in thirty-nine rhesus monkeys in a classic paper published in *Brain* in 1968 (Lawrence and Kuypers 1968a). We will not outline the time course of recovery, reserving that for chapter 3, but instead remark on the steady-state motor phenotype observed in the eight animals that had the purest bilateral PT lesions. As was observed by Tower, the animals lost their ability to individuate the digits, and this was not regained even as far out as eleven months. In contrast to what was seen by Tower, Lawrence and Kuypers state that axial stability and proximal limb movements returned, stressing that their motor deficit was restricted to an isolated deficit in fractionation of the digits. The monkeys could, for example, grip the cage bars with sufficient strength to support their whole body weight. The Lawrence and Kuypers paper is vague with respect to upper limb movements. Although the authors state that "ultimately the animals could fully extend either arm with the wrist slightly dorsiflexed and the fingers semiflexed and abducted," at another point in the paper, the authors state that "all movements were slower and fatigued more rapidly than in the normal animal." In the summary of the paper, the authors state that "the corticospinal pathways superimpose speed and agility upon subcortical mechanisms," which seems to imply that proximal limb movements, although present, are not as fast and accurate in the absence of the PT.

This vagueness with respect to proximal limb control is a problem and in retrospect can be attributed to a number of factors. First, detailed movement kinematics were not obtained. Second, as also occurs when examining patients at the bedside, failure to either individuate or perform a pinch grip with the digits is easier to detect than are reductions in the speed and accuracy of proximal limb movements, especially in the setting of concomitant weakness. Finally, Lawrence and Kuypers were trying to make a strong case for a double dissociation between the effects of PT lesions on control of the fingers and the effects of medial brainstem pathways on axial and proximal arm control. This is because of their second paper in the series, in which they lesioned the ventromedial or lateral brainstem pathways two to eight months after the bilateral PT lesions (Lawrence and Kuypers 1968b). Of most relevance to this chapter is the fact that the ventromedial lesion interrupted the reticulospinal tract (RST): the output from the medial pontine and medullary reticular formation. The phenotype generated was of flexion bias of the trunk and limbs, as well as severe impairment in axial and proximal arm movements. Critically, the authors pointed out that independent distal extremity movements were unaffected, which they attributed to intact rubrospinal pathways. Similarly, in three monkeys with ventromedial brainstem lesions but no prior pyramidotomy, there was no detectable impairment in hand dexterity using a food well task.

Thus, Lawrence and Kuypers constructed a dichotomous argument, one that is still the dominant view today, that the PT is for fractionated control of the fingers and hand, whereas the medial descending pathways such as the RST control proximal and axial muscles, with no clear distinctions between strength and control. Their formulations had a qualitative either/or quality that is somewhat at odds with the quantitative differences in speed and accuracy that they themselves described for both proximal and distal limb control after pure PT lesions.

An interesting coda to the Lawrence and Kuypers studies is a study reported in 1970 by Charles Beck and W. W. Chambers, in which a distinction was made between weakness and control for proximal movements in monkeys with unilateral pyramidotomy (Beck and Chambers 1970). Recall that in their study of bilateral pyramidotomized monkeys, Lawrence and Kuypers (1968a) state, "The most striking change after the first four to six post-operative weeks was a progressive increase in their general strength and in the speed and ease of movements." For proximal limb control, Beck and Chambers looked both at the ability of the monkeys to reach quickly to strike a key with the whole hand and make fast isolated extension movements at the elbow or shoulder. In both tasks, they reported normal performance. In contrast, when they tested flexion and extension strength at the shoulder, elbow, and wrist, as well as grip strength, and calculated a percentage relative to the unaffected side, they found significant differences. Specifically, they found that strength deficits were greater in flexion than in extension and that there was a proximal-distal gradient for flexion but not for extension. Interestingly, when they looked at the absolute strengths on the ipsilateral side, they suggested that the pattern of weakness on the contralateral side could be explained if the strength deficit is inversely proportional to the normal absolute strength around that joint. They concluded, claiming complete concordance with the PT results of Lawrence and Kuypers, "In this study, the pyramidotomized limb's fine control in accuracy of movement contrasted sharply with its profound weakness." This conclusion is confusing, however, because in the Lawrence and Kuypers study, the implication is that initial weakness after bilateral pyramidotomy fully recovers. They only comment, albeit briefly, on slowness after bilateral PT lesions, and weakness is not brought up in the general discussion at all. In their view, strength is primarily mediated by brainstem descending pathways. Most relevantly, as Stuart Baker has emphasized, monkeys could grip the cage bars and support their own weight when the RST was the only surviving descending pathway (Baker 2011).

Thus, when one reads the Tower, Lawrence and Kuypers, and Beck and Chambers papers in sequence, there is convergence on two poststroke upper limb dichotomies that continue to be debated and investigated to this day: proximal versus distal involvement and strength versus dexterity/control. Sarah Tower's view was that both arm and digit control were compromised and that the residual hemiparesis phenotype is made up of what is lost and what "extrapyramidal activity" provides (Tower 1940). Lawrence and Kuypers moved away from the pyramidal-extrapyramidal dichotomy but instead characterized descending pathways on the basis of their spinal targets (Lawrence and Kuypers 1968b). Later in the chapter, it will become clear, based on more recent research, consistent with the classic studies mentioned above, that separate descending systems differentially contribute strength and control. The seemingly contradictory results may be the consequence of the kinds of tasks used to assess impairments in the upper limb. For example, the proximal tasks used to assess accuracy might not be as difficult from the standpoint of control as the tasks used to stress individuation of the digits. Direct quantification of strength (e.g., as in Beck and Chambers) may reveal abnormalities not apparent when just observing strength-requiring behaviors (e.g., as in Lawrence and Kuypers)—that is, some accuracy requiring tasks may not require the limits of strength detected on confrontational testing. Thus, differences in proximal versus distal task difficulty could manufacture dissociation in control that is not really present, whereas measuring maximal voluntary strength may identify weakness that would otherwise not be appreciated during behaviors that do not require high levels of strength. Finally, it is interesting to note that by the time we get to Lawrence and Kuypers and the work that has followed it up to the present day, positive signs, or spasticity, are decreasingly emphasized. For example, in the now classic 400+ page monograph *Corticospinal Function and Voluntary Movement* by Robert Porter and Roger Lemon, spasticity and reflexes each get only one page reference in the index (Porter and Lemon 1993). This is a far cry from Sherrington and Walshe. We shall have more to say about spasticity and its relevance to voluntary movement later in the chapter.

2.4 Weakness

Loss of force generation (weakness) is a significant component of the hemiparesis pheno-type and the one that neurologists most emphasize in their evaluation. For example, in the National Institutes of Health Stroke Scale (NIHSS), all the motor measures pertain almost exclusively to strength rather than control. One reason for this is that weakness often is the dominant contributor to the paretic deficit in the first month after stroke. In the chronic state, contractures and synergies become more prominent. The physiological mechanisms of weakness remain to be fully elucidated, but there has been careful empirical character-ization of poststroke weakness and assessment of its impact on function.

In a classic, albeit fairly low *n* study, James Colebatch and Simon Gandevia characterized weakness in the arm after what appears to be stroke—there is some vagueness with re-spect to the precise diagnosis for the patients in this study (Colebatch and Gandevia 1989). They tested upper limb strength in fourteen patients with hemiparesis of varying severi-ties. There was some heterogeneity in the time after onset of hemiparesis that the patients were tested, but in the paper, it is stated that thirteen of the fourteen patients in the study were "recovering" from acute weakness, which suggests that it was within months post-stroke. The muscles tested were the shoulder abductors and adductors, as well as the ex-tensors and flexors at the elbow, wrists, fingers, and thumb. Handgrip was also assessed. Results were expressed as torques around the isolated joints using the unaffected side as a reference (the authors were aware that this assumes that the unaffected side is truly spared and also examined for this). A fixed hierarchy of involvement of the different mus-cle groups was not seen across patients, but there was still a hint of a gradient with spar-ing of relative weakness at the shoulder and elbow compared to the wrist and fingers. More specifically, the flexors of the wrist and fingers were most affected, with relative weaknesses of 32 percent and 36 percent, respectively, compared to 47 percent for elbow flexion and 57 percent for shoulder abduction. Wrist extension was at 43 percent and fin-ger extension at 50 percent—percentages are calculated with respect to the unaffected side, reported here for ten patients (Colebatch and Gandevia 1989). In terms of *absolute* force, extensors were weaker than flexors at all joints, which was also true of the controls.

A more recent study by Justin Beebe and Catherine Lang largely corroborates the results of Colebatch and Gandevia, although they did not find evidence for even a slight proximal to distal gradient in strength. The study was performed in thirty-three patients within two to three weeks of stroke (Beebe and Lang 2008). Dynamometry, which was obtained for flexion and extension of the shoulder, elbow, wrist, and index finger, showed no difference proximally to distally in the percent reduction in strength compared to the unaffected side, with average strength percentages across the four joints ranging from 30 percent to 39 percent. Although there are differences between the Beebe and Lang and the Colebatch and Gandevia studies in terms of the muscles studied, the methodology used, and time after stroke (earlier in Beebe and Lang, although not chronic in either case), overall both show that the whole limb is affected and that this is true for flexion and extension. Thus, it is perhaps time to retire the idea of a proximal to distal gradient—it has likely persisted because of the nature of how clinicians examine hemiparesis in the poststroke patient, which exaggerates the impression that the hand is more impaired than the arm. That the flexors are as weak as extensors may seem contradictory to the reader's experience, but as we shall see in chapter 3, one explanation is that the more typical pattern of flexors being stronger than extensors emerges over time, especially at the wrist and fingers. One should also note that at the bedside, one tends to compare absolute rather than relative forces.

There are two potential neural mechanisms for limiting force production across a joint—lack of drive to the agonist and unwanted contraction of the antagonist (i.e., excessive coactivation). Although peripheral factors such as muscle fiber shortening, changes in motor unit properties, and contractures can also limit movement around joints after stroke (Jakobsson et al. 1992; O'Dwyer et al. 1996; Kallenberg and Hermens 2009), especially in the chronic state, evidence suggests that central neural mechanisms are the main contributors to weakness (Levin et al. 2000; Ng and Shepherd 2013). Derek Kamper and colleagues sought to determine the relative contributions of neural and peripheral factors to passive and voluntary rotational movement across the metacarpophalangeal joints of the four fingers using a rotary actuator (Kamper et al. 2006). Patients were divided into moderate and severe hand paresis based on the Chedoke-McMaster Stroke Assessment (CMSA). Essentially, severe means that the patients have some mass finger flexion but not extension, whereas moderate means some mass finger extension but not finger individuation. It should be emphasized that the CMSA does not directly test for strength but for movement. Normalized isometric extension torque was significantly smaller than normalized grip strength, which again indicates that the symmetric weakness across flexors and extensors seen in the acute state becomes asymmetric in the chronic state. Flexor spasticity was determined by deriving velocity-dependent torques, voluntary isometric strengths were estimated by measuring torque generated from the metacarpophalangeal joints in the neutral position, and coactivation was determined with electromyography (EMG). The main result was that the difference in the CMSA in the two groups was mainly attributable to loss of voluntary drive and not to coactivation, although the latter was more abnormal in the severe group (Kamper et al 2006). In another recent study, M-wave EMG recordings were used to determine whether abnormal torque generation around the elbow joint by the biceps brachii poststroke is primarily due to a voluntary activation deficit or to loss of peripheral neuromuscular capacity (Li et al. 2014). The approach was to see whether there was a difference in the EMG values when the biceps was voluntarily activated versus when it was externally triggered by electrical stimulation. The finding was that weakness correlated most closely with voluntary activation rather than with peripheral changes.

Where does drive on spinal segmental circuitry come from? In humans, we know that pure primary motor cortex (M1) lesions do cause weakness, as do isolated pyramidal lesions in monkeys and humans. Lesions in the corona radiata and posterior limb of the internal capsule are harder to interpret because many indirect pathways, such as the corticoreticular tract, are also present, along with corticospinal projections. That said, it is notable how similar the motor phenotype is when strokes occur in these locations in humans compared to the deficit after pyramidal lesions in monkeys. In a classic experimental paradigm in monkeys, Edward Evarts first identified cells in the arm area of the precentral motor cortex that projected axons to the pyramids—pyramidal tract neurons (PTNs) (Evarts 1968). He did this through antidromic stimulation of the motor cortex with electrodes inserted in the pyramidal tract. He then showed that the discharge rates of PTNs correlated with static flexion and extension forces around the wrist after controlling for position. Many subsequent studies have corroborated this finding that there are neurons in M1 that code for static force (Evarts 1981). A common finding across these studies was that most modulation of these M1 neurons occurs at low force levels (less than 1N), which is consistent with a role for them in the fractionated control of forces associated with precise fine movements like precision grip. Support for this idea comes from an experiment in which recording was done from a single PTN that was known from spike-triggered averaging to project monosynaptically to the first dorsal interosseus muscle, a muscle that contracts in both precision and power grips (Muir and Lemon 1983). The striking finding was that the PTN fired far more strongly in the setting of the precision grip even though significantly less force was required. This result is consistent with the fact that in monkeys, the size of the excitatory postsynaptic potentials (EPSPs) from single PTNs is greatest at synapses of motorneurons innervating intrinsic hand muscles and that postspike facilitation (PSF) of EMG is also largest for intrinsic hand muscles (Lemon 1993). PSF is much weaker or absent for proximal arm muscles in the monkey. In their book, Porter and Lemon state that "relatively few M1 neurons are found to be recruited specifically high force levels, . . . and relatively little work has been done to explore the contribution, if any, of M1 to the generation of large forces" (Porter and Lemon 1993). Putting all this together would suggest that at least in the case of power grip, force is not being generated by fast-conducting monosynaptic PTNs but instead may have a subcortical origin (figure 2.1). Such a conclusion would be consistent with the classic observation by Lawrence and Kuypers reported above—namely, that power grip was relatively spared after bilateral pyramidotomy. Relative preservation of power grip is also often seen at the bedside in patients after stroke. A secondary question is whether the generation of large forces around the shoulder and elbow can be considered the proximal analogs to power grip and therefore subserved by the same subcortical mechanisms. As we have pointed out above, recent evidence does not suggest that arm and hand weakness dissociates to as large a degree as conventional clinical testing has taught.

A subcortical (brainstem) origin for drive onto spinal circuits for the generation of large forces does not of course mean that these brainstem nuclei are not themselves driven by cortical inputs. Indeed, this would have to be the case given the profound weakness that results from strokes above the level of the brainstem. Thus, we need to ask which

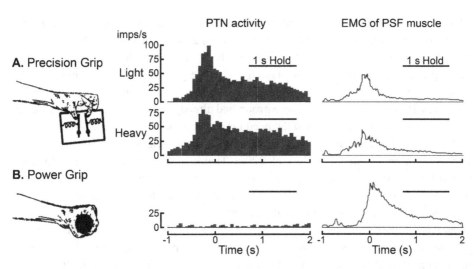

Figure 2.1
Task-related activity for a pyramidal tract neuron (PTN) producing postspike facilitation (PSF) in the first dorsal interosseous (FDI) muscle. (A) PTN discharges for sixteen repetitions of the precision grip task showing an increase in activity before and during the light and heavy force conditions. Rectified and summed electromyography (EMG), recorded at the same time (shown at right). (B) Corresponding task-related analysis for the same PTN and muscle during the power grip. Adapted from Muir and Lemon (1983).

brainstem pathways are relevant to force output in humans and which cortical areas project to these nuclei. For the rest of this chapter, the focus will be on the corticospinal tract (CST) and the RST as evidence suggests that these are the two descending pathways of most importance in humans, as the rubrospinal tract is vestigial in adults (Nathan and Smith 1955; Onodera and Hicks 2010). The RST is a brainstem pathway that originates from the pontomedullary reticular formation (PMRF) and is organized bilaterally: it descends in the medial and lateral regions of the ventral funiculus, and single reticulospinal axons terminate bilaterally in the spinal cord, mainly onto interneurons but also monosynaptically onto motor neurons, over multiple segmental levels with minimal ipsilateral predominance (Kuypers 1964; Matsuyama et al. 1997; Sakai et al. 2009). Both PMRF stimulation and spike-triggered averaging studies in monkeys have shown, in the case of shoulder and arm muscles, a double reciprocal output organization: facilitation of ipsilateral flexors and contralateral extensors, as well as simultaneous suppression of ipsilateral extensors and contralateral flexors (Davidson and Buford 2004; Davidson et al. 2007; Sakai et al. 2009). Cortical input to the PMRF originates bilaterally from primary, premotor, cingulate, and supplementary motor cortices (Keizer and Kuypers 1989). The availability of a more diffusely distributed corticoreticular projection has potentially important implications for recovery from strokes affecting the more cortically focused and predominantly contralateral CST.

Recent studies in humans and nonhuman primates make a case for the RST as a contributor, possibly the main one in severe hemiparesis, to residual motor function after stroke. With respect to force, an interesting study in humans by Jules Dewald and colleagues provides evidence that the ipsilateral RST is the generator of maximum voluntary torques in elbow flexors and shoulder abductors on the affected side after stroke (Ellis et al. 2012). Previous work by the same group had already shown that involuntary flexor torques are generated at the elbow if the need for more motor drive for shoulder abduction is induced by providing progressively decreasing levels of antigravity support to the arm during a reaching movement (Ellis et al. 2009). That is to say, the need to generate more force to lift the arm unmasks coupling of shoulder abduction with elbow flexion (flexor synergy). In a more recent study, the same group showed that loading of shoulder abduction during a reach also leads to involuntary flexion of the wrist and fingers. This is very interesting as together these results suggest that the RST might explain patterns of residual force generation after stroke, wherein preservation of power grip is the distal equivalent of coupled shoulder abduction and elbow flexor torques. These results in humans are entirely congruent with the results discussed previously in monkeys showing modulation of M1 neurons for precision but not power grip, as well as preservation of handgrip after bilateral pyramidotomy. It remains unknown which cortical areas generate the command to the reticular formation to generate power grip other than it not being new M1. Old M1 and premotor areas are potential candidates, but this remains to be elucidated.

2.5   Residual Motor Control and the Idea of Synergies

Despite the importance of weakness to hemiparesis, it is clear that there are other independent contributors to the phenotype. A major part of the motor abnormality after stroke that cannot be attributed to weakness is the presence of motor synergies. The idea of synergy in the context of stroke has its origin in the longstanding clinical observation of obligate coordination patterns across joints that limit the ability of patients to precisely control the proximal and distal limb. Unfortunately, the notion of synergy has become somewhat vexed and confusing because not only is it a descriptive term used by the clinical community to refer to *abnormal* movement patterns after stroke, but it is also a principle used by the motor control community to explain the planning and control of *normal* movements. The term *synergy* refers to either systematic coupling/coarticulation across different joints or to a fixed pattern of coactivation of muscles (McMorland et al. 2015; Roh et al. 2015). For any given task in healthy subjects, there are many joint configurations that could be used to accomplish it (i.e., there is *redundancy*). So, for example, if you are asked to trace a vertical line on a wall with a laser pointer, this could be done using any of one of an infinite number of combinations of rotations about the shoulder, elbow, and wrist joints (figure 2.2). The actual combination chosen, a synergy, could reflect optimization of a cost function that trades off between effort and accuracy.

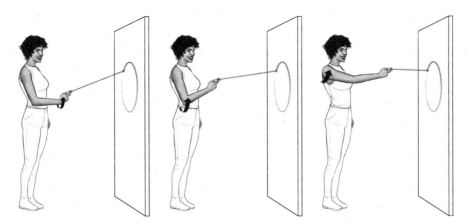

Figure 2.2
Redundancy in arm joint coordination for the same kinematic goal. The goal here is to draw a circle on the wall using a laser pointer. This can be accomplished by rotating about the wrist, elbow, or shoulder. An almost infinite number of combinations are possible.

Similarly, redundancy can exist for the muscles used to actuate joints. The wrist, for example, contains five muscles—more than the minimum four that is needed to be capable of generating a force in all directions. The single solution adopted in this instance can also be explained by minimization of a cost, for example, the summed squared activity of all the muscles (Fagg et al. 2002). The crucial point is that in the case of healthy subjects, coordination patterns are often thought to result more from task and biomechanical constraints than from limitations in neural capacity: one *could* trace the vertical line with the laser pointer just using your shoulder, and one *could* generate a directed isometric wrist force with another combination of wrist muscles. But in either case, the optimal combination of joints and muscles is selected rather than alternative solutions.

Another way that synergies have been conceptualized is as modules or building blocks that simplify the nature of the descending commands required from motor cortex and elsewhere to generate movements. Here the emphasis is not on how the synergies are derived but rather on how they might make planning and control easier by limiting the degrees of freedom problem introduced by redundancy, i.e., how to pick a movement when there are so many to choose from. One proposal has been that the problem is made easier if the brain only needs to control a small number of synergies rather than a large number of individual muscles. Here most of the work has focused on the analysis of EMG data whereby the activity of multiple muscles can be approximated by a linear combination of a smaller number of fixed activation patterns across muscles: *muscle synergies*. Thus, in this view, synergies represent a form of dimensionality reduction to simplify the degrees of freedom problem. These synergies need not be derived through an optimization process; instead, they could be hardwired in neural circuitry or be habits converged upon during learning and development. For many reasons, it remains controversial whether muscle synergies, regardless of how they are constructed, are useful elements for motor planning and motor control. The details of this debate need not detain us too much here (de Rugy et al. 2013), but it is worth mentioning those aspects that are pertinent to both how hemiparesis is characterized and how motor recovery is conceptualized.

A major objection to the derivation of muscle synergies through application of dimensionality reduction techniques to EMG data is that it is just an exercise in data compression. For instance, jpeg compression offers a good way to represent images in a lower-dimensional space, but the details of the jpeg representation do not provide us any meaningful information about the contents of the scene that generated the image. A more pertinent question, perhaps, is not whether the EMG data can be reconstructed using a small number of synergies but whether this reconstructed activity can explain task achievement. The problem is that even if the unaccounted for variance in muscle space is small, large kinematic errors in task space may remain. Indeed, this has recently been shown for wrist data in healthy subjects (de Rugy et al. 2013). This finding has serious implications for performing similar analyses on EMG in patients after stroke. Identifying that patients either abnormally recruit normal muscle synergies or have abnormal muscle synergies is of questionable use unless these findings are relevant in task space. It should be remembered that the classic poststroke synergies familiar to clinicians refer to joint kinematics, not muscle activations. Given the poor mapping between muscle synergies and kinematics in healthy subjects and the fact that what ultimately matters to the patient is task completion, it would seem more logical to examine patient performance in movement or task space and tailor therapies accordingly. To design therapies with the goal to return patients to a normal muscle synergy structure rests on two questionable assumptions. The first is that making muscle synergies after stroke look more normal will translate to kinematics looking more normal. The second is that alternative muscle synergies poststroke cannot lead to adequate task performance. We conclude that interest in muscle synergies is misplaced and is based on the controversial assumption that synergies are the building blocks the brain uses to generate behavior. The idea that muscle synergies simplify motor control in healthy subjects has, confusingly in our view, been applied to the empirical observation of abnormal joint synergies poststroke.

In our view, the more useful way to think about synergies in the context of stroke is to think about them as a loss of degrees of freedom or *repertoire* due to neural damage. That is to say, unlike in the case of healthy subjects, where the capacity of the nervous system is assumed infinite and synergies are proposed as a means to both exploit and solve the redundancy problem, the nervous system after stroke loses capacity. To illustrate this point, consider again the laser pointer task but now consider the case in which your elbow has been locked into position by a splint so that the only degrees of freedom left are around your shoulder and wrist, which means that the only way to accomplish the task of drawing a vertical line requires coordination across these two remaining joints. This recourse to a new synergy to accomplish the task can be considered a form of compensation and is directly analogous to what patients do after stroke, except that after stroke, it is central nervous system (CNS) damage that has limited the available repertoire of joint motions.

An observed synergy can reflect the combination of a loss of repertoire (no elbow movement) and adoption of a compensatory pattern with what remains. To extend this example further, imagine that you get so used to this new way of drawing the vertical line that it becomes a habit and so you fail to notice when the elbow splint is slowly relaxed. Precisely such sticking to a habit despite the presence of an energetically more efficient movement has been demonstrated in studies of healthy subjects when they make reaching movements in force fields that favor curved over straight reaches—subjects continue to reach in a straight line even if more effort is required (Kistemaker et al. 2010). In the case of stroke, Amy Bastian and colleagues have shown that patients will revert to an inefficient asymmetric gait even after split-belt treadmill adaptation has made their gait more symmetric (Reisman et al. 2007). Thus, task-optimized solutions and habits (*soft* synergies) can coexist with constraints imposed by remaining neural circuitry (*hard* synergies). Disambiguating these two kinds of synergy can be a challenge—is the patient performing poorly out of habit, or is that the best the patient can do with the circuits he or she has left? Constraint-induced movement therapy (CIMT) is predicated on the idea that the unaffected arm is being used out of habit despite latent remaining capacity on the affected side. We shall have more to say about CIMT in chapter 6.

2.6   The Dissociation between Strength and Motor Control

Any given motor task requires a combination of strength and control. An example for the hand would be playing the piano *fortissimo*; the keys still need to be pressed precisely but with more force. For the arm, reaching to press an elevator button requires overcoming the weight of the arm and precisely aiming the finger. An example of how these two aspects of movement are dissociable is the patient with cerebellar ataxia, who will likely miss the button even though he or she has no weakness (pure control deficit). A patient with proximal myopathy will have trouble lifting an arm but will not miss the button (pure strength deficit). Hemiparesis is a challenge because both kinds of deficit coexist. Poststroke synergies can be thought of as the behavioral readout of the residual capacity of the damaged nervous system when attempting tasks that require both strength and control. Studies that parametrically alter the control and strength requirements of a task can provide insights into the nature of the systems contributing to the hemiparesis phenotype.

One of the first studies to carefully examine the relationship between strength and dexterity in the upper limb following stroke was performed by Louise Ada and colleagues in 1996 (Ada et al. 1996). Specifically, they designed a paradigm to measure elbow dexterity within the strength capabilities of patients with stroke. The arm was supported in a frame on a horizontal surface for antigravity support, and friction was minimized. Dexterity was assessed with a cursor-tracking task that had to be performed at a slow and a fast speed using elbow flexion and extension around a mean position of 60°, which is the optimal mid-range position for the flexors and extensors. There was no significant correlation between strength and dexterity, or any correlation between the ability of the patients to generate fast discrete flexion and extension movements—all the patients could move faster than what was required to track the target—and dexterity. Interestingly, the patients and the age-matched controls showed the same maximum speed during the fast-tracking task but widely diverged in dexterity. The authors concluded that when dexterity can be examined separately from strength, they are not related. The authors speculate that although slowness of movement is largely attributable to weakness, it may sometimes be a compensation for loss of dexterity at faster speeds. The problem with going more slowly is that some tasks of everyday life have been learned at a particular tempo and the skill is lost at lower speeds. An example of this comes from Alfred Brodal, the Norwegian professor of anatomy who published an article in 1973 entitled "Self-Observations and Neuroanatomical Considerations after a Stroke" (Brodal 1973). He described the trouble he had tying a bowtie about two months out from his stroke. He did not attribute the difficulty only to loss of speed or strength but to his "fingers not knowing the next move. It felt as if the delay in the succession of movements (due to paresis and spasticity) interrupted a chain of more or less automatic movements." This is very interesting as it suggests that his bowtie control policy depended on a particular speed of transition through a sequence of states and did not generalize to slower speeds. Alternatively, the policy depended on his original way of generating the finger movements and failed to be triggered by the compensatory ones he learned after his stroke.

To investigate synergies in the context of upper limb reaching movements, Jules Dewald and colleagues have developed a novel reaching paradigm, previously mentioned in the context of weakness earlier in section 2.4 (Ellis et al. 2009). The paradigm consists of having patients try and cover as large an area of a planar workspace with their hand in the horizontal plane at 90° of shoulder abduction. The group's main hypothesis, which propagates through a series of their experimental stroke studies, is that flexor synergies are increasingly expressed during arm movements the more the weight of the arm has to be supported by shoulder abduction. The flexor synergy, as we discussed in chapter 1, was first defined by Twitchell and consists, in part, of coactivation of shoulder abductors and elbow flexors when attempting to lift the arm. To test the hypothesis, the authors quantified the effects of progressively greater shoulder abductor activation on the total work area covered by the paretic arm during planar reaching movements. An admittance robot was used to both create a haptic horizontal plane and provide forces along its vertical axis to alter the amount of active limb support a patient was required to generate. For example, in the 0 percent active support condition, the robot provided a force equal to that of the subject's relaxed limb such that no net abduction torque was required for the subject to maintain his or her arm in a position above the table. Support levels were randomized and titrated in increments of 25 percent of limb weight; they ranged from 0 percent to 200 percent of

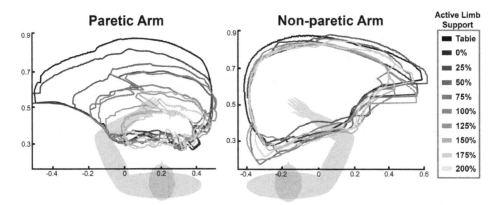

Figure 2.3
Decrease in work area of the arm as a function of various levels of limb support. Work areas in the left paretic limb are inverted to facilitate the comparison with the right nonparetic limb. Data from a single patient; axes units are in meters. Adapted from Sukal et al. (2007).

required active limb weight support (figure 2.3). The main result was that the planar movement area covered decreased as loading on the arm increased and was correlated with a large reduction in elbow extension and shoulder flexion degrees of freedom. This result is consistent with earlier work by the same group, in which they found abnormal patterns of muscle coactivation when patients attempted to generate isometric torques at the elbow and shoulder (Tang and Rymer 1981; Bourbonnais et al. 1989; Dewald et al. 1995).

Dewald and colleagues interpret the loss of elbow extension and shoulder flexion to be a consequence of loading-dependent expression of the flexor synergy. This view is consistent with an elegant earlier study by Kathleen Zackowski and colleagues, in which they compared two kinds of fast-reaching movement to a suspended Styrofoam ball in patients with hemiparesis (Zackowski et al. 2004). In both cases, the patients were seated. In one movement ("reach up"), the patient had to reach using 40° of shoulder flexion and 40° of elbow flexion. In the other kind of movement ("reach out"), the reach required 40° of shoulder flexion and 40° of elbow extension. The main result was that the reach-up movements were straighter and faster than the reach-out movements (figure 2.4). This difference in the quality of the control in the two kinds of reach was not related to weakness. The authors interpreted the difference in trajectory deficits for the two kinds of reach to the fact that the reach-up movement was congruent with a flexor synergy while the reach out required breaking out of the flexor synergy as it required elbow extension. Support for this interpretation came from a separate individuation assay, in which the patients had to flex the shoulder, flex the elbow, and extend the wrist joints in isolation; the most common problem was unwanted flexion at the shoulder and elbow. Individuation was quantified with an individuation index, which captures how well a joint is able to move while keeping the

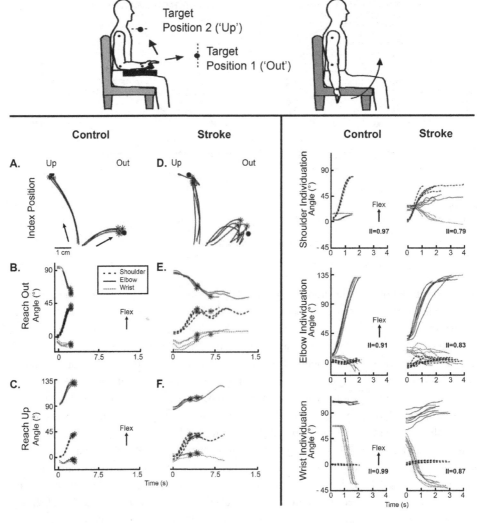

Figure 2.4
Trajectory and joint kinematics during two kinds of reach. Top Pane: Schematic of (A) reaching condition, target position 1 for reach "out" and target position 2 for reach "up" and (B) shoulder individuation task (elbow and wrist individuation tasks not shown). Left Pane: Overlaid single trials for index finger paths and associated excursions of the shoulder, elbow, and wrist joints from both reaching conditions (up and out). (A) Control index finger paths for reaches out and up. (B) Control joint angle plots during the reach out. (C) Control joint angle plots during the reach up. (D) Hemiparetic subject index finger paths. (E) Hemiparetic subject joint angle plots during the reach out. (F) Hemiparetic subject joint angle plots during the reach up. Solid black circles: target. Asterisk: index end-point position or time of index end point prior to corrective submovements. Angular excursions are graphed as follows: bold dashed traces, shoulder angular excursion; narrow solid traces, elbow angular excursion; narrow dashed traces, wrist angular excursion. Right Pane: Shoulder, elbow, and wrist individuation movements graphed over time from start of movement to maximum angular excursion in the instructed joint. For a healthy control (left) and a hemiparetic subject (right). Bold dashed traces, shoulder angular excursion; narrow solid traces, elbow angular excursion; narrow dashed traces, wrist angular excursion. Adapted from Zackowski et al. (2004).

others still. Notably, the average individuation index accounted for 50 percent of the variance in reach curvature. It is of interest that a study that took a similar approach in patients within the first two weeks of stroke found that weakness was the best predictor of the speed, accuracy, and curvature of a forward reach (Wagner et al. 2006)—that is, the opposite result to what was found in chronic patients. As we shall see, weakness and control are dissociable components of the hemiparetic deficit both cross-sectionally and longitudinally. Early after stroke, a state analogous to spinal shock likely dominates, causing weakness severe enough to mask other deficits. It would be of interest to compare kinematic performance early and late after stroke in patients matched for weakness to see if the predictors still differ.

Several important points are raised by the reaching studies described above. First, they reveal the value of careful quantitative movement analysis of hemiparesis in general and in particular of studying kinematics/kinetics and not just EMG under isometric conditions. This is because the behavioral consequences of stroke in task space cannot be inferred from abnormal muscle coactivation patterns. Second, in the unloading studies, the loss of planar workspace area cannot be attributed to weakness of the elbow extensors because none of the vertical robot-induced forces act about the horizontal axis of elbow joint rotation. If weakness were the cause of a reduced workspace area, then this reduction would be expected to be invariant across loading levels until patients are unable to produce enough shoulder abductor torque to lift the paretic arm. Similarly, weakness was not the explanation for the kinematic abnormalities in the Zackowski et al. (2004) study. Third, impaired movement is partly the result of dependence on a flexor synergy, which can only partially compensate for the loss of CST-based control. Indeed, as we shall see in chapter 6, therapeutic studies have been conducted that attempt to train away these unwanted synergies, an emphasis that brings us back to earlier approaches to the rehabilitation of hemiparesis, which focused on a return to more normal joint coordination patterns (see chapter 1).

In a recent study, Jing Xu and colleagues used a novel ergonomic keyboard that measures isometric forces produced by each finger (Xu et al. 2017). The goal was to examine the poststroke paretic hand from the acute to the chronic periods and identify abnormalities in control that cannot be attributed to weakness by titrating strength requirements. In this sense, this is the hand version of Dewald and colleagues' arm unloading task. Two kinds of behavior were tested: maximal voluntary force (MVF) and individuation. During each MVF trial, subjects were asked to depress one finger at a time with maximum strength—it did not matter if the other fingers exerted some degree of force too. For individuation, subjects were asked to depress each finger at parametrically varying degrees of sub-MVF (20, 40, 60, and 80 percent) but with the additional instruction to exert no force with the other fingers. Unsurprisingly, the patients exerted less absolute MVF than controls. The critical finding, however, was that patients showed more enslaving of passive fingers for any sub-MVF (figure 2.5). Thus, even when normalizing for

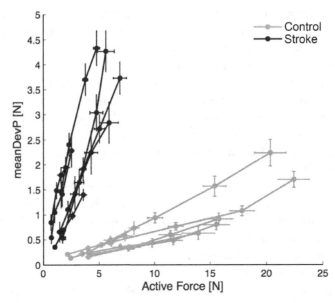

Figure 2.5
Isometric finger individuation task. Mean deviation from baseline in the passive fingers plotted against the force generated by the active finger. The higher the slope of the relationship, the worse the individuation ability. Data from seven patients and seven healthy controls. Adapted from Xu et al. (2017).

their weakness, they still had markedly less control. This demonstration of a finger flexor control abnormality that is not attributable to weakness is directly comparable to the results of Dewald and colleagues for arm reaching.

Despite the clear dissociation, for the both the proximal and distal limb, between the ability to generate large forces and perform dexterous movement, it should be stated that the distinction between strength and control blurs when it comes to low-force levels. For example, we have already discussed how fast-conducting monosynaptic corticomotor-neuronal connections in nonhuman primates have their largest effects on the intrinsic muscles of the hand and extensors of the wrist (see section 2.4). Fingers are very light and so, for example, extending them is about properly focused recruitment of motor neurons to generate the low levels of force required. In this case, control and force generation converge. Thus, the dissociation between strength and control, when observed, may be more about anatomical separation between neural systems that generate large load-bearing forces versus those that make small force adjustments.

2.7   The Contributions of the Corticospinal and Reticulospinal
Tracts to Strength and Control

The studies described above for the hand and arm show that after stroke, there is a qualitative change in the trade-off between strength and control compared to healthy subjects. In the case of the hand, an attempt to generate normal levels of force in one finger leads to excessive enslaving of the others. In the case of the arm, an attempt to lift it against gravity triggers a flexor synergy that precludes making accurate reaches. The challenge is to begin to map these behavioral changes after stroke onto the underlying anatomy and physiology of both the lesion and of the residual undamaged CNS. Dewald and colleagues have been the main proponents of the idea that the appearance of the flexor synergy in patients after stroke is the result of an incremental substitution of corticospinal by ipsilateral reticulospinal projections as force requirements increase. Indeed, there is experimental evidence that the RST may be particularly important for generating higher force muscle contractions as outputs from the PMRF facilitate the ipsilateral biceps and deltoid during reaching in monkeys (Davidson and Buford 2004).

Further support for invoking the RST as a driver of the hemiparesis phenotype comes from a recent study of mirror movements by Naveed Ejaz and colleagues. Mirror movements in the context of poststroke hemiparesis refer to involuntary movements of the fingers of the passive hand while attempting to move a single finger of the paretic hand. Mirror movements increase with paresis severity and when more force is generated by the finger of the paretic hand (Ejaz et al. 2016). In their study, Ejaz and colleagues provide evidence that mirroring cannot be attributed to either increased ipsilesional or contralesional motor cortical activation. The conclusion is that mirror movements poststroke are best explained by a subcortical mechanism, and they make a case for the RST given its known bilateral monosynaptic connections to finger flexors (Baker 2011).

Stuart Baker and colleagues have also made a convincing case for the RST as the main driver of the hemiparesis phenotype in a fascinating study in macaque monkeys (Zaaimi et al. 2012). Unilateral pyramidal tract (PT) lesions were made and then the monkeys were allowed to recover over six months. At this point, the magnitudes of synaptic responses for motor neurons innervating hand and forearm muscles were examined after stimulating either the unlesioned ipsilateral PT or the ipsilateral medial longitudinal fasciculus (MLF), which carries mainly fibers of the RST. The main finding was that, compared to unlesioned control animals, EPSP amplitude only significantly increased with RST stimulation, and this occurred in motor neurons innervating forearm flexor and intrinsic hand muscles but not in forearm extensor motor neurons (figure 2.6). Baker and colleagues point out the similarity of this pattern of RST stimulation responses to the arm paresis phenotype after stroke, which often shows

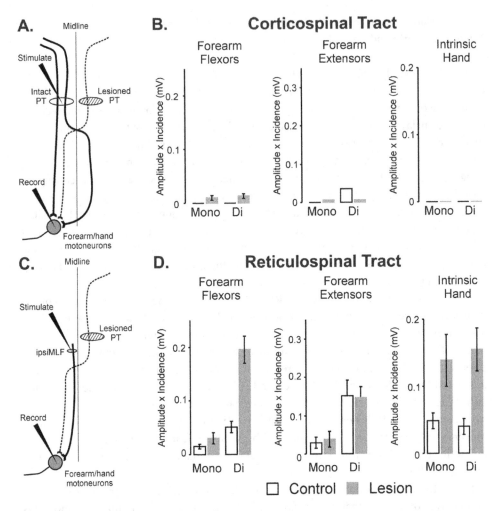

Figure 2.6
Upregulation of the ipsilateral reticulospinal tract after stroke. Responses of identified motor neurons to ipsilateral pyramidal tract stimulation. (A) Schematic showing the lesioned pyramidal tract (PT, dashed line) and potential pathways descending through the ipsilateral (intact) PT that might contribute new connections. (B) Quantitative measurements of input to motor neurons from the ipsilateral PT. Data are presented separately according to the muscle group that the motor neuron innervated: forearm flexors, forearm extensors, and intrinsic hand muscles. Light-shaded bars relate to control and dark to lesioned animals. Bars are grouped depending on if the responses are monosynaptic or disynaptic. The product of amplitude and incidence measures connection strength. There were no significant differences between control and lesioned animals. (C) Schematic showing the lesioned pyramidal tract (PT, dashed line) and potential pathways descending through the ipsilateral (intact) reticulospinal tract (medial longitudinal fasciculus [MLF]) that might contribute new connections. (D) Same quantitative measurements of input to motor neurons from the reticulospinal tract. There were significant differences between control and lesioned animals for forearm flexors and intrinsic hand muscles for both monosynaptic and disynaptic responses Adapted from Zaaimi et al. (2012).

persistent extensor weakness but preserved ability to flex all the fingers together in a power grip.

A core idea then, extending from the classic studies of Tower, Lawrence, and Kuypers to contemporary studies in humans and nonhuman primates, is that the hemiparesis phenotype is attributable to a combination of residual CST capacity and an upregulated RST. Despite the weight of evidence supporting this dualistic framework, it raises something of a puzzle. If it is indeed the case that finger enslaving, mirror movements, and flexor synergies all represent intrustion phenomena caused by RST activation and that the RST contributes force in the healthy CNS, why are these intrusions not also seen in healthy subjects when large forces are required for the arm and hand? While the answer to this question is not yet known, there are interesting clues. Finger enslaving and mirroring are indeed seen in healthy subjects but to a lesser degree (Armatas et al. 1994; Zijdewind and Kernell 2001; Sehm et al. 2009). It could be speculated that when stroke disproportionately affects the CST, which is likely to be often as it is heavily unilateral and has more focal cortical origins, then, in addition to loss of control over effectors, it also causes an imbalance in the interaction between the CST and RST. That is to say, the phylogenetically newer CST does not just exert finer control over effectors but also exerts control over older descending pathways, either in the brainstem or at the level of spinal cord interneurons.

There is evidence in the primate that the RST and the CST have largely overlapping effects on spinal interneurons (Riddle et al. 2009). Disynaptic effects on motor neurons are not normally observed with PT stimulation in macaques (Maier et al. 1997; Alstermark et al. 1999), which suggests that CST input onto spinal interneurons might instead be playing a modulatory role. Bénédicte Schepens and Trevor Drew, on the basis of a study of reaching movements in the cat, suggest that one such role of the CST is gating of RST input onto spinal interneurons (Schepens and Drew 2006). A key finding in this study in cats was that the relationship of the RST to reaching behavior was asymmetric, even though the RST projections are bilateral. This is important, as skeptics of the upregulated RST explanation of flexor synergies in the paretic limb might wonder why such synergies are not seen in both limbs. An explanation might be that the contralesional motor cortex can gate out compensatory upregulated RST effects in the unaffected limb via its normal CST. John Buford and colleagues have started to directly address the question of interactions between the RST and CST in the control of the upper limb in awake, behaving monkeys. In a very interesting recent study in which they jointly stimulated cortex and the PMRF, they found a pattern of response consistent with an indirect inhibitory effect of the cortex onto an interneuron blocking disynaptic excitation by the RST (Davidson et al. 2007).

2.8   Poststroke Resting Posture

Anyone who has interacted with patients after stroke will know that these patients frequently adopt abnormal limb postures at rest (i.e., when they are not moving). Indeed, in
the classic 1952 paper by Thomas Twitchell titled "The Restoration of Motor Function following Hemiplegia in Man," which we will discuss in depth in chapter 3, he devoted a
whole section to abnormalities in resting limb posture. The treatment approaches advocated by Berta and Karel Bobath, and Signe Brunnstrom are predicated on optimizing the
interaction between posture and voluntary movement through appropriate sensory stimulation and careful positioning of the body. The classical hemiplegic posture consists of
flexion at the fingers, wrist, and elbow, along with adduction at the hip, extension at the
knee, and plantar flexion at the ankle. Derek Denny-Brown, a neurologist from New Zea-
land, who trained with Sherrington and then spent most of his career at Harvard, carried
out extensive studies on the "hemiplegic attitude" in macaques (Denny-Brown 1964). It
turns out that Twitchell was also involved in some of this work on monkeys, and Denny-
Brown's influence is apparent in his 1952 human paper. It is of great interest to read the
work by Denny-Brown in macaques and the parallel work in patients by Twitchell, for
both historical and scientific reasons. It is yet another example, as we saw earlier between
Sherrington and Walshe, of how awareness of physiological work can be fruitful for the
generation of clinical insights. Here we will summarize some of the observations they
made about the configuration of the paretic arm at rest. Both Denny-Brown and Twitchell
noted that the hemiplegic arm posture was dependent on overall body posture. For example, if a monkey, a few days after bilateral ablation of Brodmann area 4 (primary motor
cortex), was suspended in the air by the body or pelvis or seated in a chair, the arms were
observed to be "held in a flexed adducted posture, the hands and fingers semi-flexed, and
the lower limbs, including ankles and toes, extended." In addition, the stretch reflexes at
the biceps, wrist, and finger jerks were extremely brisk. Fascinatingly, when the animal
was held upside down, the reactions were reversed (i.e., the hemiplegic posture and the
heightened reflexes switched from flexors to extensors in the arm). Similarly, Twitchell
describes how when a patient was turned over from a supine position onto his or her side,
so that the paretic limbs were upper most, the flexor posture of the arm was accentuated,
as was flexor spasticity. Thus, both in monkeys and humans, whole body posture can
modulate the abnormal resting posture of the limb.

   It can be conjectured that the flexor posture of the paretic arm is the resting version of
the flexor synergy seen during movement. The argument for this would go as follows: as
originally described by Twitchell and Denny-Brown, poststroke resting flexor posture
changes with full body orientation. Changes in body orientation trigger vestibular right-
ing and tonic neck reflexes, and part of the correcting motor response is driven by output
from the vestibular nuclei to the reticular formation, which is upregulated after stroke. Thus,

the abnormalities in arm posture at rest in patients after stroke likely represent exaggerated interaction between body-righting reflexes and upregulated brainstem descending pathways.

2.9 Abnormal Postural Control Mechanisms during Movement

While it is clear that the postural abnormalities seen in the arm at rest are telling us something important about the state of the motor system after stroke, it does not follow inevitably that abnormal resting posture will cause problems with voluntary movement. It might turn out to be true, either because they share underlying mechanisms or because they interfere with each other. Why might there be an interaction between abnormal resting posture of the arm and problems moving it voluntarily? A normal voluntary arm movement entails accurately moving from point A to point B. But it also requires being able to hold the arm still and in the right configuration at point A before starting the movement and being able to decelerate and hold still at point B after the movement is over. For example, when making a karate chop in the air, the arm moves fast and then decelerates to a precise stop. Thus, as has recently been stated by Reza Shadmehr, "Motor control is characterized by brief periods of transition between states of holding still" (Shadmehr 2017). If a voluntary movement is bookended by postures, postural abnormalities might well interfere with it. To explore this possibility, it is important, however, not to trip up on the word posture, which, somewhat confusingly, can refer to the configuration of the limbs at rest, to the end position of the arm after movement, or to the position of the whole body, for example, lying supine, sitting, or standing. *Postural control* refers to how the limb is decelerated, held still, and stabilized at the end of a movement and is contrasted with limb trajectory control, which refers to the generation of the arm's trajectory up to the hold phase. Therefore, the question being addressed really has two parts. Is there a postural control abnormality after stroke, and if so, does it relate to the abnormalities in resting posture seen in the paretic arm?

That there might be separate modes of control and separate neural substrates for moving and holding was first demonstrated in seminal work by David Robinson, David Zee, and colleagues on saccadic eye movements (Zee and Robinson 1979). Subsequent work suggests that there might be a similar separation of trajectory and postural control for other effector systems, including the head and arm—Reza Shadmehr has recently written an insightful and comprehensive review about the behavioral and neural differences between moving and holding still (Shadmehr 2017). For the arm, psychophysical experiments demonstrate the existence of a fast feed-forward trajectory controller, which shows anticipatory control of dynamics, for the reach phase of the movement and a separate feedback controller that stabilizes the arm's final position by modulating its stiffness and viscosity-like behavior. These two controllers are conjectured to combine to ensure accurate reaching movements (Scheidt and Ghez 2007; Scheidt et al. 2011; Yadav and Sainburg 2014). It is of interest that while there is evidence for the motor cortex being the trajectory controller, there is as yet no good evidence for where the postural controller is located (Shadmehr 2017). One small clue comes from recent work in the mouse, which suggests that a brainstem nucleus, the ventral part of the medullary reticular formation, may control fine paw placement at the end of a reach (Esposito et al. 2014). This is interesting because in primates, reticulospinal axons provide direct and indirect inputs to motorneurons controlling the arm and hand (Riddle et al. 2009; Riddle and Baker 2010) and so perhaps are involved in postural control.

In an elegant study of reaching movements after stroke, Robert Sainburg and colleagues (Mani et al. 2013) showed that there is a postural control abnormality that is dissociable from abnormalities in trajectory control. Patients with mild hemiparesis from either left or right hemispheric strokes were required to make planar reaching movements to targets using the contralesional paretic arm (figure 2.7). Regardless of the hand used, the hand-paths of control subjects tended to be fairly straight and directed toward the target and terminated close to the target. In general, the patients with hemiparesis made movements that were more variable and less accurate than those of control subjects. In addition, there were consistent differences in the movements of patients with left and right hemisphere damage. The patients with left hemisphere damage showed systematic and variable direction errors for all three targets, whereas the patients with right hemisphere damage made straighter movements in the direction of the targets but consistently overshot them. The bar graphs show that damage to the left but not right hemisphere resulted in significantly higher initial direction errors. Conversely, damage to the right but not left hemisphere produced higher distance errors than those of control subjects, indicating deficits in the deceleration of reaching movements, a finding consistent with earlier studies (Haaland and Harrington 1989; Winstein and Pohl 1995). Thus, there was a double dissociation between the hemisphere that was damaged and the type of movement deficit seen: while right hemispheric stroke produced deficits in the ability to stop accurately at the end of motion, left hemispheric stroke produced deficits in early trajectory features, including movement direction and movement linearity. Together, these findings support the idea that the specification of movement trajectory features and the ability to achieve and maintain stable and accurate steady-state positions at the end of motion are mediated by independent neural mechanisms. It should be made clear that this clean dissociation was seen on average in patients with mild to moderate hemiparesis. In any given patient, and as severity increases, deficits in both kinds of controller can be seen in the same arm after stroke. Thus, the primary importance of the study is the demonstration that there are abnormalities in two kinds of control in the arm after stroke. The laterality point is interesting but secondary for the current discussion.

The natural next question is to ask whether there is any relationship between abnormal postural control during voluntary movement and abnormalities in resting posture. Robert Scheidt and colleagues explicitly investigated this (Simo et al. 2013). As a first

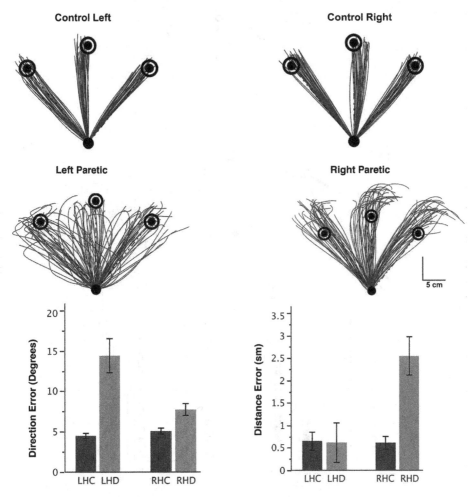

Figure 2.7
Comparison of hand paths in the left and right arms of controls and patients with mild hemiparesis (all subjects for both groups were right-handed). Left hemiparesis is associated with end-point errors, whereas right hemiparesis is associated with directional trajectory errors. LHC: Left Healthy Control; RHC: Right Healthy Control; LHD: Left Hemisphere Damage; RHD: Right Hemisphere Damage. Adapted from Mani et al. (2013).

step, a robot manipulandum was used to hold the paretic arm in place and patients instructed to relax. Steady-state posture forces were measured at the hand throughout a horizontal workspace. The robot moved the hand smoothly between the sample locations and waited five seconds before sampling to avoid velocity-dependent effects (spasticity). The average hand force vector at each spatial location was plotted and a "posture map" of hypertonic bias forces was interpolated between sample points (figure 2.8). As can be seen in figure 2.8A, the force vectors were predominantly directed into elbow flexion, which is

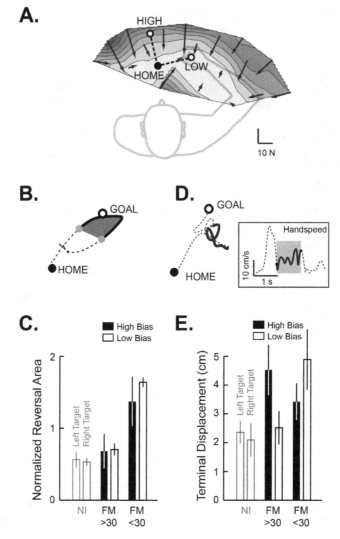

Figure 2.8

Differential impact of posture-dependent hypertonia on the control of limb trajectory and end-posture stabilization after stroke. (A) Topography of posture-dependent flexor forces acting on the hand of one stroke survivor instructed to relax as a planar robot held the hand at various sampled locations. Each stroke survivor yielded a unique "posture map." Also shown are target locations for cued movements from the home position into regions of HIGH (dark gray areas) and LOW (light gray areas) postural bias forces. (B) Schematic view of an out-and-back reversal movement produced by a stroke survivor reaching without concurrent visual feedback into a HIGH bias force region of workspace. Gray dots: trajectory locations of hand speed maxima during the outward and return phases. Shading: reversal area. (C) Average reversal areas for neurally intact (NI) control subjects, less-impaired stroke survivors (Fugl-Meyer Motor Assessment [FMA] > 30) and more impaired survivors (FMA < 30). Reversal areas were normalized by the square of the target movement distance to account for individualized differences in active range of motion and consequent differences in target distance. (D) Overhead view of an out-and-hold reach hand path produced by a stroke survivor without concurrent visual feedback. Inset: corresponding hand speed profile during stabilization against small, 3N peak-to-peak robotic force perturbations applied after hand speed dropped below 20 percent peak speed. The ability to stabilize the hand at the target was quantified using the sum of measured hand displacements for the one-second period after the onset of perturbations. Solid line: portion of the trajectory used to compute terminal hand displacement for this movement. (E) Average terminal hand displacement for the same NI control subjects, less-impaired stroke survivors (FMA > 30), and more impaired survivors (FMA < 30). Error bars: ± 1 SEM. Adapted from Simo et al. (2013).

consistent with the classic paretic flexor posture. This "high-bias force" was not uniform throughout the workspace and in fact diminished with increasing flexion. This nonuniformity in the resting bias allowed Scheidt and colleagues, in a second step, to compare reaching movements into high- and low-bias regions of the workspace. Notably, patients with moderate hemiparesis were able to stabilize the end points of their movements in a manner comparable to controls in areas of low flexor resting bias, but end-point stabilization was markedly compromised in areas of high flexor bias. In contrast, resting posture abnormalities did not have an effect on trajectory control. There are three important conclusions to be drawn from this study. First, as we saw in the study by Sainburg and colleagues, deficits in trajectory and postural control are dissociable. Second, resting posture abnormalities are also dissociable from trajectory and postural control abnormalities during movement. Third, resting postural abnormalities can interact and interfere with postural control during voluntary movement. This last point is one of the long-held core assumptions of the neurophysiological approach to rehabilitation.

2.10 Spasticity

Spasticity is confusing both with respect to terminology and phenomenology. The term has been subject to a great deal of historical and conceptual zigzagging but we are stuck with it, and so we must try to be as clear as possible when we use it. As we saw at the beginning of the chapter, the history of hemiparesis can be seen as arguments about the importance of negative and positive signs, the relative contribution of pyramidal and "extrapyramidal" pathways, the difference between the corticospinal tract and brainstem descending pathways, and, as stated by Denny-Brown, "the nature of the loss of movement in relation to each [component of spasticity]" (Denny-Brown 1964). This last quote is critical as it suggests a way to think fruitfully about spasticity—as a set of dissociable component phenomena elicited at rest, each of which may or may not causally interfere with voluntary movement. Unfortunately, this more neutral multicomponential view was (unintentionally) set back by what has come to be the most lasting definition of spasticity, coined by James Lance in the closing synopsis of the proceedings of a symposium on spasticity published in 1980. Lance stated, "Spasticity is a motor disorder characterized by velocity-dependent increase in tonic stretch reflexes ('muscle tone') with exaggerated tendon jerks, resulting from hyperexcitability of the stretch reflex, as one component of the upper motor neuron syndrome." (Lance 1980). There are many things that are unsatisfactory about this statement. First of all, it calls spasticity a motor disorder, which immediately implies that it has an impact on movement and therefore should be a target for treatment. Second, it attributes the disorder monocausally to the stretch reflex, echoing Sherrington's belief that the stretch reflex was the fundamental sensorimotor unit or movement primitive of the CNS. Third, it makes spasticity a component of the upper motor neuron syndrome rather than making it, as we prefer, the term for an overall syndrome that is itself made up of components. Fourth, it implies that increased tone and hyperexcitable tendon jerks should coexist when in fact they regularly dissociate. For example, in a study of the biceps brachii of patients with hemiparesis, an increased tendon jerk response developed progressively over a year, whereas increased tone reached a peak at one to three months and then decreased (Fellows et al. 1993).

Here we will instead use spasticity as a blanket term to refer to phenomena elicited at the bedside while the patient is at rest: velocity-dependent resistance to passive stretch (hypertonia), exaggerated phasic stretch reflexes (hyperreflexia), the clasp-knife phenomenon (a powerful inhibitory reflex elicited in the knee in humans mediated by stretch sensitive free nerve endings located in muscle interstitial tissues and aponeurotic surfaces [Ashby and Burke 1971]), clonus (oscillation in a reflex feedback loop involving distal limb musculature), and the Babinski response (toe extension is recruited into the flexor reflex withdrawal response). All of these phenomena can present together or separately after stroke. Some might argue, however, that the term *spasticity* would not be used if only one of the phenomena were present. As we have already seen, in the early twentieth century, studies of hemiparesis in humans and nonhuman animal models placed considerable emphasis on the positive or extrapyramidal components of the hemiparesis phenotype. Some current approaches to rehabilitation continue to consider spasticity reduction a way to improve voluntary movement after stroke (Das and Park 1989; Bobath 1990). The hypothesis that links spasticity to voluntary movement is that hyperexcitable reflexes will lead to intrusive activation of the stretched antagonist during contraction of the agonist, which will result in reduction in the movement's amplitude and/or speed. But does this actually happen? One study that set out to answer this question examined the ability of patients with chronic stroke to make fast planar 90° elbow flexion and extension movements against loads (to simulate the fact that the elbow usually is operating against gravity) with their paretic arm. The first finding was that the observed slow and abnormal movements made by the patients was more related to degree of maximal isometric torque that the agonist could generate than the level of passive muscle hypertonia in the antagonist. The second, critical, finding was that antagonist activity *during* the movement was at or below normal levels. This was true even in those patients with the most hypertonia at rest in that muscle (Fellows et al. 1994).

In a more recent study, Michael Ellis and colleagues sought to compare the contributions of flexor synergy-related biceps activation and reflex-mediated bicep coactivation to decreased reaching velocity due to their effects on elbow extension. Patients with chronic stroke made planar reaches with varying degrees of antigravity arm weight support, which was standardized as percentages of isometric shoulder abduction maximum strength (figure 2.9) (Ellis et al. 2017). As we have described earlier, Dewald and colleagues have shown that increased involuntary flexor torques are generated at the elbow if more force is required to lift (abduct) the arm. The key finding was that the great majority of bicep EMG activation opposing elbow extension was generated by the flexor synergy rather than reflex bicep activation. The small contribution of reflex-mediated bicep activation to extension opposition became even smaller with increasing abduction loading, which it could be argued more closely approaches real-life arm use. Thus, these two studies showed no evidence for exaggerated reflex activity in antagonists opposing movements related to voluntary activation of agonists. These results are consistent with a study of fast multijoint reaching movements in patients with hemiparesis, previously mentioned in the context of weakness (see section 2.4), which showed that performance impairments were not related to the presence of spasticity (Beer et al. 2000; Zackowski et al. 2004). The observation that stroke patients can have hypertonia at rest without movement interruption caused by exaggerated stretch reflexes in the antagonist muscles is likely because of differences of transmission through spinal pathways during movement compared to at rest. Zev Rymer and colleagues have shown in several studies that there is no evidence for increased gain of the stretch reflex in voluntarily activated spastic muscle (Powers et al. 1989). A more recent study by John Burne and colleagues has come to the same conclusion— reflex amplitude and joint mechanical resistance were linearly related to active contraction in the *same* way for patients and healthy controls (Burne et al. 2005).

The causal mechanisms for spasticity continue to be debated. The main assumption is that there are immediate and delayed effects in spinal cord segmental circuitry due to loss of descending input, particularly onto the elements that make up the stretch reflex arc: spindle efferents (gamma motor neurons), spindle afferents (Ia and group II), associated interneurons, and alpha motor neurons. In recent work, the two main ways that hypertonia has been posited to occur is either through an increase in the gain between afferent input and motor neuron output or through an increase in motor neuron excitability. Despite the multiplicity of spinal pathways that could be implicated in a gain increase, at the current time, only decreased homosynaptic depression (loss of the normal reduction in transmission at the Ia afferent-motor neuron synapse with repetitive stimulation) and increased group II spindle afferent excitation, likely due to loss of gating by descending monoaminergic pathways, are considered likely candidates. There is no evidence for increased gamma drive (Pierrot-Deseilligny and Burke 2012).

The possibility of increased motor neuron excitability being a part of the spasticity mechanism derives in part from what is known about decerebrate rigidity in the cat, which, although it differs in many ways to human spasticity (Burke, Knowles, et al. 1972), has features that remain relevant over a century after Sherrington first described it (Sherrington 1898). In particular, following brainstem transection at the mid-collicular level in the cat, there is a sharp increase in excitatory drive to motor neurons innervating antigravity muscles, giving rise to enhanced stretch reflexes in these muscles or even to spontaneous muscle activation in some cases. Although these phenomena are described as

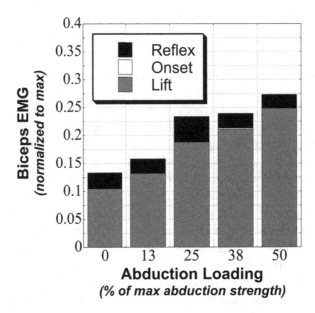

Figure 2.9
Flexor synergy-related activation of the biceps rather than reflex coactivation is a major impediment to elbow extension velocity poststroke. Biceps electromyography (EMG) was partitioned by movement kinematics (elbow angular velocity). The "Lift" window represents the EMG activity taking place when the participant lifts the arm at a specific abduction load and maintains the end point in the starting home position. This represents the "synergy-related" biceps activation prior to elbow extension movement. The "Onset" window represents the EMG during the first 25 ms of elbow joint extension. This represents the change in EMG that occurred following the onset of elbow extension but prior to the potential contribution of stretch reflex. Note that in general, there was no measurable change. The "Reflex" window represents the EMG beginning 25 ms after the onset of elbow extension up to peak elbow extension angular velocity. Reflex activation was statistically significant, but its contribution in comparison to synergy-related EMG was very small. Biceps EMG increased and elbow extension angular velocity declined (not shown) as a function of abduction loading. Adapted from Ellis et al. (2017).

being "decerebrate rigidity," the length and velocity dependence of these stretch responses are strikingly similar to those responses recorded in intact human subjects and even in spastic muscles of hemiparetic stroke survivors. Thus, some have argued that there is no reason to see this increase in stretch reflex responses as being radically different in origin between these different systems (Pierrot-Deseilligny and Burke 2012). Burne and colleagues have shown that the increased tone seen in patients at rest is attributable to incomplete relaxation, that is, they are unable to reduce background muscle contraction to normal levels (Burne et al. 2005). When background contractions were matched to normal levels, no evidence of exaggerated reflex activity or mechanical resistance was found. There was also no evidence for increased reflex activity in antagonist muscles. Thus, both controls and patients showed similar velocity-dependent effects when actively contracting, but patients also showed it at "rest" because they were not really at rest. It appears then that both spastic and intact muscles show similar dependence on initial muscle length, but spastic muscle just manifests hypertonia at shorter starting muscle lengths because of a different degree of baseline motor neuron depolarization, which presumably is the result of changes in the influence of descending brainstem pathways.

The decerebrate preparation has been examined extensively to determine which pathways are necessary to sustain the increase in motor neuron excitability and the associated increase in stretch reflex responses. Using the tonic vibration reflex (TVR) as a surrogate probe, David Burke, James Lance, and colleagues showed some years ago that selective lesions of brainstem and above were useful in determining which pathways were critical to maintaining these spinal reflex excitatory responses. They reported that the lateral vestibular nuclei and reticular pathways were both likely to be involved since lesions of these centers sharply reduced the level of motor neuron excitation in the spinal cord (Gillies et al. 1971; Burke, Andrews, et al. 1972). So why does brainstem transection increase excitation of motor neuron pools in the spinal cord? It appears that in the intact animal, there is substantial tonic inhibitory drive from cortex to many brainstem nuclei, so that when the upper brainstem is transected, this inhibition is sharply reduced, allowing brainstem nuclei to discharge in an unrestricted manner. The resulting increase in excitatory drive to spinal motor neuron pools gives rise to enhanced stretch reflexes, to heightened tendon jerks, and also to clonus, which is oscillation in a reflex feedback loop involving distal limb musculature. The precise pathways involved in this central inhibition are unclear, and we do not know, for example, whether the descending inhibition is mediated by dedicated projections from cortex or whether branches of other cortical projections mediate the inhibition.

It is worth noting, however, that the standard decerebrate cat preparation does not replicate other key features of spasticity in humans after stroke such as abnormal cutaneous reflexes (e.g., the Babinski) and the clasp-knife reflex response. However, if a mid-thoracic dorsal spinal hemisection is superimposed on the decerebrate animal, then many of the clinical features of spasticity do emerge (Rymer et al. 1979). These include a range of segmental cutaneous reflex patterns similar to the Babinski response and clasp-knife inhibition. The spinal hemisection appears to interrupt key dorsolateral reticular pathways relaying inhibition of segmental spinal circuits, while leaving excitatory vestibular projections intact. The combination of these lesions does then replicate many features of human spasticity. The combined loss of inhibition to brainstem circuits, including vestibular and reticular nuclei, coupled with release of segmental inhibitory systems controlling cutaneous and muscle afferent reflexes, does replicate most of the cardinal features of human spasticity. Thus, it is possible that spasticity is due to loss of modulation of brainstem and segmental circuits with the result that there are changes in reflex gain and changes in motor neuron excitability—unfortunately, there is no method at this time to selectively assess alpha motor neuron excitability in humans (Pierrot-Deseilligny and Burke 2012). Spasticity after stroke in humans, unlike immediate induction of decerebrate rigidity in the cat after intercollicular transection, takes place over days, weeks, and months. These different time courses in the human likely reflect loss of descending modulation of spinal circuits followed by slower plastic changes at interneuronal and motor neuronal synapses within these circuits.

Spasticity only develops in about a third of patients with hemiparesis, and severe motor impairment is seen almost as much in patients with and without spasticity (O'Dwyer et al. 1996). Nicholas O'Dwyer and colleagues examined stretch-evoked muscle activity with EMG (hyperreflexia) and resistance to passive stretch (hypertonia) in twenty-four patients recruited unselectively from rehabilitation units within one year of their stroke. There were two main findings. First, hypertonia was associated with contracture rather than reflex hyperexcitability; indeed, the latter was rare. Second, there was no relationship between hypertonia and either weakness or loss of dexterity. Thus, as interesting as the components of spasticity and their underlying physiological mechanisms are, the emerging consensus is that they are readouts of the state of the motor system after stroke rather than the cause of impairments of voluntary movement. In this view, damage to motor areas in the brain leads to the emergence of brainstem (flexor posture) and spinal cord (spasticity) phenomena at rest, but attempting to therapeutically target these emergent phenomena may be no more beneficial for voluntary movement than switching off an alarm would be for putting out a fire. Indeed, most evidence suggests that treating spasticity in the upper limb does not improve voluntary control (Sheean 2001). The time has probably come for those interested in the physiology of poststroke hemiparesis and its relation to motor control to exercise the lingering ghost of Sherrington and turn attention away from abnormalities of the stretch reflex at rest and toward a broader view of how descending pathways interact with spinal circuitry during voluntary movement. From the clinical standpoint, poststroke spasticity should take a backseat from here on out.

2.11   The Relationship between Lesion Location and Arm Paresis Phenotype after Stroke

At the current time, almost all attention on neuroanatomy in the context of hemiparesis is focused on the integrity of the CST, mainly for the purposes of predicting recovery (chapter 3). In a sense, this can be considered the current incarnation of the belief dating back to Walshe and others that pyramidotomy in monkeys could serve as a model of hemiparesis in humans. The primate physiologist Roger Lemon in a recent review has stated, "Indeed it is remarkable that clinically, capsular lesions give rise to the classical pyramidal signs of contralateral weakness, and especially those affecting the distal musculature despite the fact that CS fibers probably constitute only a few percent of the total number of ascending and descending fibers within the capsule" (Lemon 2008).

The concern that pyramidal lesions do not cause significant spasticity is mitigated by the fact that spasticity is not the relevant factor when it comes to loss of voluntary motor control in the arm and hand after stroke, as we have outlined in detail above. It is not our purpose to go through all the cortical and/or subcortical lesion locations that can cause arm paresis. In fact, our null position is that the initial paresis phenotype is similar regardless of the location where the stroke interrupts the CST (i.e., an upper extremity Fugl-Meyer Motor Assessment [FMA] score of 25 from a cortical lesion, a capsular lesion, or a pontine lesion reflects the same hemiparetic deficit). In support of our view, a recent pooled individual data systematic review found no relationship between severity of arm paresis, defined as a FMA score of 31 or less, and stroke volume, location, side, and corticospinal tract asymmetry index (obtained from diffusion-weighted imaging) (Hayward et al. 2017).

In a recent study that compared the best anatomical predictors of the paretic deficit in the arm for cortical versus subcortical lesions in patients with chronic stroke, the finding was that while damage to the CST was the best indicator of deficit in patients with subcortical stroke, this was not the case for patients with lesions involving both the CST and any of five prespecified ipsilesional motor cortical areas (Park et al. 2016). This is interesting, but it does not mean that the underlying *mechanism* of the paretic deficit is any different in the two groups because all the cortical regions chosen for the analysis contribute to the CST. Thus, this study does not in fact imply that the hemiparesis phenotype is qualitatively different for subcortical versus cortical strokes. Of the four behavioral measures, the Nine-Hole Peg test was worse in the cortical stroke group, but in this group, the lesion load was greater and so there could have been more CST involvement overall. What the study *does* tell us is that simply quantifying the degree of involvement of the CST with a subcortical CST region of interest (ROI), when there is concomitant lesion burden in motor cortical regions, will underestimate the severity of hemiparesis.

Our anatomy-agnostic position with respect to arm paresis does not preclude the possibility of additional layered-on anatomy-dependent abnormalities such as spasticity, sensory loss, differential impairments in trajectory and postural control, impaired sequencing, neglect, and apraxia. The importance of these additional abnormalities will ultimately be determined by whether they multiply on the basic paretic deficit. For example, does a patient with sensory loss for the same degree of CST damage have a worse FMA score? In contrast to our claim for invariance between where a lesion interrupts the CST and paresis phenotype, brain reorganizational processes and potential for recovery do depend more on the specifics of anatomy and will be discussed in chapter 3. It is important to appreciate that the neural correlates of motor performance at any given time are not necessarily related to the neural correlates of subsequent motor recovery.

2.12   Remote Physiological Effects: Diaschisis

The idea that a focal brain lesion can have effects on adjacent and remote uninjured brain regions has a history that dates back to the neurologist Charles-Édouard Brown-Séquard, who suggested that the functional deficit after a brain lesion could be the sum of the effects of the lesion itself and its effects on other connected undamaged brain areas (Brown-Séquard 1875). In one way this is obvious; the effects of a brain lesion, at least when it comes to the motor system, are always remote: a motor cortical stroke exerts its behavioral effect through undamaged brainstem and spinal circuits. Thus, it is not a huge leap conceptually to posit that a focal brain lesion can similarly affect other undamaged brain regions, with behavioral consequences. In 1914, Constantin von Monakow, a Russian-Swiss neurologist, formalized and gave a name to these remote effects: *diaschisis* (von Monakow 1914). von Monakow's complete definition of diaschisis, slightly paraphrasing from an excellent recent review article by Carrera and Tononi (2014), which we will return to often in this section, can be summarized as (i) presence of a focal brain lesion, (ii) sudden shock-like reduction in neuronal activity in a remote region due to loss of input, (iii) interruption of the connections between the lesion and the remote region, and (iv) reversibility of the effect on the remote region, presumably through homeostatic regulatory mechanisms.

In humans, diaschisis can be captured through measures of metabolism and cerebral blood flow (CBF) using positron emission tomography (PET) imaging. The most studied form of diaschisis has been *crossed cerebellar diaschisis*, a reduction in glucose metabolism and CBF in the cerebellar hemisphere contralateral to a cortical stroke (Baron et al. 1981). Similar subcortical depression has been seen in the ipsilateral striatum and thalamus after cortical stroke (Carmichael et al. 2004). Notably, however, these diaschisis phenomena have not been shown in any consistent way to be behaviorally relevant, that is, they do not add to the effect of the lesion itself (Carrera and Tononi 2014). Thus, despite the intuitive appeal of the concept of diaschisis, evidence for its behavioral relevance, especially in humans, is sparse at best, at least in von Monakow's conception of it. That said, after a century of relative neglect, there has been a surge of interest in characterizing and intervening upon remote effects of lesions, an interest partly attributable to the availability of new techniques for making reversible lesions in the brain in humans and animal models, and new approaches for imaging and quantifying changes in whole networks of connected regions. This recrudescence of interest has been attended by significant changes in the how diaschisis is conceptualized, with a switch in emphasis to *dis*inhibition at remote sites and a focus on *functional* diaschisis, which refers to a change in the physiological responsiveness of a remote region to a stimulus or its activation during behavior (Carrera and Tononi 2014).

Here we shall focus on a form of cortico-cortico diaschisis called *transhemispheric diaschisis* because this phenomenon may be relevant to the new interest in interhemispheric interactions—specifically the *interhemispheric competition model*, which posits that stroke leads to disinhibition of the contralesional hemisphere, which in turn inhibits the ipsilesional hemisphere and thereby contributes to the hemiparesis phenotype and impedes recovery (Ward and Cohen 2004). There is good evidence in animal models of rapid lesion-induced disinhibition of the contralesional hemisphere, which results in it showing increased task-evoked activity and increased responsiveness to stimuli, sometimes even to those that are usually only responded to by the ipsilesional hemisphere (Buchkremer-Ratzmann et al. 1996; Mohajerani et al. 2011; Imbrosci et al. 2014). The evidence for analogous hyperexcitability in the contralesional hemisphere in humans after stroke is much more questionable, with a recent meta-analysis concluding that there is no neurophysiological evidence for contralesional M1 hyperexcitability either within the first three months after stroke or in the chronic stage (beyond six months) (McDonnell and Stinear 2017). Why this difference from the animal models? The answer is not known, but the time frames that are described in the animal models (hours, days) are much shorter than those for investigations in patients (weeks, months). It is possible that with more time, homeostatic mechanisms normalize excitability in the contralesional hemisphere. The idea of contralesional excitability has become an important issue because of the interhemispheric competition model, discussed further in the next section, which refers to effects operating in the opposite direction to diaschisis, with the overactive contralesional hemisphere now inhibiting the ipsilesional hemisphere via transcallosal connections. Thus, in this model, transhemispheric diaschisis as disinhibition could make behavior worse, although indirectly and not in the way that was originally envisaged by von Monakow. This difference is the likely explanation for why the burgeoning literature on hemispheric interactions after stroke rarely makes use of the term *diaschisis* but instead invokes imbalance or competition.

Overall, it seems that diaschisis itself is not a significant contributor to severity of hemiparesis. This does not mean that diaschisis is not a useful concept or relevant phenomenon for other behaviors or for recovery from hemiparesis. One of us has shown in a rat model of focal stroke that there was overlap between a large area of cortical hypometabolism extending around the infarct and areas where reorganization took place (Carmichael et al. 2004). That areas of axonal sprouting and ipsilesional diaschisis were coextensive suggests that the latter may have an enabling effect on the former. Another important point is that brain areas that are connected are not necessarily engaged in the same behavior—a lesion in a motor area may have an acute effect on an *independent* function of a connected area (i.e., a function that does not relate to what the lesioned area does). In this scenario, one would expect some other behavior but not hemiparesis to get worse. Finally, as Carrera and Tononi delineate in their review, the concept of diaschisis has been expanded to include the effects of a lesion on changes in functional connectivity, for example, between homotopic motor areas—*connectional diaschisis*—and more globally on the connectome, captured with graph theory metrics—*connectomal diaschisis* (Carrera and Tononi 2014). Importantly, changes in connectivity between areas can occur without concomitant mean changes in activation in these areas. Thus, an acute focal lesion can lead to rapid rerouting and propagation of signal throughout latent networks, which remains true to the fast-onset notion of diaschisis. A nice example of rapid connectional diaschisis comes from an experiment done in mice (Mohajerani et al. 2011). A small stroke in forelimb somatosensory cortex led to enhancement, in under an hour, of responses of the contralesional somatosensory cortex to stimulation of either forelimb. Notably, this result was not reproduced by pharmacological lesioning with tetrodotoxin, which suggests that ischemia may have unique connectional diaschisis effects.

2.13 Remote Physiological Effects: The Interhemispheric Competition Model and the Use of Noninvasive Brain Stimulation for Treatment of Hemiparesis

As introduced above, the interhemispheric competition model posits that the hemiparesis phenotype is made worse by inhibition from the contralesional hemisphere to the ipsilesional hemisphere (i.e., the ipsilesional hemisphere is "double-disabled"). Is there any compelling evidence for this? To answer in the affirmative, it is necessary to (1) show that there is indeed increased transcallosal inhibition of the ipsilesional hemisphere in patients after stroke and then (2) that reducing it improves impairment measures of hemiparesis. To our knowledge, no single study has convincingly demonstrated these two things. In 2004, Leonardo Cohen's group at the NIH published a highly influential study of patients with chronic stroke, which reported abnormally high inhibitory drive from the contralesional to ipsilesional hemisphere prior to movement onset. In addition, the degree of release from this premovement inhibition positively correlated with motor performance measures (Murase et al. 2004). It is this paper, along with others like it (Cicinelli et al. 2003), that launched the interhemispheric competition model and subsequent attempts to treat poststroke hemiparesis by using noninvasive brain stimulation to inhibit the contralesional hemisphere and/or excite the ipsilesional hemisphere. Many small n studies ensued, using either repetitive transcranial magnetic stimulation (rTMS) or transcranial direct-current stimulation (tDCS) to excite the ipsilesional hemisphere and/or inhibit the contralesional hemisphere—and sometimes both (Fregni et al. 2005; Khedr et al. 2005; Mansur et al. 2005; Takeuchi et al. 2005; Fregni et al. 2006; Kim et al. 2006; Schlaug et al. 2008). Not surprisingly, some results were positive and others were negative. All this culminated in an ill-fated industry-sponsored (Nexstim) phase III clinical trial called NICHE (Navigated Inhibitory rTMS to Contralesional Hemisphere Trial). It was halted following the recommendation of their own Data and Safety Monitoring Board to stop due to prespecified criteria for futility being met and therefore the trial being unlikely to meet its primary end point. The trial is no longer recruiting but data have not yet been released. At the Nexstim website, the reason given is that the group that got sham stimulation over the contralesional hemisphere also did surprisingly well—so well in fact that the company states on its website, "The unexpected sham treatment data have led Nexstim to file a patent application on this novel stimulation method" (Nexstim 2016). It would be hard to make these things up.

In addition to the failure of the NICHE trial, there have been two recent Cochrane reviews on the therapeutic efficacy of poststroke noninvasive brain stimulation: one for rTMS (Hao et al. 2013) and the other for tDCS (Elsner et al. 2016). Both reviews conclude that there is no evidence that these interventions benefit upper extremity function after stroke.

Clearly, these sobering facts should give us pause. It is time to examine why so many people in the field got so excited about the promise of the interhemispheric model in the first place. Sadly, part of the answer is that this situation is all too common in neurorehabilitation— act clinically now, think scientifically later. We will begin by revisiting the study that arguably started it all—the Cohen group's highly cited paper published in 2004. We will argue that a conceptual mistake was promulgated right here at inception. The main finding in the paper was that interhemispheric inhibition (IHI), determined during a simple reaction time task using a double-pulse TMS paradigm, was similar at rest and just after the Go Signal in patients and controls but then failed to switch to facilitation just before movement onset in patients. Two results are reported: (1) a mean difference in premovement IHI at the group level and (2) premovement IHI was negatively correlated both with finger tapping speed and the Medical Research Council (MRC) scale for muscle power. The authors then conclude, "It is conceivable that modulation of such interactions might influence motor disability in these patients" (Murase et al. 2004). Unfortunately, this is problematic—two variables that are strongly correlated will not necessarily continue to covary when one of them is changed (Moreau and Conway 2014), in this case manipulating IHI with TMS. Now this is just a variant of the old saw that correlation is not causation, but what seems to have happened is that the *combination* of the correlations with the presence of a mean difference made a causal link appear more plausible. This is spurious, however, because means and correlations are statistically independent of each other. Thus, an intervention was born based on two statistical misunderstandings. We should stress that the work is interesting, but the findings alone did not merit the jump to noninvasive brain stimulation interventions, which continue apace. In addition, rodent stroke experiments directly inspired by the human model are ongoing (Dancause et al. 2015). Results from these animal experiments are and will continue to be interesting, but we suspect that they will ultimately find explanations other than hemispheric imbalance for the purported benefits of contralesional hemisphere inactivation. In support of our view, a recent study in humans by Winston Byblow and colleagues showed that contralesional cathodal tDCS in patients with mild hemiparesis reduced impairment and improved function in the paretic arm (Bradnam et al. 2012). Importantly, however, the authors concluded that the benefit was not attributable to a reduction in transcallosal inhibition because they found no correlation between the improvements and ipsilateral silent periods, which are indicative to some degree of transcallosal inhibition.

Positing that transcallosal inhibition contributes to the hemiparetic deficit implies that it should also be present in the acute and not just the chronic stroke period, as it is in the acute period when the paretic deficit is at its most severe and increased contralesional activation has been most consistently reported. Addressing precisely this issue, a recent study by Jing Xu, Pablo Celnik, and colleagues tracked premovement IHI in twenty-two patients from the first two weeks after stroke up to one year (Xu et al. 2016). Notably, premovement IHI was *normal* early after stroke and only became abnormal in the chronic

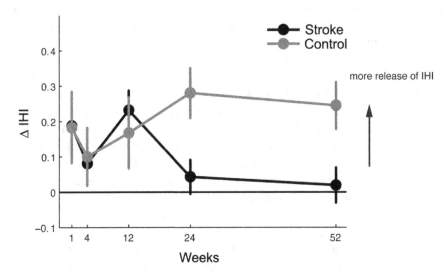

Figure 2.10
Evolution of premovement interhemispheric inhibition (IHI) for hemiparetic patients and healthy controls over a one-year period. IHI was normal early after stroke and only became abnormal in the chronic phase (week 24 onwards). Adapted from Xu et al. (2016).

stage (figure 2.10). In other words, the IHI premovement abnormality emerged as patients' hemiparesis improved, hardly what one would expect for a mechanism contributing to the paretic deficit. Cathy Stinear and colleagues have also reported that they could find no evidence for asymmetric IHI early after stroke (Stinear et al. 2014). Thus, an alternative explanation needs to be sought for the correlations between the premovement IHI abnormality and clinical variables. One possibility is that the causality runs in the opposite direction—the nature of the systems mediating recovery in the paretic arm lead to the change in premovement IHI.

There are many lessons to be learned here. A preliminary finding in a small number of patients led to a questionable model that in turn led to many subsequent low *n* studies that sought to either excite the ipsilesional hemisphere or inhibit the contralesional hemisphere. Some of these studies reported positive outcomes and others negative ones, which is precisely what is to be expected from low *n* studies, which unavoidably yield contradictory findings, overestimates of effect size, and unreproducible results (Button et al. 2013). The results of low *n* (pilot) studies are often used to gain an impression of the efficacy of an intervention, but they are in fact absolutely unsuited for this purpose. Rather than questioning the model or raising concerns about the dangers of small *n* studies, reasons given for inconclusive results thus far have focused instead on issues of heterogeneity of severity of hemiparesis, different outcome measures, and varying stimulation protocols (Di Pino et al. 2014). The implication is that the intervention will show efficacy once we get better at patient stratification, which would be a perfectly laudable conclusion if it had been based on adequately powered large studies. In the meantime, the best that we can hope for are meta-analyses of small positive and negative studies, which, as we state above, have concluded that there is no benefit from noninvasive brain stimulation in the treatment of upper extremity stroke. Thus, overall, the conclusion must be that *if*, and it remains a big if, noninvasive brain stimulation has an effect on motor impairment, it is a small effect at best and is not likely attributable to effects on IHI. It remains possible that one day we will discover, with properly conducted and powered clinical trials, that a subset of patients exists that benefits—for example, patients with very mild hemiparesis or patients very early after stroke. The possibility that brain stimulation might help at the ICF level of activity by enhancing learning of compensatory strategies will be discussed in chapter 6.

2.14 The Ipsilateral "Unaffected" Arm

In one section of his self-report on his stroke, the anatomist Alf Brodal describes how he became aware that although he had suffered a right subcortical stroke, the quality of his writing with his right hand had deteriorated (Brodal 1973). A multitude of studies have since investigated motor control abnormalities in the "unaffected" arm and hand after stroke. Here we shall first summarize what has been found with respect to motor control and strength in the arm and hand. We will then briefly touch upon apraxia as this condition is often unappreciated because it is more apparent in the ipsilateral limb, which is often not tested, and is frequently masked by paresis on the contralateral side. As we shall see, apraxia can contaminate certain clinical assessments that are considered tests of dexterity.

2.14.1 Strength

In section 2.4, we cited the Gandevia and Colebatch study on weakness in the affected upper limb early after stroke (Colebatch and Gandevia 1989), in which they tested maximum voluntary strength in proximal and distal muscles. Notably, the patients had weakness, although less, in almost all the same muscles on the ipsilateral side as they did on the paretic side, with the shoulder adductors particularly affected and only the thumb extensor being spared. Similarly, Sunderland and colleagues found reduced grip strength in the ipsilateral side in patients who were tested within one month of their stroke (Sunderland et al. 1999). In a more recent study that tracked recovery in the paretic hand, maximal voluntary contraction force for flexion in each finger was reduced in the ipsilateral hand compared to age-matched controls, especially in the first three months after stroke (Xu et al. 2017).

2.14.2   Control

In addition to weakness, control (or dexterity) has also been examined in the ipsilateral arm and hand. As we have described previously, these two aspects of hand and arm function can be dissociated after stroke. One way to isolate control deficits from the effects of weakness in the arm is to have patients perform planar reaches with full antigravity support and with friction minimized. Several studies have analyzed the kinematics of ipsilateral planar reaching movements and found that they are abnormal compared to controls (Mutha et al. 2012). Interestingly, the kind of ipsilateral deficit seen depends on the side of the stroke. Patients with left hemisphere stroke made curved movements with accurate final positions, and patients with right hemisphere damage stroke made straight movements with poor final position accuracies. These results are consistent with the hemispheric dominance hypothesis for these two kinds of control proposed by Sainburg and colleagues, which we discussed in section 2.9. Indeed, the deficits on the ipsilateral side look like subtler versions of those that we presented earlier for the contralateral side in patients with mild paresis (see figure 2.7).

Abnormalities in the ipsilateral limb suggest that both hemispheres contribute to accurate control of the upper limb, a conclusion consistent with several imaging studies in healthy subjects, although these have mainly examined individuated and sequential finger movements rather than reaching (Salmelin et al. 1995; Haslinger et al. 2002; Filippi et al. 2004; Poldrack et al. 2005; Diedrichsen et al. 2013). Another important implication of these ipsilateral results is that they suggest that arm control is hierarchical—the deficit on the contralateral paretic side is the result of loss of bihemispheric coordination of trajectory and positional controllers plus a more generic deficit caused by unilateral interruption of the CST. The ipsilateral abnormality itself reflects more than loss of one of these high-level cortical controllers, because, as we have seen above, it also manifests as weakness and loss of finger dexterity, deficits also seen with isolated subcortical strokes below the level of the corpus callosum (Noskin et al. 2008). Thus, the anatomical basis of ipsilateral deficits likely reflects a combination of (1) loss of one cortical controller or another, which may either be damaged directly or affected indirectly by being disconnected from its output targets or from its thalamic input, and (2) loss of cortical projections to brainstem nuclei that project bilaterally to the spinal cord. For example, ipsilateral responses to magnetic stimulation of the contralesional hemisphere have characteristics consistent with excitation of a corticoreticulospinal pathway (Benecke et al. 1991).

What the ipsilateral deficit is *not* likely due to is interruption of the ipsilateral CST. Anatomically, most ipsilateral CST terminations in the spinal cord do not come from ipsilateral descending fibers but come instead from contralateral CST projections that cross the midline at the segmental level via the spinal commissure and then synapse on lamina VIII (Kuypers 1964; Lemon 2008). Physiological evidence for a lack of a contribution of the ipsilateral CST to control of the ipsilateral forearm and hand comes from two very important studies by Stuart Baker and colleagues in monkeys. Their first study was conducted in healthy monkeys (Soteropoulos et al. 2011). Recording from motor neurons when stimulating the ipsilateral PT revealed no convincing evidence for either monosynaptic or disynaptic responses. This was corroborated by a lack of monosynaptic EMG responses to either ipsilateral M1 or PT stimulation and by spike-triggered averaging. Finally, the modulation of ipsilateral M1 cells that occurred when the monkey reached out and then grasped a lever with a precision grip was more correlated with weak muscle activity in the contralateral unmoving arm. In the second study, the question was asked whether the ipsilateral contribution to movement might only become apparent after recovery from a lesion (Zaaimi et al. 2012). Monkeys were given unilateral lesions of the PT and then given six months to recover. As was the case for healthy monkeys, however, there was no evidence for significant motoneuron responses with stimulation of the unlesioned ipsilateral PT. It should be emphasized that these results to do not mean that the ipsilateral hemisphere does not contribute to movement in health or disease, just that it is not likely occurring via ipsilateral projections of the CST. As stated above, ipsilateral responses to TMS after late hemispherectomy were more consistent with an effect on the corticoreticular tract (Benecke et al. 1991).

2.14.3   Apraxia

An aspect of limb control that gets surprisingly little attention in considerations of hemiparesis is apraxia, a notoriously confusing condition in terms of its terminology, its phenomenology, and its conceptual framework. Ideomotor apraxia, the most common after stroke, is a disorder of skilled actions not attributable to weakness, incoordination, or other low-level sensory or motor impairments and is present in approximately 50 percent of patients with left hemisphere stroke (Zwinkels et al. 2004). It is most often diagnosed on the basis of spatiotemporal errors in the production of object-related gesture pantomimes, both to the sight of objects and upon imitation of others (Goldenberg 2013). A critical aspect of apraxia is that it manifests bilaterally even though the lesion, nearly always in the left hemisphere, is unilateral. This means that it is more easily detected in the ipsilateral left arm because the signs are often masked by the paretic deficit in the right arm. The ipsilesional side, however, is often not examined in detail by neurologists in the setting of stroke because attention is focused on the paretic side. In addition, the erroneous idea that apraxia cannot be diagnosed in patients with aphasia, which is present in about 70 percent of patients with apraxia, is still prevalent. Apraxia has been referred to as the cognitive side of motor control (Goldenberg 2013), which raises interesting questions about the intersection of higher-order motor deficits, for example, in planning and sequencing of actions, as well as in tool use, with the execution deficit associated with classic hemiparesis. A critical and fascinating aspect of apraxia is that motor execution is intact. For example, if a patient with ideomotor apraxia has to reach and grasp a cylinder whose orientation is varied, he or she can do so without difficulty (Buxbaum et al. 2005). In a classic paper published in 1917, Albert Sidney F. Leyton and Sir Charles Scott Sherrington reported the results of their meticulous studies on the effects of motor cortex ablation in nonhuman primates, mainly in the chimpanzee (Leyton and Sherrington 1917). Here is a remarkable paragraph from that paper that speaks to the separation of the cognitive aspects of motor behavior and execution:

> A point which impressed us repeatedly was the seeming entire ignorance on the part of the animal, on its awakening from an ablation-experiment, of any disability precluding its performance of its willed acts as usual. Surprise at the failure of the limb to execute what it intended seemed the animal's mental attitude, and not merely for the first few minutes, but for many hours. It was often many hours before repeated and various failures to execute ordinary acts contributory to climbing, feeding, etc., seemed to impress gradually upon the animal that the limb was no longer to be relied upon for its usual services. The impression given us was that the forerunning idea of the action intended was present and as definitely and promptly developed as usual. All the other parts of the motor behaviour in the trains of action coming under observation seemed accurate and unimpeded except for the role, as executant, of the particular limb whose motor cortex was injured. And there seemed to be, and to persist for some time, a mental attitude of surprise at the want of fulfillment of that part of an act which had been expected to occur as usual. The surprise seemed to argue unfulfilled expectation, and defect in the motor execution rather than in the mental execution of the act, raising the question whether the function of part of the cortex ablated in such cases be not indeed infra-mental.

This insight into the modular organization of motor behavior is profound. Apraxia, along with other supra-executional aspects of motor behavior that may be affected by a stroke when the lesion extends into premotor and parietal regions, may be better characterized by examining the ipsilesional limb. The crucial point is that these deficits are also likely to present in the paretic limb and, although hard to detect initially, may begin to interact with rehabilitation and have an impact on what can be accomplished with both the affected and unaffected limbs in everyday life. An example of how higher-order cognitive deficits can contaminate tests presumed to capture lower-order paretic deficits comes from a study by Alan Sunderland and colleagues when they compared the ipsilateral deficit in patients with either right- or left-sided acute middle cerebral artery stroke (Sunderland et al. 1999). All the patients were tested within one month of their strokes. Both groups of patients had similarly reduced grip strength compared to controls. Hand dexterity was tested using the Jebsen Hand Function Test, which, for example, requires participants to turn over cards and use a teaspoon to spoon five kidneys beans off a ledge into a can. Critically, the patients with left hemisphere stroke made more errors than the patients with right hemisphere strokes, despite both groups having similar degrees of weakness. Interestingly, the error rate correlated with an imitation measure of apraxia in the patients with left hemisphere strokes. Thus, a task that was devised to test dexterity was in fact detecting apraxia. The implication of this study is that there are many potential causes of the ipsilateral limb abnormalities after stroke, ranging from apraxia to true dexterity loss to weakness. Careful choice of task and proper behavioral analysis are required to characterize the exact nature of the ipsilateral arm deficit after stroke, which is important both for gaining an understanding of the mechanisms of hemiparesis on the affected side and for patients who may come to depend on their nonparetic arm.

2.14.4   Clinical Relevance of Ipsilateral Deficits

Ipsilateral arm deficits are most severe and debilitating in patients with severe paresis, precisely those patients for whom the emphasis is on task-specific training of compensatory strategies for essential activities of daily living (ADLs) using the "unaffected" arm (Sainburg et al. 2016). Thus, the "unaffected" arm may need impairment-based therapy to ready it for compensatory therapy of the affected arm. The nature of the therapy for the "unaffected" arm will depend on whether it is apraxic, has loss of motor control, or is weak.

In the traditional literature, apraxia is claimed to have few or no real-world implications (De Renzi and Lucchelli 1988), which probably explains why it is so little looked for on examination and is not a component of the NIHSS. However, numerous recent studies show that patients with apraxia are more dependent on caregivers for ADLs and significantly less likely to return to work. Apraxia predicts dependency in the home more accurately than aphasia, memory impairment, or hemiparesis (Sundet et al. 1988; Saeki et al. 1995). Last, deficiency in producing symbolic gestures may significantly affect a patient with apraxia's ability to communicate (Borod et al. 1989), particularly given that apraxia often co-occurs with aphasia (Weiss et al. 2016).

2.15   Conclusions

Arm paresis refers to a motor disorder made up of distinct behavioral components. Abnormalities are present both at rest and during movement. Weakness and loss of motor control make separable contributions to impairment and activity limitation after stroke. Spasticity refers to a set of phenomena elicited at rest, but their contribution to the disorder of voluntary movement in the upper limb is minimal. Weakness early after stroke affects proximal and distal upper limb flexors and extensors fairly uniformly and then evolves to the more typical pattern of predominant extensor weakness, especially at the wrist and fingers. Motor control is compromised both proximally and distally, although the former may need more careful movement analysis to detect. The motor control deficit reflects the combination of loss of the unilateral CST and upregulation of brainstem pathways, particularly the bilateral corticoreticulospinal tract, the latter manifesting as the flexor synergy. Abnormal resting posture can interfere with successful voluntary movement.

Hemiparesis has a similar phenotype regardless of whether the lesion is in motor cortical areas, subcortical white matter, or the brainstem, given that interruption of the CST is the common denominator. At this time, neither diaschisis nor interhemispheric inhibitory imbalance has been convincingly shown to contribute to the paresis phenotype. It is unclear whether there are brain areas affected by the lesion that add to the paretic deficit in a way that goes beyond their contribution to the CST directly or indirectly via M1. Concomitant sensory loss, ataxia, neglect, apraxia, fatigue, or other cognitive deficits such as executive dysfunction (Mulick et al. 2015) might add to the arm deficit at the activity level, but it has to be determined empirically which ones worsen weakness or control per se. For an understanding of the paretic deficit at any given time, a modular and physiological approach still appears more fruitful than consideration of connectivity changes in local or in more global brain networks.

3

Acute Hemiparesis
Spontaneous Biological Recovery, the Effect of Training, Sensitive Periods, and Reorganization

3.1 Introduction

Recognition of the problem of stroke recovery goes back over 2,000 years to the Greek physician Hippocrates, who was particularly interested in prognosis, with the statement in relation to stroke (apoplexia) that "it is impossible to remove a strong attack of apoplexia, and not easy to remove a weak attack" (Lidell 1873). Here already one can see recognition of the difficulty of recovering from stroke and of the relationship of initial severity to likelihood to recover. In this chapter, we will discuss the time course, behavioral phenomenology, and underlying physiology and anatomy of motor recovery after stroke. As has already been explained in previous chapters, the emphasis will be on *true* recovery or *restitution* (i.e., recovery at the level of impairment or of strength and motor control), improvements that can largely be attributed to neural reorganization and repair. Compensation and recovery at the level of activities, which result from strategies and normal motor learning processes, will be discussed but mainly to contrast them with restitution and to indicate how they should not be confused with it. The natural history of arm and hand recovery in humans and nonhuman primates will be placed in the context of the phenomenon of spontaneous biological recovery. Emphasis will be placed on how the endogenous post-stroke repair processes that are thought to underlie spontaneous biological recovery evolve over time and interact with training.

3.2 The Modern History of Motor Recovery after Stroke: Hughlings Jackson, Leyton and Sherrington, Ogden and Franz

Neurologically informed debate and conjecture about motor recovery from brain lesions really only began in earnest in the second half of the nineteenth century. The story is fascinating because it arose out of the dual notions of a hierarchically organized nervous system and the importance of the reflex as the fundamental building block for the control of movement. The attraction of reflexes as a concept for these early pioneers was twofold. The first was that if the spinal cord, from which reflexes are elicited, is a continuation of the brain and brainstem, then the entire central nervous system (CNS) might be subject to the law of reflex action. The second was that reflexes remained intact after severing of the spinal cord of cats and dogs, and even after decapitation, which was described by Marshall Hall in a study of a headless turtle in the 1830s (Hall 1833). These experiments provided evidence that the capacity to generate movement remains despite loss of connection to higher levels of the motor hierarchy, which meant presumably that the same would be the case with more focal damage to higher centers. The notion of a hierarchical organization to the nervous system and its role in recovery from focal lesions was first put forward systematically by the neurologist John Hughlings Jackson and arose directly out of his theory of cerebral localization (Jackson 1884). His theory of localization was arrived at through his consideration of the temporal sequence of movements evoked by unilateral seizures, which led to him to conclude that movements were represented contiguously or somatotopically in striatum and cortex (Jackson 1881). Coupled to localization was his idea that what localized became more complex as one progressed up the neuroaxis, which was an anatomical reflection of the evolution of the CNS. This second idea, what could be called evolutionary neurology, was probably adapted from Herbert Spencer's claim that human reasoning and thought evolved from the automatic responses of lower beings (Spencer 1855). However, Hughlings Jackson posited that the brain and spinal cord make up a sensorimotor machine with reflexes as the analogs of Spencer's automatic psychological responses (Jackson 1884). The sensorimotor machine had a three-tiered organization composed of the spinal cord, then the motor cortex and basal ganglia, and finally the prefrontal cortex (York and Steinberg 2011). The intriguing idea associated with this was that each level had a full representation of the level below it, which means that the highest level had a re-re-representation of all the elements of the lowest level. In this way, Hughlings Jackson was coming up with a way for the highest cortical level to generate complex behaviors by combining simple building blocks from lower levels, the nested representational structure presumably aiding in this, and would do this by excitation and inhibition of these lower levels, which would explain why cortical lesions led to two sets of symptoms, "negative" from loss of high-level contributions and "positive" from disinhibition of lower levels. Positive symptoms could be considered evolution in reverse or "dissolution" (Jackson 1881). For example, a stretch reflex for the biceps (element) would be a localized representation for that muscle and its antagonist. Higher levels, because all the reflexes were represented, would control whole movements rather than muscles. This is surprisingly prescient of today's modern notion of a hierarchy of ever more intelligent feedback loops or control policies, going from the simple short-latency segmental stretch reflex to more context-dependent long-latency transcortical reflexes (Scott 2012).

To explain recovery, Hughlings Jackson came up with his principle of compensation and the theory of weighted ordinal representation (Jackson 1884; York and Steinberg 2011). The basic idea is that after focal damage, a remaining undamaged area at that level, because it contains a full representation of the lower level, can generate the lost movement but not quite as well because it was preferentially weighted for another body part or movement pattern. Thus, for example, recovery from plegia to a flexor synergy at the shoulder, elbow, and wrist would be explained as return of a full-limb movement but driven by progressive upregulation of a remaining representation over time that is weighted toward flexors. Hughlings Jackson's notion of recruitment of remaining undamaged brain areas that get more activated over time to mediate recovery remains the driving idea behind modern-day investigation of stroke recovery, including single-unit studies of map expansion in primates and rodents, and functional imaging studies in humans.

In chapter 2, we discussed how seminal neurophysiological studies of spinal cord reflexes and of the pyramidal tract (PT) by Sir Charles Scott Sherrington and others informed neurological thinking at the time about hemiplegia in humans. An interesting coda to this story is the fact that Sherrington himself, almost in passing, made some very interesting observations about recovery of upper limb function after motor cortical lesions, which deserve some detailed consideration here. The pertinent results were first reported in a study of a single chimpanzee in 1913 (Graham Brown and Sherrington 1913) and then much more extensively in a paper titled "Observations on the Excitable Cortex of the Chimpanzee, Orangutan, and Gorilla," published in 1917, in a section titled "Experiments on Ablation" (Leyton and Sherrington 1917). Ablative lesions (note these were not ischemic lesions but lesions created with a sharp scoop-like knife) in the arm and hand regions of the motor cortex led to, as expected, impaired movements in the contralateral limb. What was not expected was the large degree of recovery that occurred over the ensuing several weeks. As the authors state, "The main object in view being to 'localize' the motor function of each cortical point."

Despite the fact that recovery was not expected and therefore obviously not the focus of the experiments, once noted, it is impressive how closely Leyton and Sherrington paid attention to its time course. They describe in detail the time course of recovery from hemiparesis in a chimpanzee, beginning the afternoon of the day that ablative surgery of the arm representation in the motor cortex was performed (Leyton and Sherrington 1917). They note that the ape seemed lively, ate two bananas, and seemed surprised that it had a problem with its right arm. They describe the deficit as a wrist drop with imperfect movement at the elbow and shoulder, although the elbow was worse. The hand could not grasp the vertical bars of the cage, and the thumb and index finger were plegic. By the next day, the authors note that the ape no longer even tried to use its right hand, and they state, "It seems to have learnt its disability with regard to that hand and do without it." So here we have a clue to learned nonuse eighty years *avant la lettre*, although it does not seem to have prevented subsequent recovery. A month later, the shoulder and elbow seemed perfect in all movements, the wrist was almost completely recovered, and the three medial fingers were able to grasp. Notably, the index finger never recovered the ability to be moved independently. At about five weeks, the authors seem to imply that recovery had reached a plateau, and they went on to make a homologous large ablation in the other hemisphere. Critically, no recrudescence of the original deficit was seen, leading Leyton and Sherrington to conclude that the contralesional motor cortex is not responsible for mediating the original recovery. In addition, both arms recovered after the large bilateral motor cortical lesions, suggesting that either remaining areas of the cortex and/or subcortical structures can mediate recovery of the arm to a large degree. Evidence for the latter possibility was provided by their demonstration that stimulation of the perilesional cortex did not elicit elbow, wrist, or finger movements. As we discussed in chapter 2 in great detail, a potential subcortical contributor to the partially recovered hemiparesis phenotype is the reticulospinal tract.

The conclusions section of the Leyton and Sherrington paper has twenty numbered points. Here is number 15, which is the only one that deals with recovery: "Ablation of the cortex of the larger portion of an arm or leg area in gyrus centralis anterior produced heavy paresis of the corresponding limb, but this paresis quickly lessened, and the limb soon regained in large measure its volitional motility, and became successfully used for climbing, picking up food, picking the teeth, etc. Ablation further of the greater part of the arm area of the second hemisphere, after previous ablation of the greater part of that area from the other hemisphere, induced no recrudescence of paresis in the already paretic and partly recovered arm. After the double lesion considerable recovery of the volitional use of both limbs somewhat rapidly ensued, the hands, for instance, being used freely in climbing, picking up food, etc."

In the same year that Leyton and Sherrington published their paper, Robert Ogden and Shepherd Ivory Franz described similar experiments in four rhesus monkeys but made larger ablative lesions that encompassed all of primary motor cortex (M1) and dorsal premotor cortex (PMd) (Ogden and Franz 1917). This work, unlike Leyton and Sherrington's, explicitly set out to investigate recovery from hemiplegia and relate it to the human case. The relevance of this paper to what will come later in this book cannot be overemphasized and, like the Leyton and Sherrington paper, deserves a close look. In the introduction to the paper titled "On Cerebral Motor Control: The Recovery from Experimentally Induced Hemiplegia," the authors state, "The effect of suitable exercises in long-standing human hemiplegics suggested that if the paralyzed segments of an animal with an experimentally produced hemiplegia were adequately dealt with the recovery would be more rapid and more complete than if the animal were permitted to recover by itself. The suggestion was tested and the results of the observations are given in the subsequent paragraphs." Thus, the authors wanted to compare spontaneous recovery with the effects of training, with the implication, which they explicitly state later in the paper, that perhaps humans recover poorly from hemiplegia because they are not adequately treated.

We will restrict ourselves to a description of the results of two experiments in a single monkey (experiments 1 and 2), as the conclusions are the same as those derived from results in the other three monkeys. The motor cortical lesion was purposely extensive: the goal was to remove any cortex from which movements could be elicited by stimulation and to make sure to destroy all tissue deep in the central sulcus. The main result reported was that the monkey's affected arm recovered back to normal by three weeks but only if the unaffected arm was constrained and the affected arm actively trained. Here are the descriptions of both these treatment components: "The left (normal) arm was strapped to the trunk by means of a jacket so constructed that the arm could not be used for any of the important operations of feeding and climbing," and "The animal was held by a strap attached about the waist and the dorsal surface of the right hand was struck with a strap; this appeared to 'anger' the animal and he endeavored to escape from the irritation (by the use of shoulder and arm muscles), and to lift the arm and hand to grasp the irritating stimulus." So here we have the idea of constraint-induced movement therapy three-quarters of a century before it was introduced by Edward Taub and colleagues in the context of deafferentation in monkeys and stroke in humans (Taub 1980). To emphasize how well the monkey recovered, the authors describe an incident about three months out from the surgery when "the animal was observed to catch with the right hand a fly that had alighted in the monkey's cage. The coordination and quickness for the performance of this act will readily be appreciated." John Darling and colleagues in an excellent recent review (Darling et al. 2011) discuss the Ogden and Franz results and caution against their claim of *full* recovery because no overt test of precision grip was performed; presumably, the fly could have been caught with the whole hand. In the fascinating next stage of the experiment, a similar lesion was made in the opposite hemisphere. This time, no treatment was given—as the authors say, the "the animal was given the chance to recover by itself without interference"—with the result that there was very little recovery even after six months. There was no recrudescence of the original hemiparesis, again suggesting that, like in the Leyton and Sherrington experiment, the contralesional motor cortex does not contribute to recovery.

These two seminal, near century-old, papers deserve continued close study as both offer important lessons for human stroke recovery: the Leyton and Sherrington study because the pyramidal tract of the chimpanzee more closely resembles that of the human than any other nonhuman primate, as well as the Ogden and Franz study because the stark differences in recovery with and without early intervention suggest, as they themselves state, "that the continued paralysis of animals, and by analogy the persistence of motor incapacities in man, is due to lack of management rather than to a real inability." We very much agree with this conclusion—current rehabilitation is woefully underdosed in the critical early poststroke period. Overall, the results reported in these two early twentieth-century studies preempted, as we shall see, many findings and concerns in the current motor recovery research literature, including rapid early recovery of impairment to a plateau within three months, a suggestion of particular vulnerability to persistent deficits in finger dexterity, lack of evidence for a significant role of the contralesional hemisphere in recovery, the likely role of subcortical structures in the residual hemiparesis phenotype (especially when cortical lesions are large), and evidence for an important interaction between spontaneous recovery processes and concomitant training.

3.3   The Modern History of Motor Recovery after Stroke: Twitchell and Brunnstrom

In chapter 2, we considered the hemiparesis phenotype in some detail but in cross section, which necessarily rests on the assumption that there are shared features of the condition across patients and that an average phenotype can be described. In this chapter, the focus is on longitudinal changes in the phenotype as recovery proceeds. We will start with a detailed consideration of a hugely influential paper titled "The Restoration of Motor Function following Hemiplegia in Man," written by the neurologist Thomas E. Twitchell and published in 1951 (Twitchell 1951). In this single-author study, Twitchell provides meticulous observations in a series of patients with hemiparesis of varied severity from their time of hospital admission to when they reached a stable condition. The challenge of natural history studies is attested to by the fact that although 121 patients were originally included in the study, only 25 patients had follow-up at all the prespecified time points. Five of the patients in the study were plegic at stroke onset but went on to recover completely; Twitchell states that they made voluntary movements "as rapidly and as dexterously as those in the uninvolved limbs." Twitchell's overarching idea is that motor recovery progresses through a fairly regular pattern of voluntary behaviors and associated reflex changes and that the recovery process can stall at any point along this recovery trajectory. It is notable that he devotes as much attention to spasticity as to voluntary motor behavior, which is quite different from the absence of discussion of spasticity in the nonhuman primate cortical lesion studies we have described above. As we discuss in chapter 2, there are many reasons why loss of voluntary motor control and spasticity are inextricably linked in past and present considerations of hemiparesis and approaches to its rehabilitation. As we point out, there is surprisingly little evidence showing that treatment of spasticity leads to an improvement in voluntary motor control of the arm and hand. Twitchell, however, states that "it has often been assumed that if spasticity could be abolished, willed movement could more effectively be performed. . . . The present study indicates that the first movements to appear following hemiplegia are themselves facilitated stretch reflexes. The problem at that stage is not so much to abolish the spastic reaction, as to harness its diffuse hyperactivity." Later he states, "Movement was gravely defective in the presence of spasticity." Thus, Twitchell considered spasticity to have a causal role in abnormal voluntary movements after stroke.

Here we will discuss Twitchell's descriptions of the return of voluntary movement and largely ignore the concomitant reflex changes because we take the view that he was wrong to think that the evolution of reflex responses has a causal role in improvements in voluntary motor control after stroke (detailed discussion about spasticity and whether it contributes to voluntary movement can be found in section 2.10). Twitchell described the following progression of recovery stages from hemiplegia to full recovery, although he stressed that it could show considerable variation in both phenotype and timing: (1) Weak flexion at the shoulder. (2) Weak flexion at the elbow that could not be performed separately from shoulder flexion. (3) Weak flexion at the wrist and fingers but only as an addition to the shoulder-elbow flexor synergy. (4) Appearance of some extension at the shoulder and elbow at around the time that flexion is appearing in the wrist and fingers, the full extensor synergy present within about forty-eight hours of the appearance of the flexor synergy. (5) Flexion and extension of joints out of synergy. In the case of the fingers, they could be flexed together and to a lesser degree extended, but finger individuation could not be performed. (6) Flexion and extension of a single finger, beginning with the index finger. Opposition of the thumb with the index finger but with most excursion done by the thumb, and with finger flexion occurring at the metacarpophalangeal joint rather than the interphalangeal joint. (7) Peeling away of the above deficits, movements becoming more rapid and more dexterous but dependent on having vision of the limb. (8) Normal performance of tasks, with eyes either closed or opened, such as buttoning and unbuttoning a shirt and tying shoelaces. (9) Very slight residual weakness and liability to fatigue.

Some further points should be made about Twitchell's classic motor recovery study. First, his view that a patient's recovery can come to a halt at any stage along the progression is profound because it implies that two patients can be on the same recovery trajectory up to a particular point but then diverge. The crucial question is whether this divergence is attributable to differences in the recovery process itself, for example, capacity to undergo plastic change or because of a neuroanatomical constraint imposed by the lesion that cannot be overcome. Second, he delineated the timing of recovery after stroke in a table dramatically titled "The Restoration." In the five patients who recovered fully, none of whom had movement of the hand and wrist at the time of admission and three of whom had no limb movement at all, voluntary hand movement was observed within two weeks, and recovery was complete by three months. In those patients who showed incomplete recovery, return of any voluntary movement in the hand was delayed another one to two weeks. This is interesting as it again suggests that good and poor recoverers may differ with respect to the repair processes underlying recovery, for example, in their capacity to trigger post-stroke plasticity or in the extent of the damage that needs repairing. We shall have more to say about this distinction later in the chapter. Last, Twitchell did not focus primarily on weakness, although it is mentioned throughout the paper, which makes the critical point that recovery from hemiparesis is as much about regaining motor control as it is about return of strength.

In 1970, the pioneering physical therapist, Signe Brunnstrom, published a book titled *Movement Therapy in Hemiplegia: A Neurophysiological Approach*, which was based both on lectures she gave at the College of Physicians and Surgeons at Columbia University in the 1960s and a student manual she prepared in 1965 (Brunnstrom 1966). In the book, she first takes the sequence of recovery from hemiparesis initially described by Twitchell and turns it into six stages. The stages for the arm can be summarized as follows: stage 1, flaccidity; stage 2, spasticity and abnormal posture; stage 3, voluntary control of synergies and increased spasticity; stage 4, out of synergy movements and decreasing spasticity; stage

5, synergies minimize and movements begin to normalize; and stage 6, disappearance of spasticity and close to normal coordination. The second thing that Brunnstrom does, going well beyond Twitchell's observations, is that she lays out detailed training procedures. The core scientific principle underlying Brunnstrom's therapeutic approach is that synergies seen in patients are no different from the stereotypical reflexes that are present in healthy people but they have been released from control by higher centers as the result of the stroke, which she sees as the CNS reverting to an earlier developmental stage. Thus, as in the case of Twitchell, Brunnstrom considers spasticity to be central to the paresis phenotype, with no mechanistic distinction made between hypertonia and hyperreflexia at rest and the synergies seen during movement. With this assumption, it follows that spasticity needs to resolve before normal movement can occur. Furthermore, it explains Brunnstrom's reasoning with respect to therapy: as reflexes are motor responses elicited by afferent input from changes in joint configuration and body posture, then movements should be trainable by careful sculpting of sensory input through movement guidance to facilitate the transition from synergistic movement patterns to more normal ones. In essence, she falsely concluded that reflexes must be trained to respond properly again.

It is important to follow these early twists and turns about the conjectured interplay between spasticity and voluntary motor control in stroke recovery because they are the origin of confusions that persist to this day, manifest, for example, in the continued use of botulinum toxin to treat patients with arm paresis and spasticity (Das and Park 1989; Bhakta et al. 1996; Mayer 1997; Bakheit et al. 2000; Brashear et al. 2002; Rosales et al. 2016). Twitchell causally relates abnormal resting posture of the hemiparetic limb with the onset of spasticity: "However, resistance to passive movement had to reach a certain intensity before abnormalities of posture resulted." One source of confusion is the failure to distinguish between abnormal resting posture and abnormal voluntary movement, which we point out in chapter 2 are probably not causally related. Similarly, we cite the result that during movement, the heightened reflex responses that were present at rest disappear during movement (Bennett 1994; O'Dwyer et al. 1996; Bakheit et al. 2000). Patients without spasticity are not less impaired in voluntary control than those with spasticity, nor do they recover more (O'Dwyer et al. 1996).

The other confusion is the perennial one of mistaking correlation with causation. For example, Twitchell points out that proprioceptive facilitation predicts recovery of voluntary movement and that spasticity disappears in patients with full or near full recovery of voluntary movement. Similarly, Brunnstrom assumed that recovery of ability to make individual joint movements depends on the disappearance of spasticity. It may well be the case that if abnormal reflex responses of a particular type are present, then this will correlate with a particular stage of recovery of voluntary motor control. As we discuss in chapter 2, voluntary movement of a limb by cortical areas requires control not just of its muscles but also of other descending systems that project to it. To the degree that synergies and spasticity are just another manifestation of abnormal descending control, then they would be expected to covary with impaired voluntary movement. The critical implication of this view is that treating spasticity will not improve voluntary movement. That is to say, a pretherapeutic correlation does not imply that there will be therapeutic covariation (Moreau and Conway 2014). Twitchell, perhaps not fully cognizant of this pitfall, stated that his results provide "a rationale for proprioceptive and contactual exercises in the re-training of movement." This conceptual mistake continues to influence stroke rehabilitation over half a century later.

3.4 The Natural History of Recovery of Hand and Arm Impairment

It has long been established that initial severity of upper limb paresis is the best predictor of good functional outcome and that almost all recovery occurs within the first three months after stroke (Bonita and Beaglehole 1988; Duncan et al. 1992). As we shall see, however, there is a considerable amount of complexity hiding in this simple statement. Given that the focus in this book is on true motor recovery or *restitution*, which is best captured by clinical impairment scales and quantitative measures of motor control, we will mainly discuss those studies that longitudinally tracked these types of measure. Pamela Duncan and colleagues conducted the classic longitudinal study of recovery at the level of impairment using the upper extremity Fugl-Meyer Motor Assessment (FMA) (Duncan et al. 1992). They made the case for tracking recovery with the FMA rather than activities of daily living (ADL) measures because the latter can be contaminated by compensatory strategies. Fugl-Meyer and colleagues made exactly this point when they devised the FMA (Fugl-Meyer et al. 1975). Duncan and colleagues also made the argument that many studies prior to theirs had drawn patients from rehabilitation facilities and thereby had missed early spontaneous biological recovery. The rationale of the Duncan et al. study in many ways exemplifies the ethos of this book, and it stands as a laudable exception to the trend, which we outline in chapter 1, toward an emphasis on the activity level of the International Classification of Functioning, Disability and Health (ICF) and motor learning concepts, and away from spontaneous recovery and notions of repair. The goal of the study was to identify the earliest measures that would predict the level of motor recovery at six months. In total, 104 patients were assessed within twenty-four hours of admission and at 5, 30, 90, and 180 days after admission. The main finding was that the FMA at thirty days explained 86 percent of the variance at six months. The authors stated,

> The most dramatic recovery in motor function occurred over the first 30 days, regardless of the initial severity of the stroke. However, the moderate and most severe stroke patients continued to experience some recovery for 30–90 days.

The authors then report that 37 percent of the patients with severe hemiparesis at baseline recovered to at least moderate levels, if not better, at six months. All the patients with mild to moderate hemiparesis at baseline were rated as independent in ADLs at six months.

Other longitudinal studies of arm paresis have reported similar findings. For example, the Copenhagen Stroke Study tracked upper limb paresis using the Scandinavian Stroke Scale subscores for the hand and arm, which mainly assess strength but also some control, and found that the best possible upper extremity function was achieved by 80 percent of the patients within three weeks after stroke onset: those with mild paresis reached their best possible level within three to six weeks and those with severe paresis within six to eleven weeks (Nakayama et al. 1994; Jørgensen et al. 1999). Full upper extremity (UE) function was achieved by 79 percent of patients with mild UE paresis and only by 18 percent of patients with severe UE paresis. To summarize what was found in these two studies: for all levels of initial severity, most recovery from hemiparesis occurred in the first thirty days, although a small amount of recovery extended beyond this period for severe paresis. All patients with mild to moderate hemiparesis recovered substantially, but only a subset of patients with severe hemiparesis did. As we shall see, considered in retrospect, results such as these were hiding an important clue about spontaneous recovery from impairment.

3.5   The Proportional Recovery Rule

In the longitudinal study of recovery discussed above by Duncan and colleagues, the main result was that the FMA at thirty days explained 86 percent of the variance at six months. As the authors themselves realized, this is because almost all recovery is over by thirty days (i.e., there is high correlation precisely because nothing much else happens to the FMA from thirty days onward). In contrast, the FMA at baseline explained only 53 percent of the variance at thirty days. Thus, for the period within which almost all recovery occurs, about half of the variance was unexplained. Importantly, unexplained variance is not necessarily noise and may in fact contain informative structure, especially when measurement uncertainty is low, as is the case with the FMA (Gladstone et al. 2002; See et al. 2013). Regardless of its strength, a correlation between baseline FMA and the FMA at thirty days is not informative about recovery per se, because recovery is best conceptualized as the biological process that leads to a change (or delta) in the FMA from baseline to thirty days later. The end point or outcome provides a measure of the *result* of the recovery process but not of the process itself. By analogy, a skin laceration and a graze may both fully heal, but the laceration engages the biological mechanisms of full-thickness wound healing, whereas a graze activates superficial epithelial repair.

In 2008, in a paper reporting on a prospective study of spontaneous motor recovery after stroke (Prabhakaran et al. 2008), we defined biological recovery as the *change* in impairment in the FMA from the first twenty-four to seventy-two hours poststroke ($FMA_{initial}$) to three months later ($FMA_{3\,months}$). Initial impairment was defined as $FMA_{ii} = 66 - FMA_{initial}$, where 66 is the maximum score for the upper extremity FMA. We found that up to a high value of FMA_{ii}, ΔFMA was approximated by the proportional recovery relationship:

$$\Delta FMA = FMA_{3\ month} - FMA_{initial} = \beta FMA_{ii}$$

with $\beta = 0.70$. For example, if a patient has $FMA_{initial} = 46$, then he or she will have a ΔFMA of $0.70 \times (66 - 46)$, which equals 14, and so $FMA_{3\ months} = 60$. At higher values of FMA_{ii} (more severe paresis), this relationship failed quantitatively, with a proportion of patients with severe FMA_{ii} showing a much smaller ΔFMA that would be expected from proportional recovery. This led us to distinguish between two populations of patients, one that showed proportional recovery, which encompassed all patients with mild and moderate hemiparesis and about 50 percent of patients with severe hemiparesis, and one whose patients did not show proportional recovery, consisting of the other 50 percent of patients with severe hemiparesis. In a larger cohort, Gert Kwakkel and colleagues corroborated both the proportional recovery rule, with a $\beta = 0.78$, and the finding that a subset of patients with severe hemiparesis fails to follow the rule (figure 3.1). They called the former "fitters" and the latter "non-fitters" (Winters et al. 2015).

In 2011, we used a cohort of sixty-four patients to estimate the FMA_{ii} demarcation, that is, the level of initial severity at which patients bifurcate into proportional and nonproportional recoverers—fitters/nonfitters (Zarahn et al. 2011). To do this, R_j, the cumulative correlation coefficient between ΔFMA and FMA_{ii}, was computed using the lowest value of FMA_{ii} up to $FMA_{ii,j}$ and plotted against $FMA_{ii,j}$, where j is an index of the ordered FMA_{ii} values. That is to say, the correlation was computed beginning with mild patients, and then more affected patients were added cumulatively. If proportional recovery were true throughout the entire range of FMA_{ii}, the resulting R_j versus $FMA_{ii,j}$ plot would tend to increase and then stabilize at some asymptotic value, as $FMA_{ii,j}$ increased. If, however, a level of $FMA_{ii,j}$ was reached at which patient behavior dichotomizes, then the asymptote would begin to destabilize at some value of FMA_{ii}. Figure 3.2 shows the latter scenario, with a breakdown of proportional recovery beginning at an initial FMA of 10 and below. It is important to state that the fact that R_j was consistently high for patients with mild and moderate paresis, not excepting the low values when the cumulative n was still low, corroborates the claim that proportional recovery is indeed an excellent predictor for the mild to moderate group and alleviates any concern that the high correlation coefficient is artificially created by removal of outliers (Kwah and Herbert 2016).

There are two important lessons to be learned from the existence of the proportional recovery rule and the situations under which it fails: (1) the high correlation between ΔFMA and FMA_{ii} implies that rehabilitation received between these two time points is having no impact on recovery from impairment. That is to say, proportional recovery is capturing spontaneous biological recovery processes that are not impinged upon by existing rehabilitation approaches. We make this strong statement because there is great heterogeneity in the care given to patients, and if this care were having an impact on impairment, then this would dilute the strength of the correlation. (2) There is something

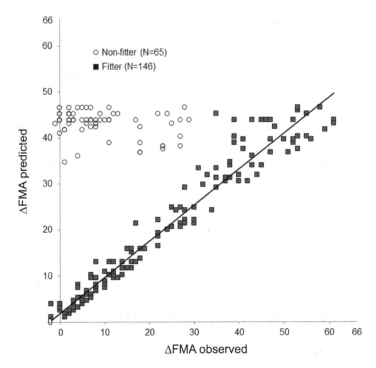

Figure 3.1
The proportional recovery rule. Predicted versus observed ΔFMA. The squares represent fitters and the circles nonfitters. The line represents the least squares regression for the fitters with slope of 0.78. FMA: upper extremity Fugl-Meyer Motor Assessment. Adapted from Winters et al. (2015).

categorically different about patients with severe hemiparesis who do show proportional recovery and those who do not. Recent studies have begun to address both (1) and (2), but before we discuss them, it is necessary to examine the proportional recovery rule itself more closely to make sure that it is capturing true biological processes and is not an artifact created by the mathematical form it takes, outlier exclusion, or the FMA measure itself.

There is always a concern that one might create spurious correlations when the same variable is correlated with itself plus another variable. In this case, the correlation is between $(FMA_{3months} - FMA_{initial})$ and $FMA_{initial}$ (the conversion of $FMA_{initial}$ to FMA_{ii} is not shown because it does not change the argument). The presence of measurement error ε on both $FMA_{3\,months}$ and $FMA_{initial}$ leads to:

$$\Delta FMA = (FMA_{3months} + \varepsilon_{3months}) - (FMA_{initial} + \varepsilon_{initial})$$

which would be correlated with:

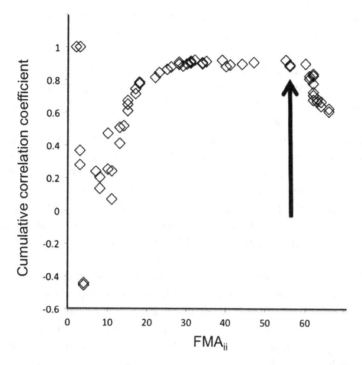

Figure 3.2
Breakdown of the proportional recovery rule at high severity levels of motor impairment as measured by the
FMA. The cumulative correlation coefficient between ΔFMA and initial impairment (FMA_{ii}) is plotted versus
FMA_{ii}. The latter defined as $FMA_{ii} = 66 - FMA_{initial}$. The arrow marks the severity threshold ($FMA_{ii} = 56$)
just before where the proportional recovery rule starts to fail. FMA: upper extremity Fugl-Meyer Motor
Assessment. Adapted from Zarahn et al. (2011).

$$FMA_{initial} + \varepsilon_{initial}$$

Note that initial measurement error ($\varepsilon_{initial}$) is present in both terms and therefore could
induce positive correlations between $FMA_{initial}$ and ΔFMA, even when there is no true
correlation. This spurious positive correlation, however, will be low if the FMA has low
measurement noise, which it does as it has high reliability (Duncan et al. 1983; Gladstone
et al. 2002; Prabhakaran et al. 2008), and when the true variability in $FMA_{initial}$ is large.
In addition, given low ε, one can interpret ΔFMA as true biological change. Thus, there
is good reason to believe that the high positive correlation between $FMA_{initial}$ and ΔFMA
is real.

Another concern with proportional recovery is the possibility of it being an artifact of
ceiling effects—namely, that it is the result of including patients with minimal to low impairment, for whom ΔFMA necessarily has to be smaller because they are already close
to maximal performance. This is not the case, however, as such effects have not been

observed; substantial room for improvement remains for all but the least affected patients. Moreover, the proportional recovery relationship was identified just as strongly in subjects with low $FMA_{initial}$ (moderate to severe impairment), who therefore had plenty of room to improve (Prabhakaran et al. 2008). A related concern is that the measurement properties of the FMA itself transform a latent linear recovery relationship into a proportional one. For example, a change from $FMA_{initial} = 56$ to $FMA_{3\ months} = 61$ ($\Delta FMA = 5$) may need as much plasticity as a change from $FMA_{initial} = 36$ to $FMA_{3\ months} = 51$ ($\Delta FMA = 15$). If this were the case, then the amount of true recovery is constant rather than proportional to initial impairment. This concern, however, even if it were true, does not invalidate proportional recovery as a rule for predicting changes in FMA. It also would provide an explanation of how the proportional relationship may arise whereby the same or more change in substrate is required for ever-smaller changes in performance. That the proportional recovery rule for paresis has now been shown to apply to aphasia (Lazar et al. 2010) and perhaps to neglect (Winters et al. 2015) makes it less likely that it is an idiosyncrasy related to the FMA.

Proportional recovery is a particular form that spontaneous biological recovery appears to take, but this by no means had to be the case. If the relationship relating recovery (R) to initial impairment (I) is $R = mI + b$, where m and b are the free parameters estimated by regression, then proportional recovery is the special case when $b = 0$. This is strictly an empirical question; spontaneous recovery could be driven by a process for which b takes on a nonzero value, in which case recovery would not be proportional (in our original paper, however, we found no evidence for a significant b term; Prabhakaran et al. 2008). For the same reason, exclusion of outliers could not itself cause a proportional relationship but only improve the regression fit to $R = mI + b$. It should be added that m does not have to be 0.7; it could take on any value, and recovery would remain proportional as long as $b = 0$. To the degree that a scale other than the FMA would change m does not mean it would change the zero intercept.

We can now return to the original study by Duncan et al. and see why they obtained the results they did (Duncan et al. 1992). All the patients with mild to moderate hemiparesis showed substantial recovery, which is because proportional recovery operates in all these patients. Recall also that the FMA at baseline accounted for only 53 percent of the variance at thirty days and that 37 percent of patients with severe initial hemiparesis made good recoveries. These results are to be expected if some severe patients show proportional recovery and others do not. Proportional recovery suggests that the failure to account for more of the variance at thirty days was not due to biological or measurement error but to the presence of distinct subpopulations of recoverers and non-recoverers (Prabhakaran et al. 2008; Winters et al. 2015).

As we have pointed out previously, the high quality of the fit for the proportional recovery rule for patients with mild to moderate hemiparesis between the first few days after stroke and at three months implies that intervening rehabilitation is not having an influence on spontaneous recovery from impairment. This was shown to be the case empirically in a recent study by Winston Byblow and colleagues, who compared how well the proportional recovery rule predicted the upper limb FMA in two groups (Byblow et al. 2015). One group received an extra thirty minutes of upper limb therapy five days per week for four weeks between two and six weeks poststroke. The other group essentially received regular therapy. The result was that the proportional recovery rule held equally well for both groups, which is exactly what one would expect if spontaneous recovery were impervious to alteration by regular rehabilitation.

The natural next question is whether there is a way to predict which patients with severe hemiparesis will go on to show proportional recovery and which will not. Two main forms of contribution to recovery after stroke can be envisaged. The first is availability of residual descending pathways, such as the corticospinal tract (CST) and the corticoreticulospinal tract, and the second is reorganization in cortical and subcortical areas to facilitate the use of residual descending pathways. In recent years, the emphasis has been mainly on using either single-pulse transcranial magnetic stimulation (TMS) or diffusion tensor imaging (DTI) to determine the residual integrity of the CST after stroke (Stinear et al. 2007; Summers et al. 2007; Schulz et al. 2012; Byblow et al. 2015). In the case of TMS, the approach has been to ascertain the predictive value of being able to elicit a motor evoked potential (MEP) early after stroke before the patient can make a voluntary movement. For DTI, CST injury is usually quantified by the loss of various white matter water diffusion metrics such as fractional anisotropy (FA) and axial diffusivity (AD). It should be noted that there are concerns about how reliable these DTI metrics are as predictive biomarkers in the acute stroke period (first week) because of edema and delayed Wallerian degeneration (Kim and Winstein 2016). Here we shall discuss those few studies that have specifically looked at early prediction of upper limb FMA in patients with severe hemiparesis, using either TMS or DTI.

In a study already mentioned above, Winston Byblow and colleagues specifically set out to distinguish between those patients with severe hemiparesis who would show proportional recovery and those who would not (Byblow et al. 2015). They obtained the upper limb FMA in a cohort of ninety-three patients at two, six, twelve, and twenty-six days after a first-time ischemic stroke. MEPs and resting motor threshold (RMT) were recorded from the extensor carpi radialis on the paretic side. FA was measured within the posterior limb of each internal capsule and used to quantify structural integrity of the corticospinal tract as an asymmetry index (AI). The main finding was that patients with detectable MEPs at two weeks all showed proportional recovery of 0.7; this was even true in patients with initial FMA scores as low as 5. Conversely, patients without MEPs and with FA AI less than 0.15 made a small but variable recovery, whereas those with FA AI more than 0.15 made essentially no recovery. Thus, this study was indeed able to use the presence or absence of an MEP at two weeks to predict those patients with severe hemiparesis who would show proportional recovery (figure 3.3). DTI only distinguished poor recoverers from complete non-recoverers. A final interesting finding was that patients' RMT also showed proportional recovery—with the recovered RMT on the paretic side reaching its stable value at twelve weeks, the same time they reached their predicted proportional level of impairment. This is an important result as it suggests that there is an underlying biological process that leads to the proportional recovery rule rather than an idiosyncrasy of the FMA.

A study by Wuwei Feng and colleagues sought to determine whether a measure of CST integrity obtained early after stroke (within two to seven days) would correlate with the upper limb FMA at three months better than initial FMA (Feng et al. 2015). Instead of FA, they calculated a weighted CST lesion load (wCST-LL) for each patient by overlaying the patient's diffusion-weighted imaging (DWI)–based lesion map with a probabilistic CST derived from DTI scans of age-matched healthy subjects. There were four main results. First, for patients with nonsevere arm paresis (FMA > 10), wCST-LL correlated no better with the FMA at three months than did initial FMA. This is to be expected because initial impairment is going to be largely dependent on the degree of CST damage, so they are essentially measuring the same thing. Second, the nonsevere patients obeyed proportional recovery. Third, for patients with severe hemiparesis (FMA < 10), the FMA at three months correlated better with wCST-LL than initial FMA; we shall discuss this result further below. Finally, the authors suggested that beyond a specific CST lesion load, patients no longer showed meaningful (proportional) recovery. This lesion load threshold could perhaps play a similar role to the presence or absence of an MEP in the study by Byblow and colleagues. Unfortunately, in a follow-up study, the prespecified lesion-load threshold, when feed-forward applied to a second cohort, failed to distinguish between recoverers and non-recoverers (Doughty et al. 2016), which suggests that physiological measures will consistently prove superior to anatomical measures.

It might seem puzzling at first that a measure of CST integrity in patients with severe hemiparesis correlates with both the initial FMA and the three-month FMA, but the correlation is higher with the three-month than the initial FMA (the third finding in the study above). To solve this puzzle, we first need to deal with a paradox that arises when measures of CST integrity are used to predict recovery: consider a patient with severe hemiparesis, presumably due to extensive damage to the CST, who undergoes proportional recovery; in absolute terms, the patient will have a large ΔFMA. But this large ΔFMA cannot also be attributed to a measure of CST integrity because any such measure will be low—the patient had severe hemiparesis. A way out of this paradox is to posit that the relationship between CST integrity and initial FMA is noisy and saturates; the function can be considered sigmoidal. That is to say, after a certain degree of severe CST damage, the initial FMA takes on the same low value—it hits a floor. The additional damage is to latent descending pathways and does not contribute further to the initial

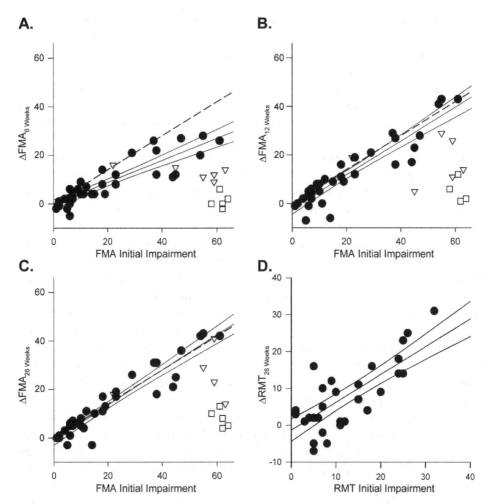

Figure 3.3

Expression of proportional recovery from impairment depends on corticospinal tract integrity. Plots of
ΔFMA against FMA initial impairment. (A–C) MEP+ patients (filled symbols) make a proportional recovery,
while MEP+ patients (open symbols) do not. For MEP+ patients, recovery is incomplete at six weeks and
complete at twelve weeks. For MEP– patients, recovery is more variable and slower for those with FA < 0.15
(triangle) compared to MEP+ patients. There is essentially no recovery from UE impairment for patients
without MEPs and FA > 0.15 (squares). (D) The dynamics of ΔFMA and ΔRMT are similar (only twenty-six
with data shown). By definition, RMT is only known for MEP+ patients. FA = fractional anisotropy; FMA =
Fugl-Meyer Motor Assessment; MEP = motor evoked potential. Adapted from Byblow et al. (2015).

impairment, but these latent pathways are relevant to future recovery. This model can now solve our puzzle above: a measure of CST lesion load can capture damage to *both* active and latent descending pathways and therefore would correlate better with the FMA at three months than the initial FMA. Similarly, two patients with the same degree of severe paresis can diverge with regard to proportional recovery because the non-recoverer had damage to latent descending pathways compared with the recoverer. The presence of early MEPs and decreasing RMT is consistent with the presence of latent descending pathways and their gradual coming online, respectively.

As we have been discussing, an interesting feature about the positive predictive value of TMS-evoked MEPs in patients with initially severe hemiparesis is that it is observed to act at a delay, sometimes up to weeks, such that patients can be MEP positive while still having their recovery ahead of them (Swayne et al. 2008; van Kuijk et al. 2009). This delay suggests that other processes, perhaps cortical reorganization, are required to make effective use of residual but latent descending pathways. Those patients with severe hemiparesis who undergo such reorganization might show proportional recovery. This possibility was directly addressed in a study by Eric Zarahn, John Krakauer, and colleagues, who tested to see whether functional magnetic resonance imaging (fMRI) performed within the first two days after stroke might distinguish recoverers from non-recoverers (Zarahn et al. 2011). In the scanner, patients were asked to attempt hand closure, even if they could not do so in the case of plegia. An fMRI-based scalar measurement was obtained by converting each patient's task-based activation map into degree of expression of a whole-brain multivariate recovery pattern (obtained by correlating all the other patients' activation maps with their ΔFMA). This fMRI-based predictor was then combined in a regression with initial impairment to attempt better prediction than proportional recovery in patients with severe paresis (defined as a FMA of 10 or below). The result was a nonsignificant increase in predictive explanation from 16 percent to 47 percent of the total sum of squares of ΔFMA explained in patients with severe hemiparesis. Even though this result was negative, the authors provide convincing arguments as to why this might be a false negative: there was a significant correlation between patients' expression of the fMRI recovery pattern and ΔFMA, the predictive model they applied was quite conservative, and the study was underpowered. Overall, the studies discussed in this section are intriguing because they suggest that spontaneous recovery, expressed as proportional recovery, arises as a result of an evolving interaction between remaining cortical motor areas and residual descending pathways.

3.6   The Recovery of Motor Control in the Arm and Hand:
Kinematic and Kinetic Measurement

As we have discussed in detail above, longitudinal studies of the paretic upper limb show that recovery from impairment is largely complete at three months (Skilbeck et al. 1983; Duncan et al. 1992; Nakayama et al. 1994; Jørgensen et al. 1999; Duncan et al. 2000; Kwakkel et al. 2006), whereas the time course of recovery of ADLs can extend further to six months (Skilbeck et al. 1983; Jørgensen et al. 1995; Jørgensen et al. 1999; Duncan et al. 2000). It is readily apparent that the precise time course of recovery depends on the outcome measure chosen, which is attributable to the fact that different processes underlie improvements in them. For example, activity limitation assessments, such as the Action Research Arm Test (ARAT), can improve via restitution or compensation (see chapter 1) (Roby-Brami et al. 2003; Yozbatiran et al. 2008; Levin et al. 2009; Krakauer et al. 2012). The Fugl-Meyer assessment of the upper extremity (FMA), which measures impairment, is largely immune to compensatory strategies but has significant antigravity strength requirements (Fugl-Meyer et al. 1975; Gladstone et al. 2002), which means it cannot fully disambiguate improvements in motor control from those in strength. Kinematic and kinetic analysis, combined with specially designed tasks that minimize strength requirements and prevent use of compensatory strategies, provides a way to properly isolate and quantify recovery of motor control (Beer et al. 2007; Krebs et al. 2014; Cortés et al. 2017).

Before focusing on those kinematic studies that sought to isolate and characterize the time course of recovery of motor control, mention should be made of how kinematic analysis can also be used to provide finer grained detail about recovery of ADLs. One study placed retro-reflective markers on the hand, arm, trunk, and face to analyze patients while they reached to pick up a glass of water, drank from it, and then put it back down (Alt Murphy et al. 2012). The goal of the study was to characterize those kinematic changes that could be deemed clinically meaningful as determined by their association with accompanying changes in the ARAT. This is interesting since very few studies have directly compared clinical scales with kinematic assessments, although the goal was not to separate recovery of motor control from compensatory strategies. Indeed, the authors state, "We speculate that trunk displacement reflects primarily the component of compensation, movement smoothness the recovery, and movement time both compensation and recovery."

Kinematic analysis of functional tasks can also be used to explicitly capture the transition from compensatory strategies to more normal movements as patients recover. Surprisingly, there seems to be only one published study that has done this, and it was a case report (van Kordelaar et al. 2012). The patient was a forty-one-year-old man who had a left lacunar infarct causing paresis in his dominant right arm. He was assessed with the FMA, the ARAT, and an instrumented three-dimensional reach and grasp task at weeks 1, 3, 5, 8, 12, and 26 poststroke. The task required that he sit at a table, reach out and grasp a block, and then move it across his midline to another position on the table (figure 3.4).

Motion sensors were attached to the thorax, scapula, upper arm, lower arm, hand, thumb, and index finger. At one week, the patient's FMA was 23 and ARAT was 3. These reached their ceiling values of 66 and 57, respectively, at eight weeks, whereas certain kinematic measures were more sensitive than the clinical scales and continued to show improvements beyond eight weeks. For example, elbow angle and peak hand speed also reached a plateau at eight weeks, whereas forward trunk displacement and movement time continued to improve and approach the values of the control subject at twenty-six weeks. Thus, as measures of motor control recovered, there was concomitant reduction in some compensatory responses. It should be emphasized, however, that although certain compensatory strategies can diminish as motor control improves, others might persist and be refined through learning and then combined with residual motor control capacity. The concern that premature adoption of compensatory strategies may impede a patient's full potential for recovery of motor control early after stroke could be addressed by using the kind of kinematic analysis described here to monitor rehabilitation of ADLs (Levin et al. 2009).

In a follow-up to their case study described above, Gert Kwakkel's group used the same task, motion sensor placement, and assessment time points to test a cohort of forty-four patients as they recovered from hemiparesis (van Kordelaar et al. 2014). The focus of this second kinematic study, however, was on recovery of motor control, specifically the smoothness of both the initial movement of the hand from reach to grasp and of grasp aperture. The interesting finding was that smoothness for both components reached a plateau at eight weeks (figure 3.4). The authors concluded that this represented improvement in motor control, rather than compensation, and was mostly explained by spontaneous biological recovery.

Juan Camilo Cortés, Tomoko Kitago, and colleagues tracked recovery in eighteen patients (Cortés et al. 2017), beginning early after stroke, using a two-dimensional reaching task that allows precise measurement of reaching kinematics, clamping of movement speed, separation of motor control from antigravity strength requirements, and minimization of compensation (Kitago et al. 2013; Kitago et al. 2015; Goldsmith and Kitago 2016). The time course of motor recovery assayed with the planar reaching task was compared to recovery time courses for impairment using the FMA, activity using the ARAT, and strength using dynamometry of the biceps. Assessments were made 1.5, 5, 14, 27, and 54 weeks poststroke. The quality of reaching movements was characterized using functional principal components analysis (FPCA), a generalization of traditional PCA to kinematic time-series data. This technique dramatically reduces the dimension of the analysis problem while retaining the major patterns that differ across movements (Kitago et al. 2015; Goldsmith and Kitago 2016; Cortés et al. 2017). The approach also precludes *a priori* choice of specific kinematic variables, such as directional error, smoothness, or end-point accuracy, which can bias the analysis toward specific components of motor control and also lead to a multiple-comparisons problem. Instead, FPCA compares distributions of movements at a global level and is sensitive to changes in overall movement quality. It is

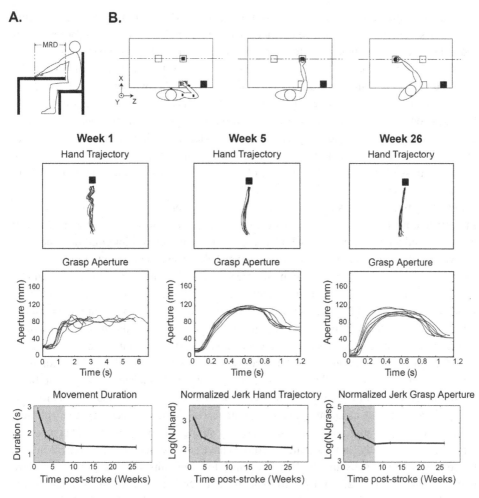

Figure 3.4
Motor recovery of the upper-extremity after stroke measured with three-dimensional kinematics. Top Pane:
(A) Determination of the MRD (maximum reaching distance). (B) Illustration of task. During task execution,
the participant starts in the initial position (left panel). Then the participant reaches for the block (small black
square) at the block position (middle panel) and places the block at the end position (right panel). The small
rectangles on the participant (left panel) indicate the positions of the sensors. The large black square at the
side of the participant indicates the position of the electromagnetic source. The dashed line represents the
maximum reaching distance of the arm. Middle Pane: Reaching trajectories of the hand and grasping profiles
for a single stroke patient in weeks 1, 5, and 26 poststroke. The black square represents the position of the
block. Each trace represents one repetition of the reach-to-grasp movement. Note that the improvements in
smoothness of the reaching trajectories and grasp aperture profiles are larger between week 1 and week 5
compared with week 5 and week 26 poststroke. Bottom Pane: Change in movement duration and hand
trajectory and grasp aperture smoothness (normalized jerk) as a function of time poststroke. All three
kinematic parameters reached a plateau at week 8 poststroke (gray area). Adapted from van Kordelaar et al.
(2012).

possible to compute a squared Mahalanobis distance (MD^2) with respect to the reaching trajectories from the dominant arm of a normal control population. Thus, the approach provides a scalar quantity that indicates how close to normal a patient's movements are.

The main result was that improvements in motor control in the paretic arm were observed up to week 5 but then stopped, with no significant improvements seen beyond this time point. In contrast, improvements in the FMA, ARAT, and biceps dynamometry continued beyond five weeks, with a similar magnitude of improvement between weeks 5 and 54 as between weeks 1.5 and 5 (figure 3.5). The shorter time window of motor control recovery (five weeks) compared to the three-dimensional study described above (eight weeks) is possibly due to contamination of the three-dimensional task by concomitant strength increases.

Several aspects of this study, and quantitative movement approaches in general, merit further comment. First, although the greatest rate of improvement was seen between weeks 1.5 and 5 for all measures, which is consistent with previous longitudinal studies of recovery in the arm after stroke, there was nevertheless robust continued improvement in the clinical scales when there was none for the motor control measure. This is an example in which the more sensitive kinematic measure was the one that did not show a change, which is to be expected if it successfully isolated and detected a component that is usually mixed in with other components that continue to improve. Second, the most parsimonious explanation for the dissociation in time courses is that strength improvements can contribute to both the FMA and the ARAT, a conclusion consistent with the various examples given in chapter 2 showing that strength and control can dissociate in the paretic limb. Third, measurement of recovery of motor control with specially designed assays provides a more accurate and clean measure of spontaneous biological recovery than clinical scales, because the latter will be contaminated by strength improvements and compensatory strategies. Finally, some kinematic studies have suggested that improvement in motor control can go on substantially longer, even up to six months, especially in more severe patients (Semrau et al. 2015). How to explain the discrepancy between kinematic measures that show prolonged recovery versus those, as in the studies described above, that reach plateau within five to eight weeks? We speculate that there are two answers to this question. The first is that, as we have already outlined, kinematic measures can be used to isolate motor control, but they also can be used to quantify performance of ADLs. Therefore, some kinematic measures, intentionally or inadvertently, may parallel the time course of clinical scales. For example, if compensatory postures are not controlled for and speed is not clamped, then some kinematic measures will improve even if motor control has not. This is because practice will lead to decreases in movement time and increases in accuracy within a *fixed* motor performance envelope (Hardwick et al. 2016). The second possible contributor to discrepancies across kinematic studies is the long-known finding from longitudinal studies that patients with the most severe paresis have more prolonged recovery (Skilbeck et al. 1983; Duncan et al. 1992; Nakayama et al. 1994;

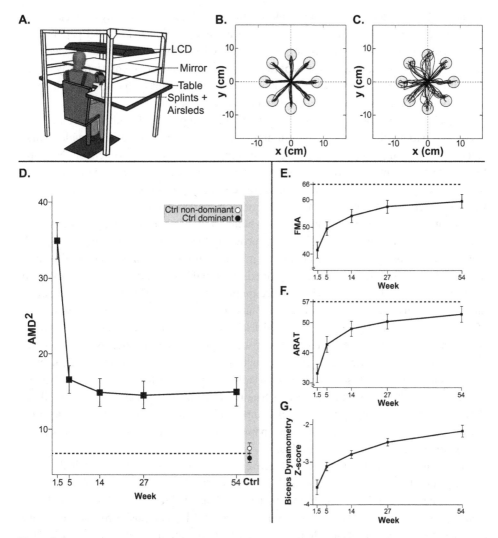

Figure 3.5
Motor recovery of the upper-extremity after stroke measured with two-dimensional kinematics. (A) Planar-kinematics apparatus and sample trajectories for (B) the dominant arm of a healthy control and (C) the paretic dominant right arm of a mildly hemiparetic patient (FMA = 58). (D) Time course of recovery of motor control in the paretic arm of 18 patients, measured with the squared average Mahalanobis distance (AMD^2) from the distribution of healthy controls' trajectories. The dotted line indicates the average AMD^2 of dominant and nondominant arms in healthy controls. Hemiparetic patients showed a clear improvement in motor control between weeks 1.5 and 5. No further improvement in motor control was seen beyond week 5. Time course of clinical scales: FMA (E), ARAT (F), and biceps dynamometry z score (G) of the same stroke patients. The dotted lines in **E** and **F** indicate the maximum possible scores for ARAT = 57 and FMA = 66. Clinical improvement was observed throughout the first year poststroke. ARAT = Action Research Arm Test, FMA = Fugl-Meyer Motor Assessment, Ctrl = Healthy Controls. Adapted from Cortés et al. (2017).

Jørgensen et al. 1995; Jørgensen et al. 1999; Duncan et al. 2000). This may be because the same spontaneous biological recovery processes are in operation as in milder patients but are more drawn out in time; by analogy, a deep cut takes longer to heal than a superficial graze. Alternatively, additional types of recovery (strength and compensation) may play a more prominent role in more severe patients. The general point is that a kinematic variable may show improvement for more than one underlying reason.

All these considerations come into play in a recent longitudinal study of motor recovery in 116 patients by Jennifer Semrau and colleagues, in which patients made planar reaches while being provided with antigravity support by a robotic exoskeleton. Notably, in patients who were most affected (moderate hemiparesis), the whole temporal spectrum was observed, with some kinematic measures continuing to improve up to six months, whereas others did not improve at all (Semrau et al. 2015). For example, in the most affected patients, the speed maxima count, which is the number of speed peaks between movement onset and offset, continued to improve up to six months. Other measures such as initial directional error, movement time, and maximum speed improved significantly up to three months, whereas a measure called the initial distance ratio only improved up to six weeks. Finally, a feedback measure quantified as the difference between speed maxima and minima did not improve at all. Thus, it is readily apparent that the time course of recovery varied greatly depending on which kinematic measure was looked at. In a sense, we are right back where we started, but now with differential time courses for kinematic measures instead of clinical measures. The authors do not speculate as to why their individual kinematic measures showed such variation in recovery time course. As we discuss above, they likely represent a mixture of true recovery of motor control (which is likely slower the more affected the patients are), compensation, strength improvements, and practice effects. Kinematic measures are not a panacea unless the meaning of each measure is prespecified and the measure controlled.

Surprisingly few studies have tracked recovery of motor control and strength in the hand using quantitative measures. Catherine Lang and colleagues evaluated thumb and finger extension against gravity in twenty-four patients with hand paresis at three weeks and thirteen weeks after stroke (Lang et al. 2009b). They used a composite rotation measure across all three finger joints and found, on average across all the digits, recovery of < 20°. They comment that this is not a lot of recovery given that the full range possible is 100°. They also state that they would not anticipate that patients would show any further changes in digit extension beyond thirteen weeks. For our purposes, the study is notable because it suggests that a compensation-proof measure of hand motor control reaches a plateau within the first three months after stroke. There were no time points between three and thirteen weeks in this study, and so the motor control plateau was likely reached even earlier.

Jing Xu, Joern Diedrichsen, and colleagues (2017) conducted a longitudinal study of hand recovery in over fifty patients, with poststroke assessments within the first two weeks, four to six weeks, twelve to fourteen weeks, twenty-four to twenty-six weeks, and

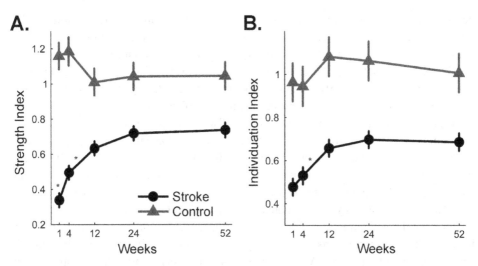

Figure 3.6
Time courses of recovery for finger strength and individuation in stroke patients and healthy controls. Almost all recovery for finger strength and individuation occurred within the first 12 weeks after stroke. Adapted from Xu et al. (2017).

fifty-two to fifty-four weeks. At each of these five visits, hand function was tested using an ergonomic device that measures isometric forces produced by each finger. Two separate aspects of finger function were tested with two reliable measures: (1) strength was assessed as the maximum voluntary force in flexion for each finger normalized to the nonparetic hand. (2) Individuation, for which patients were instructed to press down one finger while attempting to exert no force with the other fingers, was assessed with an index that represents the ratio between the force exerted by the active finger and the force exerted by the others. This index gives a measure of individuation that controls for force level (i.e., is independent of strength). The first result was that almost all recovery for both hand strength and individuation occurred in the first three months (figure 3.6). It appears, therefore, that motor control in the hand reaches a plateau at a later time point compared to recovery of motor control in the arm, which, as we saw above, is largely complete by five weeks. This is interesting and potentially good news, as it suggests that there may be a longer time window available for therapeutic intervention for the hand.

The second result of the longitudinal study was that a novel recovery function was identified: strength and individuation were tightly correlated up to a strength level of approximately 60 percent of estimated premorbid strength; beyond this threshold, strength improvement was not accompanied by further improvement in individuation. This relationship between individuation and strength took the same form at any time after stroke up to one year (i.e., it was time invariant). Any additional improvement in individuation was attributable to a second process that superimposed on the mean recovery function. Notably, this second process added to the first in the first four weeks after stroke. These results suggest that spontaneous recovery of the hand is attributable to two processes: one contributes almost all of strength and some individuation; the other contributes additional individuation. This is very interesting as these two processes, identified by behavioral analysis alone, are consistent with the interplay between the residual corticospinal tract and the upregulated reticulospinal tract described for motor recovery in nonhuman primate stroke models and discussed in chapter 2.

In summary, there is a time window for spontaneous biological recovery of motor control in the arm and hand that at best is open for three months and overlaps to a large degree with recovery of strength. This means that there is a need to intervene rapidly with new therapies to have an impact that goes beyond what is expected from spontaneous biological recovery. It can only be speculated as to the reasons for the difference in the recovery time for the arm and hand. One might be that recovery of individuation in finger flexors is different from extensors; the latter might follow a similar recovery time course to planar reaching for the arm. A second reason is that restoring individuation ability may require time to practice for any given level of spontaneous recovery. We shall see later in the chapter that evidence in nonhuman animal models suggests that spontaneous biological recovery can interact favorably with training, although such evidence is sparse in humans. Another reason is that there may be differences in the time courses of recovery for different descending systems. It is possible that the contribution that the residual CST makes to control after stroke reaches a plateau earlier than that made by other descending systems such as the reticulospinal tract (RST). This interpretation would be consistent with the same short time window Cortés and colleagues found for the arm (five weeks) and what Xu and colleagues found for the second pure finger individuation process (four weeks). Thus, spontaneous recovery may not all be the same; it may represent the sum of repair processes occurring in parallel in different substrates and that interact with training to differing degrees.

3.7   The Interaction between Spontaneous Recovery and Training: Repair versus Learning

Up until this point, we have discussed the time courses of spontaneous biological recovery at the levels of impairment and motor control. There has been some objection to the word *spontaneous* because surely patients are always doing *something* behaviorally after a stroke, with perhaps the exception of when they first transition from flaccid hemiplegia to some minimal amount of arm or hand movement. The question then is, how much recovery early after stroke is attributable to endogenous repair processes and how much to behavioral experience? As we have outlined above, both on theoretical and empirical grounds, spontaneous recovery after stroke in humans does not seem to be driven by or impinged upon by current rehabilitation approaches. The notion of stages of recovery, as first described by Twitchell and Brunnstrom, implies a stereotypical sequence that plays out as a repair process, a process that can perhaps be modulated by training but is not caused by it. It is common to see patients begin to get hand function back in the weeks following a stroke, and they often report that they watch it happen—neither they nor the therapists are making it happen. The predictable change in the FMA between the first week after stroke and three months later (proportional recovery) that we have already described captures this. The FMA was specifically designed to quantify the stages of recovery as patients regain the ability to isolate joints and to make multijoint movements in and out of synergy. As the FMA does not have functional components it is never used for training because of the switch in emphasis in neurorehabilitation to task-oriented training and away from synergies. Nevertheless, the FMA can dramatically improve. This is spontaneous biological recovery as an endogenous repair process, analogous to a wound healing. The wound can be dressed and kept clean, but these interventions are not causing healing; they are modulating it. Spontaneous biological recovery can therefore be thought of an endogenous repair process that presumably relies on residual intact neural architecture as a template for reorganization.

What complicates the picture is that the cellular and physiological conditions that allow spontaneous recovery may also increase responsiveness to intense training. In this scenario, training is causing the improvement in a task, and heightened plasticity is the modulator (i.e., the driver and modulator roles have been switched as compared to spontaneous biological recovery). This can be quickly understood in the case of normal motor learning—we would never say that someone spontaneously learned tennis. The challenge is to disambiguate these two ways to improve: spontaneous biological recovery/endogenous neuroplasticity and training-induced recovery. These are likely happening simultaneously and interacting with each other. For example, finger individuation can improve spontaneously, but a patient can also practice at an individuation task. There is no avoiding the fact that things can get murky, but the distinctions between repair and learning, as well as between driving and modulating, can help interpret studies in this area. What is clear is that in the first three months after stroke, endogenous plasticity is the main driver of recovery. It remains possible, however, that some minimal amount of activity is required to kickstart endogenous repair processes and that more intense training experiences can interact with and augment them.

Here we shall begin by considering work in animal models, primarily focusing on nonhuman primate work as we consider it most relevant to human stroke when it comes to the question of how spontaneous recovery of motor control in the arm and hand may interact with specific forms of rehabilitative training. Rodent studies are less relevant for this discussion because spontaneous recovery, especially of prehension, has been little studied and does not appear to happen to an appreciable degree (Ng et al. 2015).

3.7.1 Nonhuman Primate Studies

At the beginning of this chapter, we discussed the seminal work by Ogden and Franz in macaques in which they showed that after a large cortical lesion that included M1 and PMd, there was near-complete recovery of arm function by three weeks but only if it was trained and the unaffected arm restrained. Thus, this was a clear case in which spontaneous recovery alone was inadequate. More contemporary studies in monkeys suggest considerable variability in the effectiveness of spontaneous recovery alone. The variables that determine the need for concomitant training remain to be elucidated, but they seem to include initial severity, extent of the lesion, and the kind of behavior being investigated. The matter is complicated by the fact that some degree of spontaneous recovery will lead to more use of the limb, and so a form of self-training will ensue, at which point the boundary between spontaneous recovery and training blurs. Studying the interaction is important for two reasons. First, from a scientific standpoint, it can provide insights into the degree to which neural mechanisms overlap for recovery due to spontaneous repair processes and that due to training. Second, from a rehabilitation standpoint, it can inform about the dose, intensity, timing, and type of training that should be given for limb paresis after stroke.

In the 1990s, Randolph Nudo and colleagues conducted a series of seminal studies on motor recovery of the hand in squirrel monkeys. In one study, monkeys were first trained to retrieve pellets from food wells of different diameters (Nudo et al. 1996; Friel et al. 2007). Successful retrieval from the smallest well could only be accomplished through skillful use of one or two fingers. Training consisted of two titrated thirty-minute sessions per day and continued until 600 pellets were retrieved from the smallest well on each of two consecutive days. Note already that success at the small well task required skill (i.e., training beyond baseline levels). After training was complete, small subtotal infarcts were then induced in the hand representation of the primary motor cortex (M1). At postinfarct day 5, intense rehabilitative training began that was identical to the pretraining and continued until the monkeys reached their preoperative skill level on the smallest well, which occurred after three to four weeks. The same group had conducted a previous study in which monkeys received no preoperative or postoperative training, as the goal was to investigate spontaneous recovery (Nudo and Milliken 1996). The monkeys' retrieval performance took about two months to return to preoperative levels, which is twice as long as it took in the case of concomitant training, but nevertheless full recovery did occur. It should be noted, however, that the preoperative levels were obviously not matched in the pretrained versus the spontaneous studies. An additional experiment could therefore be envisaged: postinfarct training in the absence of preinfarct training. This would better approximate the real-life situation after stroke; we do not pretrain people on a specific task before they have a stroke and then rehabilitate that particular task. Such an experiment would also allow direct comparison to the classic Ogden and Franz result described above and others like it. Humans who suffer strokes in the cortical "hand-knob" area present with isolated hand paresis. In one study of twenty-nine patients, three-quarters were asymptomatic at a mean follow up at twenty-five months (Peters et al. 2009). This is an interesting result as hand-knob infarcts can be considered analogous to the small infarcts in the Nudo et al. studies, with recovery likely the sum of spontaneous recovery and practice.

It has been suggested that the reason why the monkeys in the Ogden and Franz experiment needed training to recover whereas those in the experiments by Nudo and colleagues did not (although they recovered faster with training) is that the lesions were much larger in the former case. Yumi Murata and colleagues directly tested this idea in macaques that were given extensive ibotenic acid lesions that encompassed all of the M1 digit representation (Murata et al. 2008). They found that precision grip, tested on a Klüver board with five cylindrical wells of different diameter, could recover to prelesion performance levels in one to two months if the monkeys were given highly intensive training (one hour per day, five days a week) that began one day after the lesion. In contrast, spontaneous recovery (no training) led to the ability to flex and extend all the joints of the thumb and index finger, which allowed the use of compensatory grip strategies but not to the level of individuation required to consistently use a precision grip (figure 3.7). Details in the time courses of recovery for both the trained and untrained monkeys strongly suggest that spontaneous repair processes and intense training have separable effects on recovery but nevertheless interact. One of these details is that at around day 30, *both* groups of monkeys made a small proportion of precision grips (5 percent or below), but within days they diverged: the proportion increased in the trained group and reduced to zero in the untrained group. This is a profound finding as it suggests that the baseline ability to individuate the digits emerged spontaneously, albeit tenuously, in both groups at around a month, but then training was needed for precision grip to become effective and lock in. It appears that without practice, precision grip was not explored or improved upon and failed to gain a foothold, being outcompeted by a compensatory strategy. Another fascinating finding was that the trained monkeys began to increase the frequency of precision grip attempts even though initially this led to an *increase* in failure rate: at around this time the untrained monkeys were failing less at the task than the trained monkeys. This suggests that the monkeys either recognized that precision grip would ultimately be more successful, perhaps based on past experience, and/or were detecting incremental improvements with training, a form of reward prediction error.

A more recent study by John Buford and colleagues has very interesting parallels with the studies by Nudo et al. (Nudo and Milliken 1996) and Murata et al. (Murata et al. 2008). Endothelin 1 was used to induce a focal ischemic lesion in the elbow and shoulder region of left M1 in two monkeys (Herbert et al. 2015). The lesion in one monkey (monkey 1) was small and spared some elbow and shoulder representation in M1. The other monkey had a more extensive lesion that destroyed the whole arm representation. Thus, the first monkey could be considered the arm equivalent of the small M1 hand area infarct in the

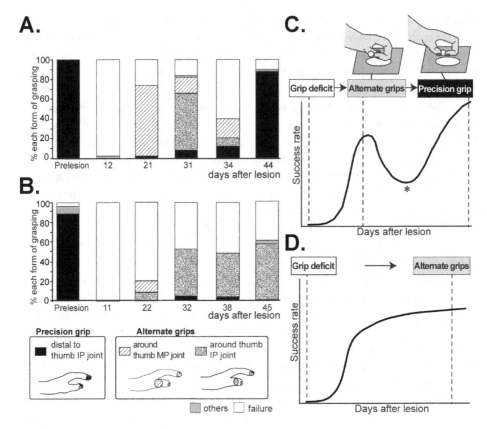

Figure 3.7
Differential recovery of the hand after a motor cortical lesion as a function of training. Although the monkeys consistently used the tip of the index finger to hold the food pellets, there were three grip types defined based on the location of the food pellet on the thumb: (1) distal to the thumb interphalangeal (IP) joint (precision grip, black), (2) around the thumb metacarpophalangeal (MP) joint (hatched), and (3) around the thumb IP joint (stippled). In trained and untrained monkeys, the precision grip, using finger-to-thumb opposition, was observed in most of the trials before lesion induction. (A) Time course of changes in the number of retrievals for each form of grasping in a trained monkey. During the transient improvement period, both monkeys often held the food pellets between the tip of the index finger and around the proximal joint of the thumb. In the trained monkey, the percentage of precision grips increased gradually during this transient improvement and in subsequent periods. By forty-four days after lesion induction, the use of the precision grip became as frequent as that during the prelesion period. (B) Time course of changes in the number of retrievals for each form of grasping in an untrained monkey. In the untrained monkey, the proportion for precision grip remained low, and alternate grips were used to hold the food pellets in most of the trials throughout the postlesion period. (C) Schematic figure showing the linkage between the change in success rate and changes in grip forms in the Klüver board task in the trained monkeys. Although the monkeys could not retrieve the food pellets at all immediately after the M1 lesion, their motor performance improved progressively during the postlesion training period. In the middle of the recovery process, they frequently used alternate grips, holding the food pellet between the tip of the index finger and around the proximal joint of the thumb. Thereafter, the precision grip use began to increase. At the point when grip strategy changed, the monkeys frequently failed to retrieve the pellets because of inadequate coordination between digits (asterisk). (D) In the untrained monkeys, many alternate grips holding the food pellet between the tip of the index finger and around the proximal joint of the thumb were observed during the recovery period. In contrast to the trained monkeys, these alternate grips were not replaced by the precision grip. Adapted from Murata et al. (2008).

Nudo et al. experiments, whereas the second monkey (monkey 2) could be considered the arm equivalent of the full M1 hand lesion in the Murata et al. experiment. Arm recovery was assessed with a task that required the monkeys to reach out to four food wells positioned in front of them at eye and waist level such that they had to make high and low lateral reaches and high and low cross-midline medial reaches. After stroke induction, monkeys were given two weeks to spontaneously recover—at the end of this period, they were both independent in their cages. In monkey 1, the right (contralesional) arm gradually improved over two weeks but did not fully return to prestroke levels. Monkey 2 had a larger lesion and a more severe initial impairment—it was not able to make any successful reaches over the spontaneous recovery period even though it did gain back enough function in the right arm to use it for general cage mobility. This distinction between the ability to move the arm versus use it for accurate reaching has direct parallels with the distinction between compensatory and precision grips in the Murata et al. study. Monkey 2 could only attain the reach performance of monkey 1 after about a month of intensive reach training of the right arm sixty to ninety minutes a day, three to four days per week, which again is very similar to the time frame in the Murata et al. study.

The results described above suggest a fundamental dichotomy between spontaneous biological recovery and training-induced recovery, although this does not imply that underlying biological and physiological mechanisms cannot overlap to a significant degree. Indeed, the heightened plasticity seen for a short time after stroke may be required for both kinds of recovery. By analogy, procedural and declarative memory may be very different, but the cellular mechanisms underlying them, such as long-term potentiation and long-term depression, may be the same. Spontaneous biological recovery can be considered a form of repair and training-induced recovery as a special form of motor learning. Spontaneous recovery appears to occur when the circuit responsible for a natural behavior suffers subtotal damage such that the remaining architecture can serve as a scaffold for repair—a hardwired "old" representation is rebuilt. This repair process is likely potentiated by behavioral experience. Training-induced recovery seems to be required when an original circuit is destroyed completely and alternative circuits need to be facilitated and trained to perform the behavior—a latent representation is upregulated by learning. Later in this chapter, we shall consider poststroke brain reorganization in the light of this repair versus learning distinction.

3.7.2 Human Studies

As we have pointed out in section 3.5, current levels of rehabilitation do not appear to have an impact on impairment beyond what can be expected from spontaneous biological recovery. Very few studies in humans have explicitly sought to test whether intense training early after stroke can either augment or add to spontaneous biological recovery in a manner analogous to the nonhuman primate studies described above. Few of the recent trials that investigated interventions within the first three months had an impairment measure as either a primary or secondary outcome. As we shall see later, there is increased responsiveness to task-specific training early after stroke, and this could lead to larger gains in activity measures such as ARAT and Wolf Motor Function Test (WMFT) without a concomitant change in impairment. Indeed, it is possible that there is a fundamental incompatibility between the task specificity of motor learning and the generalizability of repair: increased training on a particular task early after stroke may simply lead to greater gains in that task but fail to reduce impairment.

There are many reasons why there have been so few studies to date that have tried to significantly increase the dose and frequency of either current therapeutic approaches or to test new ones at high intensity. These include logistics, economics, practice biases (an emphasis of participation and activity over impairment), and the persisting scientific concern, based largely on rodent studies in the 1990s, that very early intense therapy may extend infarct volume and worsen outcome. As we suggest in chapter 4, these concerns have been exaggerated and were probably only ever relevant in the case of large cortical strokes. In addition, more recent work questions the broad applicability of the original findings.

One of the best-known early intervention studies was called VECTORS (Very Early Constraint-Induced Movement during Stroke Rehabilitation), led by Alexander Dromerick, in which fifty-two patients with stroke were randomized at about ten days poststroke to two levels of intensity of constraint-induced movement therapy (CIMT) versus standard upper extremity therapy (Dromerick et al. 2009). The study is worth discussing even though it did not target or measure impairment. The high CIMT group received an extra hour per day of CIMT: three hours versus two hours of shaping therapy per day five days a week for two weeks. The low CIMT group wore a padded restraint mitten for six hours per day, whereas the high CIMT group wore it for 90 percent of waking hours. The surprising result was that the ARAT at ninety days, the primary outcome measure, was worse for the high CIMT group. This was taken to mean that there is a U-shaped function for dosing for CIMT early after stroke (Dromerick et al. 2009; Stein 2009). A longitudinal magnetic resonance imaging (MRI) study in a subset of the patients did not show any expansion in infarct volume that could be related to intensity of treatment, so there was no evidence for infarct expansion, which was the putative explanation for intensity-related worsening in the early rodent models of early rehabilitation (Dromerick et al. 2009).

VECTORS is an impressive study, and one can only imagine how challenging it must have been to complete. We disagree, however, with its interpretation, which has led to the tendency to refer to the study in kneejerk fashion whenever an early high-dose intervention is proposed. Before suggesting alternative interpretations, we feel compelled to point out that, despite the authors stating that there were no significant differences between groups for any demographic or clinical variable, the high CIMT group differed in a number of ways from the other two groups. For example, over 60 percent of the high CIMT group had involvement of the dominant limb versus only 30 percent for the low

CIMT group and 40 percent for the control group. Women made up only 44 percent of the high CIMT group versus 68 percent of the low CIMT group and 65 percent of the control group. Finally, 6 percent of the high CIMT group had had a prior stroke versus 11 percent of the low CIMT group and 12 percent of the control group. There is always a concern in a clinical trial, especially when the n is not particularly large, that the groups will contain confounding variables despite randomization. Also, it is unclear when or where correction for multiple comparisons was employed. Nevertheless, for the purposes of the argument that follows, we will assume that there were indeed no important differences between groups before the interventions.

The most concerning omission in VECTORS was that no impairment measure was included. This is, as we have seen, fairly common in rehabilitation trials given the prevailing preference for the activity component of the ICF (World Health Organization 2002). The reason why it is especially concerning here is that we know that most spontaneous recovery occurs in the first month after stroke and that it likely drives recovery at the activity level through reductions in impairment, which is why there is such a high correlation between the ARAT and the FMA early after stroke (Rabadi and Rabadi 2006; Beebe and Lang 2009). In contrast, in the chronic stage, these two measures can dissociate (van der Lee et al. 2001; Hsieh et al. 2009), because activity measures can also improve through compensatory mechanisms. The natural question that arises is, what happens when compensatory task-based strategies are taught during the spontaneous recovery period? Can such task-based training interfere with more general recovery from impairment? That there may be complex interactions between these different mechanisms of recovery in the early period after stroke is hinted at by the varying time courses for the different outcome measures in the study. No *a priori* hypothesis was provided per measure, making it unclear whether the investigators expected them to either track each other or diverge. The investigators also did not explain why their primary outcome was ninety days after the beginning of treatment, even though CIMT was complete after fourteen days. What delayed mechanism did they have in mind? For example, why should one measure continue to improve after the treatment is over, whereas another not? The bottom line is that it is not at all clear what the VECTORS result really means. The idea of a dose-response curve would need to be replicated for CIMT and be demonstrated for other forms of early intervention. The study should not be cited as a reason to avoid studies of higher doses and intensities for new forms of impairment-focused intervention early after stroke.

A more recent and already much talked about early intervention trial examined the efficacy and safety of very early mobilization within twenty-four hours of stroke onset—A Very Early Rehabilitation Trial (AVERT) (The AVERT Trial Collaboration Group 2015). Neither its primary nor secondary outcome measures addressed the upper limb, so it is not of direct relevance to this book. Nevertheless, we discuss AVERT briefly because, like

VECTORS, it is another trial suggesting that early high-dose rehabilitation after stroke may be hazardous. The two arms of the study were very early mobilization versus usual care. What usual care entailed in this trial is vague, but this is perhaps unavoidable given its impressive scale—2,014 patients randomized at fifty-six acute stroke units across five countries. The goals of the very early mobilization intervention were (1) to get the patients out of bed within twenty-four hours; (2) to have out-of-bed therapy sessions that focused on sitting, standing, and walking; and (3) to have three more of these therapy sessions per day compared to usual care. The intervention period lasted fourteen days or until discharge from the stroke unit, whichever was sooner. The primary outcome measure was the modified Rankin Scale at three months (note the similar end point to VECTORS). The modified Rankin Scale is an ordinal and ranges from zero (no disability) to 5 (severe disability). A score of 6 means the patient died. Secondary outcomes included time taken to achieve unassisted walking over 50 meters and the proportion of patients achieving unassisted walking by three months.

The main and only significant result was that 46 percent of the very early mobilization patients had a favorable modified Rankin Scale outcome, whereas 50 percent did in the usual care group. The walking outcomes did not differ between the two groups. The median time to mobilization for the very early group was 18.5 hours and 22.4 hours for the usual care group. This difference was significant but is very unlikely to be the decisive factor. The median amount of time spent per day in out-of-bed sessions was thirty-one minutes for the very early mobilization group and ten minutes for the usual care group. The authors have little to say as to why the ninety-day primary outcome was worse in the very early intervention group, as in VECTORS, there is no speculative drilling down beyond saying that physical activity might have an effect on ischemic tissue. The problem here is that the patients must have spent more time out of bed because they had more therapy sessions, so amount of therapy is confounded with time out of bed. It is notable that death from stroke progression occurred in thirty-one patients in the very early mobilization group and nineteen in the usual care group; although this difference was not significant, it could be a clue that there might have been more infarct volume expansion (and intracerebral hemorrhage expansion) in the very early mobilization group. Consistent with this idea is a subgroup analysis that hinted at the difference between groups being greater in the patients with severe strokes and those with intracerebral hemorrhage (ICH). It is easy to get confused about AVERT because the name of the trial can lead people to think that the key factor was the difference in the time when mobilization was started, but the real difference was in the number of out-of-bed therapy sessions in the ensuing weeks. It is this second difference that makes the lead investigator, Julie Bernhardt, emphasize that this was really a trial that compared the effects of different doses of rehabilitation early after stroke. As we have stated above, however, this is not strictly correct because these patients were obviously also spending more time out of bed standing in order to engage with the increased therapy. In our view, the most parsimonious explanation is that the patients had impaired autoregulation, which can lead to infarct or hemorrhage expansion, and there is a higher probability of this happening the more time that is spent upright; nothing in the data presented can exclude this possibility. We would hazard the guess that had the patients received hand and arm therapy while remaining in bed, the results would have been reversed. Indeed, an emphasis on arm rehabilitation would have been a good way to decouple rehabilitation dose from time out of bed. In conclusion, AVERT provides an important note of caution and suggests for patients with large infarcts or ICH that there is probably a need to limit time out of bed in the first few weeks after stroke. Perhaps this could be enforced over the first five days, which is the usual length of stay in acute stroke units in the United States anyway. Unfortunately, AVERT does not answer the question of rehabilitation dose early after stroke.

Another recent trial, Explaining PLastICITy (EXPLICIT), is directly relevant to the question of how task-based training may or may not interact with spontaneous biological recovery (van Kordelaar et al. 2013; Kwakkel et al. 2016). In the arm of the trial pertinent to the current discussion, fifty-eight patients, deemed to have a favorable prognosis because they had > 10° of finger extension, were randomly allocated to three weeks of modified CIMT (mCIMT) or usual care only, both of which were started within 14 days of the stroke. Patients in the CIMT group received sixty minutes a day, while the usual care group received thirty minutes a day. Clinically relevant differences in ARAT in favor of mCIMT were found at five, eight, and twelve weeks but not after twenty-six weeks. In contrast, there were no statistically significant differences between the mCIMT and usual care groups for the FMA. Thus, while increased task-specific training led to improvement on a scale that measures activity, this was not accompanied by concomitant reductions in impairment. We agree with the authors' conclusion: "We found no evidence that we were able to influence spontaneous neurological recovery of underlying impairments based on clinical scales, suggesting that functional improvements of the mCIMT group were based on adaptation strategies to use intact end-effectors in a more optimal way." This is an important result as it suggests that increasing the dose of task-based training may not be the best way to target impairment. This conclusion is also supported by a recent robotics trial in which patients were enrolled within fifteen days of stroke but showed no additional benefit at the level of impairment compared to the usual care group (Masiero et al. 2014). Notably, however, in this study, the robot was used to mimic usual therapy rather than increase the number of movement repetitions. Finally, we should briefly mention the recently completed Interdisciplinary Comprehensive Arm Rehabilitation Evaluation (ICARE) trial, which enrolled patients on average at forty-five days poststroke (Winstein et al. 2016). The goal of this phase III trial was to both investigate the effect of dose of occupational therapy and compare two types of occupational therapy controlling for dose. There was no difference across the different interventions for any of the outcome measures. Importantly for this discussion, impairment changes were not even examined.

Several conclusions can be drawn from this survey of early rehabilitative intervention (within three months of stroke) in both nonhuman primates and humans. In the nonhuman primates, intense training at the level of the motor control deficit/impairment can augment what is expected from spontaneous recovery. In contrast, there is still no good evidence that this is true in humans early after stroke. This lack of evidence, however, should not be taken as an indication that a favorable interaction between training and spontaneous recovery cannot occur in humans. Rather, the problem is that there is surprisingly a lack of interest in targeting impairment in patients early after stroke, so the necessary studies have not been done. As outlined above, the largest trials did not even bother to measure impairment changes. Those trials that did look at impairment, such as EXPLICIT, did not have an impairment-targeted treatment. Thus, either impairment is not measured at all or therapies are not focused on it. Thus, interpreting VECTORS, AVERT, and ICARE as clues to the possibility that giving more therapy early may be of no benefit or even detrimental is, in our view, a superficial read of the studies, mainly attributable to a consistent failure both to consider impairment and to appreciate how it can dissociate from activity-level ICF measures. Overall, the current findings in humans suggest that whereas spontaneous recovery has an effect on motor activity measures via its effect on impairment, the reverse does not seem to be true—task-oriented training, even early after stroke, at best teaches compensation and has no appreciable effect on impairment. In contrast, the kind of training given in the monkey stroke studies focuses heavily on movement execution rather than goal completion and is given at much higher doses. It is an open question whether this would work in humans.

3.7.3 The Paradox of Generalization of Recovery

A way to think about the difference between the effects of training and spontaneous recovery is that the former is usually task specific and the latter is general. This leads to a paradox: how does one train general recovery or target impairment when training is always with a particular task? We do not know the answer to this question, but just being able to pose it strongly suggests that a new kind of rehabilitation needs to be developed for patients early after stroke that will interact with spontaneous recovery to maximize generalization of gains. A clue to how to do this was provided by Dale Corbett and colleagues, in an influential experiment in rats that investigated the rehabilitative effects of a more varied and enriched environment (Biernaskie et al. 2004). Rats were given a middle cerebral artery occlusive stroke and were exposed to an enriched environment (ER). The ER consisted of large cages that could house five to seven rats and contained toys, ramps, tubes and ropes, and a prehension apparatus, which was *not* overtly trained on. There were significant gains in prehension when tested on a *different* task when ER was initiated within five or fourteen days but not thirty days after stroke. In an interesting follow-up study, the same group showed that repeating the ER two more times, at a level even more intense than the initial session for two weeks at a time, beginning about three months poststroke provided no additional benefit even though there still was room for improvement. This failure of tune-ups again demonstrated that there is a window of responsiveness to ER that then closes, with no further responsiveness to even intensified versions of the original therapy.

In an attempt to devise an entirely novel treatment approach that targets impairment, an ongoing phase II pilot multicenter trial, Study to enhance Motor Acute Recovery with intensive Training after Stroke (SMARTS 2), led by John Krakauer, has patients with moderate to severe arm paresis make playful non-task-based exploratory arm movements while playing a unique videogame. The core idea is that an immersive videogame may play the role of enrichment that was described by Corbett and colleagues in the rat. The goal is to make an experience engaging enough so that patients make challenging repeated movements in the numbers usually only seen in animal studies: patient's arm movements total a distance of about a mile and half per session. The game is a full physical simulation of three cetaceans (bottlenose dolphin, Commerson's dolphin, and an orca). The patient is able to use his or her arm, which is provided weight support by an exoskeletal robot, to steer the intention of the cetacean as it swims in the ocean (figure 3.8). The goal is to provide an immersive, highly motivating experience that encourages a patient to make a large number of arm movements that are analogous to the exploratory movements young children make. Results are expected in 2018.

3.8 A Sensitive Period of Increased Responsiveness to Training

In the last section, studies in nonhuman primates show that training early after stroke can boost the gains expected from spontaneous biological recovery. Evidence from animal models suggests that the converse is also true: the conditions enabling spontaneous recovery also appear to enhance responsiveness to task-specific training. Steven Zeiler and colleagues showed that mice given a focal cortical infarction in the caudal forelimb area (CFA), the mouse analog of the primary motor cortex, fully regained their preinfarct performance level on a prehension task when poststroke training was initiated within forty-eight hours but only showed small gains if training was delayed by a week (Ng et al. 2015). Like the study by Dale Corbett and colleagues, this result strongly suggests the existence of a time-limited window of increased responsiveness to training, presumably induced by the unique poststroke plasticity milieu. A counterintuitive implication of such an ischemia-induced sensitive period is that it should be possible to reopen it with a second stroke and thereby trigger recovery from a first stroke. Steve Zeiler and colleagues tested this hypothesis directly by inducing a second focal stroke in the ipsilesional medial premotor area of mice that had only partially recovered prehension after a first CFA stroke due to delayed training (Zeiler et al. 2016). The remarkable finding was that even though the second stroke worsened the initial deficit, training initiated within forty-eight hours of the second stroke allowed prehension to return to normal levels (figure 3.9). In other

Figure 3.8
Robotic exoskeleton controlling a simulated dolphin. This setup currently used in the SMARTS 2 trial with the goal of exploiting the increased brain plasticity seen early after stroke.

words, training after a second stroke fully rehabilitated the first stroke. The benefit of the second stroke was not a nonspecific effect because no recovery was seen when it was placed in the ipsilesional occipital cortex. Clearly, inducing second strokes is not a viable rehabilitation option for patients, but it does suggest that unique molecular, cellular, and physiological changes for a short time after stroke greatly enhance responsiveness to training. Treatment with a selective serotonin uptake inhibitor prolonged the time period in which neurorehabilitative training can significantly improve recovery of forelimb use (Ng et al. 2015). This effect was associated with diminished markers of inhibitory interneurons in the cortical regions that mediate recovery, suggesting that drug and training effects may modify excitatory/inhibitory balance in connected cortical regions to benefit recovery. This conclusion is consistent with pharmacological studies in which excitatory and inhibitory signaling systems were manipulated to promote recovery after stroke (Clarkson et al. 2010; Clarkson et al. 2011; Lake et al. 2015). The molecular and cellular underpinnings of the sensitive period are discussed in detail in chapter 4. We discuss prospects for reopening of the sensitive period in chapter 8.

To our knowledge, no studies have tested whether there is a poststroke sensitive period in nonhuman primates. The best evidence to date comes from a study in monkeys with

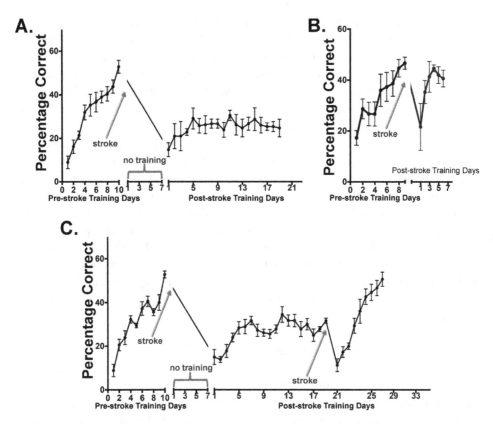

Figure 3.9
Resetting a post-stroke sensitive period window for motor recovery with a second stroke. Mice were trained to perform a skilled prehension task to an asymptotic level of performance after which they underwent photocoagulation-induced stroke in the caudal forelimb area. (A) When poststroke training was delayed by seven days, the mice hit a low plateau despite two weeks of training. (B) After a one-day poststroke delay, the mice were retrained and returned to normal levels of performance. (C) A second photocoagulation-induced stroke was then induced in the ipsilesional medial premotor cortex (agranular medial cortex). Subsequently, the mice were retrained after a one-day delay and returned to the normal levels of performance seen before the stroke. Adapted from Zeiler et al. (2016).

lateral corticospinal tract (l-CST) ablative lesions induced in the spinal cord at C4/C5 (Sugiyama et al. 2013). Eight adult rhesus monkeys were pretrained on dexterous finger movements using a Klüver board. Three of the monkeys then began the same training the day after the l-CST lesion (early-trained group), and the other five monkeys started training one month after the lesion (late-trained group). Poststroke manual dexterity was evaluated with both the Klüver board and with a second novel untrained task, which required the monkeys to use a precision grip to retrieve a small piece of food through a narrow vertical slit. The early-trained group recovered to the level of intact monkeys in the first one to two months after the lesion on both tasks. The late-trained group never recovered the precision grip in the slit task even after three months of training. The success rates for the two groups were comparable on the Klüver board, but kinematic analysis revealed that while the early-trained monkeys returned to a prelesion pattern of movement, the late-trained group adopted a compensatory strategy. The Klüver board results attest to the importance of fine-grained kinematic analysis and the pitfalls of relying on binary (success/failure) performance measures when studying stroke recovery.

At the time of writing, no human study has shown evidence for a poststroke sensitive period, either physiologically or behaviorally. Three studies, however, are ongoing. In one study, the primary outcome is the change in the MEP amplitude in a hand muscle, after the administration of a plasticity-inducing paradigm to the affected hemisphere at different times poststroke (McDonnell et al. 2015). In another study led by Pablo Celnik, both physiological measures of plasticity and motor learning capacity are being probed at two time points after stroke: the first month and beyond six months. Finally, the Critical Periods after Stroke Study (CPASS), a phase II trial led by Alexander Dromerick, is comparing the benefit of an additional twenty hours of occupational therapy when initiated within one month, at two to three months, or at six to nine months poststroke. The main outcome measure is the ARAT at one year.

3.9   Relevance of Rodent Models to Human Stroke

A potential concern with sensitive-period experiments and with rodent experiments in general is the question of their relevance to human stroke recovery. A subtext to this debate is the claim that nonhuman primate models are better than rodent models for the study of neurological deficits; rodents should be reserved for studies conducted at the levels of molecules and cells. In many rodent studies, prehension is the behavior of choice. Specifically, the tasks used are often variations on one originally introduced by Ian Whishaw and colleagues (Whishaw et al. 1992). The basic design is to have the rat or mouse reach its preferred forelimb through a slit, grasp a pellet, and then bring it to its mouth to eat. The debate about the relevance of this task hinges on the notion of homology, which is an evolutionary term and means to be derived from a common ancestor. The question of homology arises in three ways that are relevant here—with respect to skeletal structure and musculature, the neuroanatomy of the brain and the descending pathways, and the behavior itself. There is not much disagreement about limb homology, so we will focus on the other two aspects, which are more contentious—the behavior and neural structures, particularly the corticospinal tract. In a series of reviews and experimental papers, Ian Whishaw and colleagues have made, in our view, a very good case for behavioral homology between humans and rodents for reach and grasp, both in health and after motor cortical stroke (Cenci et al. 2002; Sacrey et al. 2009; Klein et al. 2012). For example, after large cortical strokes, both patients and rodents have permanent impairments in rotatory movements of the wrist, in pellet grasp and release, and make compensatory forward movements with the trunk. There are important differences, however, and the one that leads to most concern is that humans can grasp a pellet with a precision grip, whereas rodents use a power grip with all their fingers. As we have outlined in detail in chapter 2, the distinction between power and precision grips is largely attributable to the existence of a monosynaptic corticomotoneuronal (CM) connection, which is present in certain primates but not in rodents. It is this difference in the projections of the CST that has led Roger Lemon and James Griffith to question the utility of the rodent model to stroke recovery. At one point, they state, "The ability to control individual digits—not just 'skilled forelimb movement'—is a characteristic feature of primate motor behaviour that is fundamental to human activity" (Lemon and Griffiths 2002).

The debate about the utility of rodent models is of increasing importance as primate research becomes more difficult to conduct on both economic and ethical grounds (although the advent of the marmoset stroke model may change this; Virley et al. 2004). Our position with respect to the study of motor recovery after stroke is that there *is* homology between humans and rodents for both behavior and neural structures. Prehension is a skilled behavior even in those primates that lack strong monosynaptic CM connections, such as the squirrel monkey. Squirrel monkeys cannot make the precision grip seen in Old World monkeys such as the baboon and the macaque, and prehension is likely more under the control of CST inputs to the C3 to C4 propriospinal system, but this has not precluded them from being useful stroke models (as in the work by Nudo and colleagues described in section 3.7.1). There is evidence that prehension in both the rat and the mouse is supported by the corticoreticular tract (Alstermark and Ogawa 2004; Alstermark et al. 2004). If there is homology, however, at the level of the motor cortex and the limb, then the movement demands of the prehension task before and after stroke will need to be solved by these structures in a similar way despite differences in the interposed descending pathway. The additional refinement of neural structures, specifically the monosynaptic connections of the CST, the consequence of which is most apparent in finger individuation, does not mean that stroke-induced paresis should be considered a disorder of these late evolutionary refinements alone. Recent support for the translational validity of rodent stroke models comes from the demonstration that proportional recovery, first described in humans (Prabhakaran et al. 2008), also applies in rats (Jeffers et al. 2016).

A future issue in rodent models of stroke recovery is the use of fine-grained kinematic analyses that have the same resolution as those used in humans. Most current rodent skilled reach studies use simple pellet retrieval. The outcome measure is the number of pellets retrieved. This outcome measure is a rat outcome at the ICF level of activity. The actual performance of this function may indicate restitution or compensation. This is an important problem because studies that have conducted careful kinematic analysis in rodents have shown that compensation and restitution can occur simultaneously early after stroke and may even share overlapping neural substrates (Castro-Alamancos and Borrell

1995; Gharbawie et al. 2007; Moon et al. 2009). The need for kinematic analysis in rodent models is increasingly recognized and high frame rate cameras and offline automated algorithms are beginning to be used to analyze reaching trajectories (Wahl et al. 2014; Lai et al. 2015).

Finally, some commentators have expressed concern that stroke is as much a white matter as a gray matter disease in humans. In mice, white matter makes up less than 10 percent of brain volume in comparison to 60 percent in humans. Several points can be made here. First of all, as we have discussed above, behavioral homologies are as important as anatomical ones. Second, white matter lesions, even when small, can interrupt more corticofugal fibers than a sizable cortical lesion. Thus, it can be hard to compare cortical and subcortical lesions, not because one is gray matter and the other white, but because descending pathways are being involved to a different degree. Third, the ability to make small white lesions in rodent models is improving (Blasi et al. 2015; Sozmen et al. 2016), and it may soon be possible to consistently match the initial deficit for cortical and subcortical lesions. Only in this matched scenario can a meaningful comparison be made. As we point out in section 3.5, proportional recovery applies equally well for cortical and subcortical strokes. Thus, one might imagine two approaches to rodent modeling of human stroke with damage to the white matter in mind: employ a stroke model which attempts to match white matter and gray matter damage across the two species. Alternatively, one might employ multiple rodent models, with training effects studied separately for pure cortical and pure white matter damage.

3.10   Recovery and Brain Reorganization

Reorganization in the context of stroke recovery is a slippery term that sometimes refers to large-scale training-induced changes in uninjured regions remote from the focal infarct, and at other times, it is used to refer to even microstructural changes adjacent to the infarct. The term is even sometimes used to refer to learning-induced cortical changes in healthy subjects. The crux of the matter is, what kinds of change should be referred to reorganization? In our view, the term *reorganization* should be used in a specific way rather than indiscriminately to refer to any structural or physiological change in tissue occurring on any time scale after a focal infarct. Here, we will use the term *functional* reorganization to refer to changes in the peri-infarct cortex that occur in response to training and are necessary for behavioral recovery. By peri-infarct or perilesional, we mean cortical areas that lie beyond the immediate glial scar surrounding the stroke and can encompass, for example, residual M1 or ipsilesional premotor areas.

3.10.1   Cortical Map Changes
We shall make the case here that changes in or expansion of cortical maps are not *de facto* evidence for functional reorganization. Indeed, map changes have caused considerable

confusion in the field as they have come to be considered synonymous with recovery-related functional reorganization, a highly questionable assumption. Mapping is accomplished by applying a microelectrode at successive sites in a motor cortical area and delivering a low-intensity current pulse. A site is deemed part of the motor map if a fragmentary movement, usually across a single joint, is evoked. Cortical map changes occur with motor learning in healthy brains, but persistence of map expansion is not necessary for maintenance of a newly learned skill (Molina-Luna et al. 2008; Molina-Luna et al. 2009; Tennant et al. 2012). Most important for the purposes of this book is the fact that motor recovery after stroke can persist after map size reverts to normal (Nudo et al. 1996), can occur in the absence of map changes altogether (Nudo and Milliken 1996), or can occur before map changes are seen (Eisner-Janowicz et al. 2008; Nishibe et al. 2015). As this is potentially a contentious issue, we shall explore some examples in more detail. We have already discussed the seminal studies by Nudo and colleagues in the squirrel monkey in the context of the interaction between spontaneous biological recovery and training. Recall that recovery of hand function after a small subtotal M1 hand area to preinfarct levels was observed both with and without training. The critical result, which has been perhaps unappreciated, is that peri-infarct cortical map expansion was *not* required for recovery. Recovery eventually occurred in all monkeys in this study, but changes in the distal forelimb cortical representation were only seen with training (i.e., map changes were present with training but not with spontaneous recovery). In a follow-up study, the investigators allowed spontaneous recovery to occur for one month and then began Klüver board training. Notably, the monkeys made good recovery of dexterity with the training, but there was no reemergence of the digit representation in the area around the infarct. More recently, the same result has been reported for a small endothelin 1–induced lesion in the arm area of M1 in the macaque; there was good spontaneous recovery over two weeks but no evidence for map expansion in remaining ipsilesional M1 (Herbert et al. 2015).

Other studies have shown that recovery after larger lesions in M1 that covered either the whole hand or arm region occurred without evidence for increased digit or arm representations adjacent to the lesion, even with early initiation of training (Liu and Rouiller 1999; Murata et al. 2008; Herbert et al. 2015). The implication of these studies is that map expansion may be a marker of learning in residual M1, just as it is in normal M1, but it is neither necessary nor sufficient for recovery. We say not sufficient based on studies we will describe below but also based on parsimony. If a monkey or a rat can recover both with and without map changes, then either there are two different ways to recover or there is just one, which means that map changes are not the cause in either case (Nishibe et al. 2015). Changes in cortical maps most likely indicate that a region has effector-relevant representation that can be altered by training, i.e., what kind of cortical real estate it is, but the increases in map size are not the causal mediator of the recovered performance. This conclusion is congruent with more recent work showing that dendritic spine number transiently increases during learning but then normalizes (Peters et al. 2017)

3.10.2 Reorganization at Sites Remote from the Lesion: Ipsilesional Premotor Regions

Here we will operationally define functional reorganization as changes in a region that occur in response to training that can be proven to be necessary for recovered motor behavior. By proof, we mean that the original deficit is reinstated when the putatively reorganized area is reversibly or irreversibly inactivated (double-lesion approach). Weaker proof would be provided by the demonstration that either movements can be elicited from an area where previously they could not be, or a region is increasingly activated as a behavior improves but this same region is not activated or is activated much less during the same behavior in healthy controls.

Several experiments in monkeys have taken the double-lesion approach dating back to the classical study by Leyton and Sherrington in 1917 that we describe in detail at the beginning of the chapter. Recall that in their study of motor cortical lesions in the chimpanzee, ablation of the remaining arm area on the originally lesioned side did not reinstate the deficit. Thus, from very early on, it was already apparent that recovery was not being mediated by areas immediately adjacent to large motor cortical infarcts. Yu Liu and Eric Rouiller investigated recovery from total M1 hand area lesions induced by ibotenic acid in two macaque monkeys (Liu and Rouiller 1999). Both monkeys had total hand paralysis for at least a month. Slow recovery then began and was followed with daily testing of dexterity on horizontal and vertical boards containing small slots. Performance reached a plateau of 30 percent of prelesion levels in three to four months. At nine months, reversible inactivations with muscimol test infusions were then performed. As in the Leyton and Sherrington experiment, inactivation of perilesional M1 did not reinstate the deficit. The deficit was only reinstated when muscimol was infused into both ipsilesional premotor areas (PMd and ventral premotor [PMv]) but not either one in isolation. There are several noteworthy points to be made about this study. First, the two monkeys only made partial recoveries, which is different from the studies we have mentioned before by Nudo et al. and Murata et al. (section 3.7.1). As the authors state, this might be because of the size of the lesion and/or the less intensive poststroke training—they were tested on a daily basis, but training for dexterity was not the goal. Second, the authors speculate that the perilesional cortex may not have shown evidence for reorganization because there was no hand representation left in M1 and so the shift was to hand representations in the premotor cortex. Third, the authors suggest that PMv and PMd could both contribute to partial recovery because they have been shown to have projections to the cervical cord (Dum and Strick 1991).

In a study that nicely complements the one by Liu and Rouiller, David McNeal and colleagues examined recovery of prehension in macaques after an aspiration lesion that removed the arm and hand areas of both M1 and lateral premotor cortex—PMd and PMv

(McNeal et al. 2010). The rationale was that this kind of large lesion reproduces the effects of a middle cerebral artery stroke in humans, in which the medial premotor areas are usually spared because they are supplied by the anterior cerebral artery. Substantial recovery of prehension, both accurate fast reaching and finger dexterity, was correlated with an increase in synaptic connections in lamina VII and IX of the cervical cord from CST projections from the ipsilesional supplementary motor area (SMA) (figure 3.10). A second lesion in SMA partially reinstated the deficit. A very similar result has been reported recently in the mouse with reinstatement of a prehension deficit after a second lesion in a medial premotor area following recovery from a stroke in the primary motor cortex (Zeiler et al. 2013). Interestingly, in both the monkey and mouse experiments, there was a significant amount of recovery even after the second lesion, which speaks to the possibility of subcortical mechanisms of recovery.

A recent preliminary experiment specifically sought to address the possible contribution of subcortical structures to recovery (Herbert et al. 2015). A large endothelin 1 stroke was induced in left M1 of a macaque with the lesion removing all of the M1 arm area so that only prolonged training enabled any reach recovery. The interesting finding was that stimulation of the right pontomedullary reticular formation (PMRF) led to more electromyography (EMG) responses in the right arm than would be expected in normal monkeys. The result was only in one monkey, and it was a stimulation experiment rather than a double-lesion experiment. Nevertheless, it is intriguing as it is the first demonstration of upregulation of contralesional brainstem nuclei after a large cortical stroke in a nonhuman primate. The contribution of cortical-brainstem projections and the output of brainstem structures to recovery have long been suspected but have taken somewhat of a backseat to alternative cortical contributions to the CST.

The studies summarized in this section suggest a way to rethink the concept of reorganization in a way that helps reconcile studies across species. We suggest that there are three critical factors: the postischemic repair process, which radiates out from the infarct and is time limited; the nature of the residual cortical representation surrounding the infarct; and training effects. It is the interaction between these three factors that determines the degree of behaviorally relevant reorganization. For example, in the squirrel monkey experiments by Nudo and colleagues, the induced infarct was small, which means that the peri-infarct region contained residual hand representation, that is, conditions best suited for reorganization: proper representation and close to the infarct, which probably explains why recovery can occur even without training. In contrast, in the studies by Murata et al. (2008) and Rouiller et al. (Liu and Rouiller 1999), the whole hand region was removed by the lesion so adjacent areas did not have the right representation, which means that reorganization has to occur at a more distant site: the hand representation in premotor regions. To make up for peri-infarct repair processes falling off with distance and the need to upregulate premotor hand representations, more training

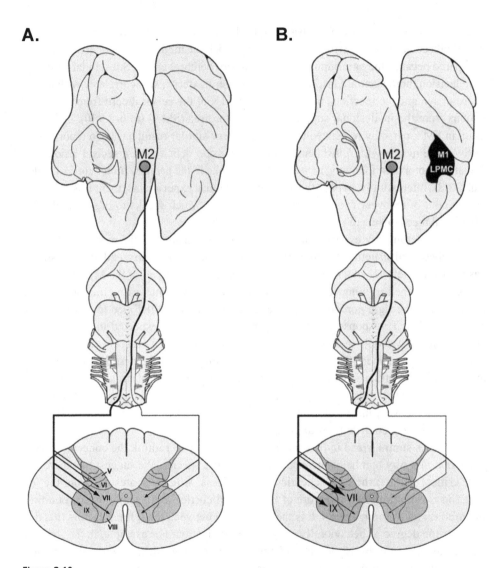

Figure 3.10
Upregulation of corticospinal projections from the supplementary motor cortex (M2) after stroke. The M2 projection originates from the medial wall of the hemisphere, crosses the midline at inferior brainstem levels (middle), and ends in the spinal cord (bottom). The relative intensity of the projection to spinal cord laminae is indicated by line thickness and arrow size. Denser terminal projections are represented by increased line thickness and arrowhead size. Progressively lighter terminal projections are indicated by progressively thinner lines and arrowheads. (A) Control monkeys. (B) Monkeys after motor recovery of dexterous upper extremity movements after large lateral motor cortical lesion. The lesion is located on the dorsal view of the hemisphere (blackened area) and involved the arm representation of the primary motor cortex (M1) and adjacent part of the lateral premotor cortex (LPMC). Extensive enhancement of the contralateral projection to lamina VII and IX occurred following the lateral motor cortical injury but not in other contralateral or ipsilateral laminae. Adapted from McNeal et al. (2010).

is required—spontaneous recovery alone is no longer sufficient. This reasoning can be captured by an equation:

$$\text{Restitution}_{[\text{magnitude}]} = \text{Behavior}_{[\text{dose}]} \times \text{Representation}_{[\text{residual amount}]} \times \text{Plasticity}_{[\text{level}]}$$

In this equation, $\text{Behavior}_{[\text{dose}]}$ refers to training relevant to the effector (e.g., prehension for the arm and hand). $\text{Representation}_{[\text{residual}]}$ presumes cortical regions with hand and arm representations of differing strengths that exist prior to the stroke and can be upregulated by repair mechanisms and training. Premotor cortex representations need more of both compared to M1 because the representations are weaker to begin with. $\text{Plasticity}_{[\text{level}]}$ refers to the conditions of heightened plasticity that fall off with distance from the infarct and normalize over time.

3.10.3   Reorganization at Sites Remote from the Lesion: The Contralesional Hemisphere

We finish the section on reorganization by discussing the vexed issue of whether the contralesional hemisphere contributes to recovery after stroke. The consensus is that it might contribute in the case of severe hemiparesis, the argument being that patients would be even more impaired without it. The implication is that the contralesional cortex only plays a role when ipsilesional areas are too damaged to contribute to better recovery. Most of the evidence for this view is indirect (Buetefisch 2015). However, double-lesion experiments have sought direct evidence for reorganization in the contralesional hemisphere after stroke. In the Leyton and Sherrington experiment, they found no recrudescence of the original deficit when they introduced a second lesion in the homologous motor cortex after having allowed the primate to make a good recovery from the first lesion (Leyton and Sherrington 1917). This result has been corroborated in recent times by several different groups in experiments done in monkeys. All demonstrate the absence of a deleterious effect of inactivation of contralesional M1 or premotor areas on the recovered arm or hand. This was even the case when the monkeys reached only 30 percent of prelesion performance, that is, they had quite severe residual hemiparesis (Liu and Rouiller 1999).

Two double-lesion studies and one viral inactivation study do, however, lend some credence to the idea that the contralesional hemisphere might contribute to recovery from severe hemiparesis. In rats, one study showed that inactivation of the contralesional motor cortex led to exacerbation of the prehension impairment that had recovered after four weeks of training (Biernaskie et al. 2004). This exacerbation was much more apparent in those rats with large infarcts and severe paresis with successful reaches being completely abolished, whereas the reduction in successful reaches was about 25 percent in rats that had recovered from small infarcts. This is a clear demonstration of an interaction effect between the contribution of the contralesional motor cortex and paresis severity. However, contralesional inactivation also had a significant effect on reach success in control rats, which suggests that bilateral cortical control of limb movements is much more pronounced in rodents compared to primates. In a second rat experiment, Martin Schwab and colleagues (Wahl et al. 2014) used a dual-virus approach to selectively inactivate corticospinal neurons in cortex contralateral to a large cortical stroke and found that doing so eliminated the recovered performance in successful pellet reach. This study indicates that the overall function of successful pellet retrieval in this study was dependent on contralateral corticospinal neurons, but the spinal viral injections are not shown and the behavior does not provide information on whether this is an effect of compensation (postural changes) or is due to contralateral sprouting in the spinal cord.

In a study of twelve patients with severe arm paresis (average upper limb FMA of 17), repetitive TMS was used to induce a reversible lesion (Mohapatra et al. 2016). The effect of TMS over contralesional M1 and PMd was to increase movement times and reduce movement smoothness. Whether these changes worsened the patients' FMA scores is not reported. Nor was it reported if the same TMS protocol had a similar effect in less affected patients. That said, the results by John Buford and colleagues referred to above showing upregulation of the contralesional reticular formation in a monkey that had recovered from large cortical stroke would suggest that there was increased contralesional cortical drive onto that brainstem nucleus. Such a view would also be consistent with the ipsilateral long-latency EMG responses that can only be elicited in poorly recovered patients using TMS (Turton et al. 1996). The origins of cortical projections to the reticular formation are much more distributed than the origins of the CST, which are mainly focused on M1. This means that double-lesion experiments that target contralesional M1 or even premotor regions may have no discernible effect because many other contralesional areas that contribute to corticoreticular projections are spared. It may be that when we talk about the role of the contralesional cortex in recovery, especially in primates, we need to consider its projections to brainstem nuclei more than either its transcallosal effects or its ipsilateral CST projections.

3.11 Conclusions

Trying to understand motor recovery early after stroke is like trying to put together a difficult jigsaw puzzle that also has many of its pieces missing. Problems arise immediately with the words *recovery* and *early*. As we have seen, the measure of motor behavior used is critical for assessing the initial deficit, tracking recovery, and choosing the treatment approach. In this chapter, we have made the case that measures at the level of impairment and motor control map best onto the neurobiology of brain repair. The term *early* is also fraught because it means different things to different people. Here we have used it roughly to mean the period of spontaneous biological recovery, but this can vary across species, and the meaning of spontaneous biological recovery is also somewhat ambiguous as it can refer to the recovery itself or it can refer to the special biological conditions that enable it. These conditions also heighten responsiveness to training, which might allow gains that are both greater and more extended in time than would be expected from spontaneous recovery alone.

Another problem in the field is that confusion can arise when concepts, techniques, and behaviors are not carefully scrutinized. The default position we take is that splitting must come before lumping; otherwise, we run the risk of being caught in an echo chamber of unexamined assumptions and half-baked ideas that mutually reinforce each other. As an example, consider the amount of time expended on the contralesional hemisphere. This has led to endless examination of how much it is activated in imaging studies, to the ubiquity of the interhemispheric imbalance model, and most recently to a great deal of emphasis on M1–M1 connectivity assessed with resting state. As we have seen both in this chapter and in chapter 2, the attention accorded the contralesional hemisphere is out of proportion to its causal relevance to the paretic deficit or to recovery. It is analogous to the excessive and unjustified interest in spasticity we discuss in chapter 2. Why do some things get disproportionate attention? We would venture that it is because it is often easier to emphasize a technique and what it can measure and then combine it with an overly simplistic model. The cost is that nuanced concepts and careful behavioral dissection fall by the wayside.

On a similar critical note, correlations of recovery with measures obtained using either resting state or task-evoked fMRI have not yet been very informative and have even been misleading. Nor are they even interesting unless a clear conceptual framework is provided beforehand that allows change in task-related mean activation or functional connectivity to be interpretable. It is also important not to conflate correlates of the recovered state with the recovery processes that lead to that state. There is also a tendency to combine models that are already tenuous on their own into ever more speculative concoctions—for example, combining the TMS-derived interhemispheric imbalance model with changes in resting-state connectivity. To show how this kind of approach can go wrong, it is worth referring to a study in monkeys recovering from a stroke in M1. The investigators performed concomitant positron emission tomography scanning, which showed activation in contralesional premotor areas in the recovered state (Murata et al. 2015). Notably, however, reversible inactivation of these areas had no discernible effect on the recovered limb. Thus, changes in an area detected by functional imaging do not themselves tell us that the area plays a causal role in recovered behavior. This is the bane of correlational imaging approaches, and causal tests are difficult to perform on functional and effective connectivity measures derived from resting-state fMRI (Grefkes and Fink 2014; Silasi and Murphy 2014). Thus, at this point, we remain skeptical about the causal relevance of functional imaging-derived metrics to recovered behavior.

What principles, even if provisional, can we extract from what has been covered in this chapter? Spontaneous recovery can be considered an endogenous repair process triggered by ischemia that can be modulated by particular forms of behavioral intervention. In animal models, the behavioral gains expected from spontaneous biological recovery can be amplified by training. Importantly, the assessment and training tasks are often the same. For example, the Klüver board is used to both assess and train hand dexterity. Although this could be seen as an example of teaching to the test, this concern is mitigated by the fact that the task is focused on improving movement quality and not just on achieving the end goal (i.e., the intention of the training is not for the monkey to get as many food pellets eaten in unit time as possible but to become better at individuating the digits). The degree to which the achieved dexterity on task A generalizes to task B is an empirical question. In contrast, task-based training, which is the norm when it comes to clinical rehabilitation, is mainly focused on compensatory movements to achieve ADLs. That is, the emphasis is on task success rather than on movement quality. This distinction was starkly demonstrated in the monkey study we discussed in section 3.7.1 by Murata and colleagues, in which a gain in dexterity was initially accompanied by a reduction in task success (i.e., a compensatory grasp was better than a precision grasp); only further training with the precision grip allowed it to overtake the compensatory strategy (Murata et al. 2008). It is doubtful that this continued training away from the compensatory response would occur on a rehabilitation ward. As we have seen, at the time of writing, there is no evidence that task-based training leads to improvements in motor impairment. To better approximate what is done in nonhuman primate studies, patients would have to be trained with a large number of repetitions of increasing difficulty on arm and hand movements using the equivalent of the Klüver board and then being assessed by the FMA test. This does not currently happen.

Although spontaneous recovery and training can add to achieve performance on a particular task and likely share molecular and physiological mechanisms, they are not the same. The difference has also been referred to as use-dependent and use-independent recovery (Murata et al. 2008). This difference is apparent in experiments showing that a small subtotal hand area infarct in primary motor cortex (M1) can be recovered from without training, but this is not the case for larger M1 infarcts involving the whole hand area. Spontaneous recovery can occur in the absence of any changes in cortical maps, whereas cortical maps can be seen to change, albeit sometimes transiently, with training. These map changes may therefore just be a marker that learning effects have been superimposed on repair effects.

Spontaneous biological recovery and increased responsiveness to training have a limited time window, which others and we have referred to as the sensitive period (Biernaskie et al. 2004; Ng et al. 2015; Zeiler et al. 2016). It lasts only weeks in rodents and maybe up to three months in primates. The hand may have more time to recover motor control than the arm. In order for spontaneous biological recovery and increased responsiveness to training, both consequences of a short-lived postischemic milieu, to optimally combine in the rehabilitation setting, it will be necessary to come up with training regimens that are very different from current ones and are perhaps analogous to enrichment in rodents. Non-task-based tasks that emphasize high intensities and high doses of complex exploratory movements will need to be devised (Krakauer et al. 2012; Zeiler and Krakauer 2013). These new behavioral therapies will likely be potentiated by pharmacological interventions, for example, selective serotonin reuptake inhibitors and anti-Nogo antibodies (Chollet et al. 2011; Wahl et al. 2014; Ng et al. 2015).

Recovery-related reorganization should only be claimed for those instances when a brain region or a change in connectivity can be shown to be required for a behavioral gain. We will also refer to such recovery-related reorganization as repair. Animal studies suggest that training on particular tasks in the sensitive period can induce changes in motor and premotor regions, strengthening their own direct connections to the spinal cord or indirect connections to other cortical regions and brainstem nuclei. Thus, recovery from impairment after stroke seems to reflect the dynamic interplay between motor cortical regions and residual corticospinal and brainstem descending pathways, in particular the reticulospinal tract, both of which can be upregulated and facilitated by training. We have proposed a model that posits that reorganization is the consequence of three factors: endogenous repair processes, preexisting effector-specific motor cortical representations, and training. All three factors exist at different levels or strengths, which means that when one is lower, the others need to be higher to induce an equivalent level of reorganization. In this model, training interacts both with the topology of peri-infarct repair and with preexisting physiological representations.

4

The Molecular and Cellular Biology of the Peri-Infarct Cortex and Beyond
Repair versus Reorganization

4.1   Introduction

Motor recovery after stroke occurs within the substrate of initial cell death, secondary damage, and early and late tissue reorganization. The next two chapters review the cellular events following ischemic damage to the brain and how they relate to motor recovery. The biology of stroke damage and tissue repair has been extensively reviewed (Terasaki et al. 2014; Rosenzweig and Carmichael 2015; Carmichael 2016a). This chapter will draw upon these reports to present a timeline of events at the site of stroke and in adjacent and connected brain regions in the context of recovery.

At the outset of a discussion of tissue changes during both initial and later recovery periods after stroke it is important to distinguish between reorganization and repair. Tissue reorganization is a descriptive term encompassing all the observed changes in the brain after stroke. Tissue repair refers to those changes in the brain that are causally relevant to behavioral recovery. Terms that also imply tissue repair include tissue restoration and tissue regeneration. These terms and their multi-syllabic cousins (neurorestoration, neuroregeneration) are omnipresent in the literature and are frequently erroneously invoked when only tissue reorganization without functional effects has been observed. However, most of the tissue reorganization that has been described after stroke has not yet been shown to be causally related to recovery. For example, angiogenesis, the formation of new blood vessels in the brain tissue around the infarct, and the neurogenesis it induces, are a dramatic case of tissue reorganization after stroke but it remains unclear if they represent actual repair. It is apparent that this field is only now exiting the foundational period of description and beginning to determine which aspects of tissue reorganization also represent repair.

The progression from stroke cell death to repair involves sequential stages of primary cell death, secondary injury events, reactive tissue progenitor responses, and formation of new neuronal circuits (figure 4.1). This progression is radial: the tissue that suffers the infarct and secondary injury radiates signals out to the surrounding tissue, including

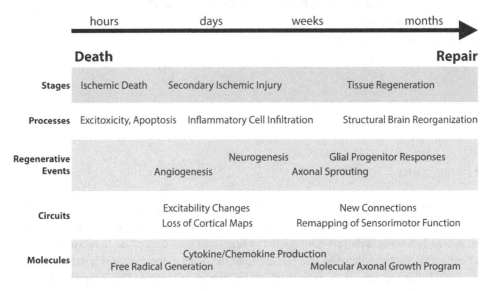

Continuum of Stroke

Figure 4.1
The continuum of stroke, from damage to repair. Below the arrow, each horizontal row progresses, from top to bottom, to an increasingly small scale of event, from stages of tissue responses down to molecules within these responses, and organizes these along a timeline. All events can be organized on a macro scale into stages of initial damage, secondary damage, and repair. At the other end of the spectrum, distinct classes of molecules participate in each point along the time continuum of stroke. Modified from Carmichael (2016a).

free radicals, inflammatory cytokines, and synchronized electrical activity. In addition to the radial diffusion of signals from the injured region to surrounding intact tissue, damage signals in stroke are relayed through neuronal connections. This is because dying tissue includes both the axons of distant neurons that projected to that tissue and the projections of neuronal cell bodies within the dying area. This means that dying neurons relay signals of their death through their degenerating connections to distant sites. . From these relayed damage signals, reactive astrocytosis, inflammatory processes, and the formation of new connections occur in distant brain areas.

4.2   Regions of Stroke Damage

Stroke can be conceived as a site of complete cellular death surrounded by zones of damage and of tissue reorganization. This concept was first suggested by the progression of early damage in acute stroke (Sharp et al. 2000), where the stroke core was the site of complete cell loss and surrounding zones at different radii from the center of tissue exhibited both injury and reorganizational responses, with the latter manifesting as distinct zones of reactive astrocytosis, dendritic spine remodeling, and axonal sprouting.

Several regions of brain can be distinguished after stroke: the infarct core, the scar, the peri-infarct tissue, and non-contiguous but connected regions. The first three regions sweep out radially from the stroke core. The non-adjacent but connected regions include, for example, ipsilesional motor cortical areas, basal ganglia, brainstem, and thalamus.

4.2.1 The Infarct Core

Ischemic stroke starts as an acute loss of blood flow to a region of the brain, which precipitates a cascade of direct damage in the distinct cell types within the brain parenchyma. The parenchyma of the brain is made up of seven principal cell types, which are best considered separately after ischemic tissue damage: neurons, oligodendrocytes, oligodendrocyte precursor cells, astrocytes, endothelial cells, microglia, and pericytes. Ischemic cell death has been mostly studied in neurons (Moskowitz et al. 2010; Lai et al. 2014). This means that most of our understanding of "cell death" in the brain after stroke is in fact limited to that of neuronal cell death. Neuronal cell death occurs quickly and prominently in cerebral ischemia, and will be described first, but injury and death to the other cell types in the brain closely follows.

The initial loss of blood supply causes the most energetically expensive neuronal function to collapse: the maintenance of the resting membrane potential. Neurons devote a substantial amount of energy to maintain ion gradients through the sodium/potassium pump, for this gradient is needed to fire action potentials. When ischemia depletes adenosine triphosphate (ATP), neurons cannot maintain this sodium/potassium gradient, which results in depolarization. This is seen as the "anoxic depolarization" in extracellular recordings: the whole tissue depolarizes. Neurons release glutamate as they depolarize, which causes further depolarization of neighboring neurons because of glutamate's actions on N-methyl-D-aspartate and α-amino-3-hydroxy-5-methyl-4-isoxazolepropionic acid (NMDA and AMPA, respectively) glutamate receptors, which let in sodium and calcium. This further depolarization, on top of the initial ischemic or anoxic depolarization, causes further glutamate release, which stimulates further depolarization, and a positive feedback cycle ensues. Glutamate, acting through the NMDA receptor, induces massive intracellular calcium entry. Calcium also enters through voltage-dependent calcium channels as the neuron depolarizes. A major calcium sink in the neuron is the mitochondrion, which swells as it buffers calcium and becomes dysfunctional. Calcium-mediated proteolysis, lipolysis, and DNA degradation lead to necrotic cell death. There are other molecular actors in this initial neuronal cell death, including free radical formation in the disordered mitochondria. In total, this is the process of neuronal excitotoxicity, first prominently described in stroke in the late 1980s and the motive force behind many of the stroke clinical trials of the 1990s (O'Collins et al. 2006) (summarized in figure 4.2). Excitotoxic cell death is quickly irreversible. But a reanalysis of animal modeling data and the experience of failed stroke

Infarct Core   Peri-Infarct Brain Tissue

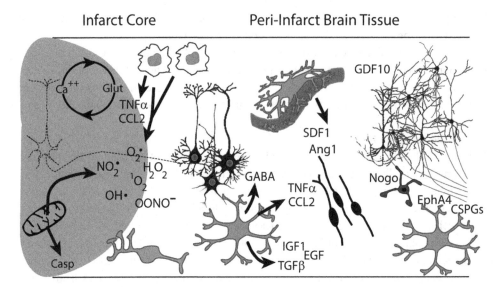

Figure 4.2
The radial progression from damage to regeneration in stroke. This schematic shows the damage processes that occur in the stroke core and the diffusible signals or axonal connection signals that relay this damage to intact tissue in the peri-infarct brain. Free radicals, damaged axonal connections (illustrated by the dotted neuron), and cytokines diffuse from the area of the core into surrounding tissue and initiate some of the processes of neural repair. Ang1 = angiopoietin 1; CSPG = chondroitin sulfate proteoglycan; GABA = gamma amino butyric acid, CCL2 = chemokine (C-C motif) ligand 2; EphA4 = ephrin receptor subtype; GDF10 = growth differentiation factor 10; IGF1 = insulin-like growth factor 1; Nogo = NogoA = reticulon 4 = neurite outgrowth inhibitor; SDF-1 = stromal-derived factor 1, TNFα = tumor necrosis factor α; NO_2^-, $^1O_2$, H_2O_2, O_2^-, OH^-, $OONO^-$ = free radicals. Modified from Carmichael (2016a).

trials indicate that the initial phase of neuronal cell death is quite likely beyond the reach of clinical medicine as a therapeutic target. In fact, clinical decision making in the emergency department in stroke seeks to define not the ischemic core but the degree of tissue around the core that has reduced blood flow but is not yet irreversibly damaged—the so-called mismatch between the densely ischemic tissue and the area of reduced perfusion but salvageable tissue (Rosso and Samson 2014).

The brain contains at least six other cell types in addition to neurons. The death and destruction of these other cell types has not been as definitively studied as it has for neurons. How damage to these cells contributes to the behavioral deficit and to partial recovery is less worked out than it is for neurons. In vitro data indicate that neurons are the most sensitive to oxygen/glucose deprivation, followed by endothelial cells, astrocytes, and microglia (Redzic et al. 2013). As with neurons, glutamate signaling may play a role in oligodendrocyte cell death after stroke, through both NMDA and AMPA glutamate receptors (Sozmen et al. 2012). In rodent stroke models, oligodendrocytes survive longer than neurons (Mabuchi et al. 2000). However, oligodendrocyte loss leads to secondary axonal degeneration and infarct expansion, particularly for white matter strokes (Sozmen et al. 2012; Hinman et al. 2015). Exposed to the same ischemic stimulus in stroke models, astrocytes die later than neurons (Chu et al. 2007; Redzic et al. 2013) and do so through mechanisms that involve swelling, ATP release, and signaling through ATP receptors—$P2Y_1$ purinergic receptors (Franke and Illes 2014). Astrocytes in a particular brain region form a network connected by gap junctions (Khakh and Sofroniew 2015), and astrocyte damage in one area is quite likely to signal and spread to adjacent areas. With this differential sensitivity in mind, hours after the infarct, all cell types in the stroke core die, although they do so on varying time scales and via different cellular mechanisms. Tissue repair and recovery depend in large part on what happens adjacent to the stroke core—the radial effects of the ischemic stimulus on surrounding brain: peri-infarct cortex.

4.2.2 Peri-Infarct Tissue

Peri-infarct is a regional term. For the biology of stroke recovery, it is crucial to have a clear description of the possible areas that are subsumed by this term. In its most general, topographic sense, peri-infarct tissue would include all of the brain adjacent to the stroke. However, there are at least two distinct biological regions within the topographic region around the infarct. The brain region immediately bordering the stroke site is a zone of intense glial reactivity and partial neuronal loss that constitutes scarring. This area contains distinct scar-related processes. It is the only zone in rodents in which there is astrocyte proliferation and bipolar or palisading astrocyte reactivity. Pericytes in this zone; cells with a close association to brain capillaries, appear to detach and form actual fibrosis and tissue contraction after stroke (Fernández-Klett et al. 2013) and other central nervous system (CNS) injuries (Göritz et al. 2011). This zone of scar formation is narrow, several hundred microns in rodents, primates, and humans. The barrier nature of this scar is contributed by both pericytes and astrocytes and serves to limit the initial damage from stroke and sequester inflammatory cells in the lesion. Just outside this zone of scar formation, is the peri-infarct cortex proper. This is the region in which neurons engage in axonal sprouting in the endogenous state. Angiogenesis, neurogenesis and oligodendrocyte progenitor cell responses also occur in peri-infarct cortex.

Peri-infarct tissue is not synonymous with the stroke "penumbra." The penumbra was originally defined physiologically as the region of electrical silence in intact but hypoxic tissue at the onset of ischemia (Astrup et al. 1981). The ischemic penumbra has come to mean the brain tissue that is not initially damaged by stroke but will progress to infarction over time if therapy is not initiated. In the emergency department and clinical stroke unit, medical care is mobilized to image the stroke penumbra and maximize resources to save it, usually with thrombolytic agents (clot busters) and stent retrieval devices (Fisher and Albers 2013; Rodrigues et al. 2016). The goal is to define the area of mismatch: a brain region that has reduced blood flow but is not yet showing magnetic resonance imaging (MRI) signals of irreversible cell death (Rosso and Samson 2014). Human imaging and rodent stroke models indicate that the ischemic penumbra will die and become part of the stroke core in roughly two days (Davis and Donnan 2014). In other words, in the acute period after stroke, the core dies and the penumbra has two fates: it will die and become part of the stroke core or be salvaged by therapy and survive. In the former case, the penumbra adds to the stroke core. In the latter, the salvaged penumbra becomes part of the peri-infarct tissue: injured but surviving. From the perspectives of the biology of tissue repair and the time course of recovery, the debate over the penumbra is moot: this is really a progression of the process of initial ischemic cell death. By the time cellular events of neural repair are activated, or by the time (in the future) a neural repair therapy might be delivered, the penumbra is gone. However, the mechanisms of cell death in the penumbra have important roles as triggers for reorganization in part via radial transmission to peri-infarct tissue.

Delayed cell injury and death in the peri-infarct tissue occur along with inflammation and free radical generation. This can be schematically collapsed into a single representation of stroke and peri-infarct tissue (figure 4.2). The partial blood flow and reduced oxygenation in the penumbra activates endothelial cells, which attract a tremendous influx of neutrophils (Schofield et al. 2013) and macrophages (Mabuchi et al. 2000). Microglia are activated within minutes in injured stroke tissue (Morrison and Filosa 2013) and, unlike in the stroke core, they survive (Mabuchi et al. 2000). These three cell types release cytokines and produce free radicals that cause secondary damage but also generate the signals for subsequent neural repair.

4.3 Triggers for Neural Repair Radiate Outward from the Infarct Core

The progression of brain ischemia from the infarct core to the penumbra releases secondary injury cascades into the tissue that will survive: the peri-infarct tissue (figure 4.3). These cascades involve reperfusion injury in damaged cells, which causes free radical production; the activation of astrocytes and white blood cells, which release cytokines and free radicals; and the production of synchronized neuronal activity. This is a radial effect of stroke, whereby these secondary injury signals activate later tissue-reorganization events. Free radicals generated in the penumbra and in peri-infarct tissue are part of the stimulus for neurogenesis (Le Belle et al. 2011; Hu et al. 2014); cytokines activate astrocytes, promote angiogenesis, and induce axonal sprouting (Brumm and Carmichael 2012; Gertz et al. 2012; Gleichman and Carmichael 2014; Li et al. 2015); and synchronized neuronal activity induces axonal sprouting and the formation of new connections (Carmichael and Chesselet 2002).

Inflammatory cytokines are released by activated microglia (early), invading neutrophils (early), and macrophages (late) (Iadecola and Anrather 2011; Benakis et al. 2015). These can directly stimulate aspects of tissue reorganization (Ekdahl et al. 2009) and also induce an activated state in brain endothelial cells and astrocytes. Activated

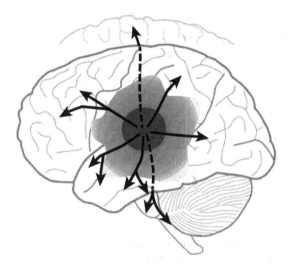

Figure 4.3
The radial and relayed nature of stroke.
Stroke damage and secondary injury is
relayed from the stroke core in a radial
manner into peri-infarct tissue. Stroke kills
not only neurons in the stroke core but also
their axons, which means that degenerating
axons and lost synaptic contacts occur in the
brain areas connected to the stroke. This is
the relayed nature of stroke damage. The
effects of this radial and relayed stroke
damage mean that inflammatory molecules
are activated in peri-infarct and more
distantly connected areas, including such
downstream areas as basal ganglia,
thalamus, and spinal cord.

astrocytes and endothelial cells then further produce molecules that induce or alter the brain's tissue response to stroke (Zhao and Rempe 2010; Brumm and Carmichael 2012; Gleichman and Carmichael 2014). For example, several cytokines from the activated astrocytes and endothelial cells influence a stem cell response to stroke (see later). Activated blood vessels in peri-infarct tissue secrete stromal-derived factor 1 (SDF-1), a cytokine that is tropic (attractive) to immature neurons as they migrate to areas of damage after stroke (Ohab and Carmichael 2008; Kahle and Bix 2013). Chemokine (C-C motif) ligand 2 (CCL2), leukemia inhibitory factor (LIF), and tumor necrosis factor α (TNFα) are also inflammatory signaling molecules that are released by astrocytes and microglia/macrophages after stroke, induce distinct stages of normal neurogenesis, and have also been implicated in poststroke neurogenesis (Liu et al. 2007; Ekdahl et al. 2009; Gonzalez-Perez et al. 2012). Other cytokines and chemokines are released by activated astrocytes and microglia and participate in poststroke neurogenesis (Yan et al. 2007; Ekdahl et al. 2009; Kahle and Bix 2013). The overall concept that emerges from these studies is that every cell type is in a state of activation because the peri-infarct tissue has in effect been given a strong poke by its neighbor (the infarct core). Through several classes of activating molecules and signaling activity from the stroke core, every cell type in the peri-infarct tissue is driven into an altered and activated state for at least days after the stroke and in some cases weeks.

Inflammatory cytokines that are released from the stroke core into peri-infarct tissue not only activate astrocytes and microglia but also participate in poststroke neuronal (axonal) sprouting. As discussed below, axonal sprouting after stroke induces the formation of new connections between peri-infarct cortex and motor, premotor, and somatosensory cortical areas. These connections causally mediate motor recovery (Overman et al. 2012; Li et al. 2015). The transforming growth factor beta (TGF-β) superfamily of cytokines contains several molecules that stimulate poststroke neuronal sprouting. Bone morphogenetic protein 7 stimulates dendritic growth in neurons (Withers et al. 2000) and promotes behavioral recovery (Ren et al. 2000). The TGF-β family member, growth differentiation factor 10 (GDF10), is induced in peri-infarct cortex after stroke and is a potent stimulant for axonal sprouting and functional recovery (Li et al. 2015). GDF10 acts by inducing a transcription factor in the cell, Similar to the Drosophila gene Mothers Against Decapentaplegic 2 (SMAD2). This transcription factor is also activated in axonal sprouting and recovery in the peripheral nervous system, spinal cord, and optic nerve and may be a common pathway for tissue repair in the nervous system (Omura et al. 2015).

Activation of astrocytes in peri-infarct cortex, in part via inflammatory cytokines, also induces molecules that block axonal sprouting, such as chondroitin sulfate proteoglycans (CSPGs) (Carmichael et al. 2005) or ephrin A5 (Overman et al. 2012). Astrocytes in the narrow region of scar around the stroke proliferate, secrete CSPGs and wall off the spread of inflammation. Astrocytes in adjacent regions show far less morphological change: they do not proliferate but do express molecules that suppress axonal sprouting (Overman et al. 2012). It should be noted that astrocytes are not the only source of CSPGs and may not really form the intense barrier to tissue repair that has become implied in the term "scar" (Anderson et al. 2016). This is an area of intense study in stroke and other CNS injuries— determining what cells actually form brain scar and whether it is the monolithic barrier to tissue repair as has been traditionally conceived.

Peri-infarct inflammatory signals control both the induction of axonal sprouting and the induction of axonal growth inhibitors that block axonal sprouting. The local interplay of progrowth and growth-inhibiting molecular systems determines the extent and pattern of axonal sprouting and recovery. Stroke generates free radicals in the infarct core and peri-infarct tissue. This process occurs through the initial ischemia itself, through reperfusion, and through inflammatory cell release (Heiss 2012). Excess free radicals have potent biological effects because they are freely diffusible, can have a long tissue half-life, and activate many downstream molecular processes. This last element of free radical action, the activation of downstream molecular processes, is a crucial action in the radial nature of stroke. In the infarct core, free radical generation produces tissue damage and cell death, in the peri-infarct cortex, however, free radical generation activates downstream signaling cascades in cells that will survive. These intracellular cascades that are activated by free radicals in the peri-infarct tissue mimic the signaling of growth factors (Sauer and Wartenberg 2005; Le Belle et al. 2011). For example, neurotrophins (nerve growth factor [NGF], brain-derived neurotrophic factor [BDNF], neurotrophin 3 [NT-3]) normally signal to a cell in a way that involves small elements of free radical generation (Sauer and Wartenberg 2005; Le Belle et al. 2011). Free radicals also directly activate neural stem cells through induction of growth factor signaling pathways (Le Belle et al. 2011).

Immature neurons in the largest germinal matrix, the subventricular zone, respond to the stroke, migrate to the area of damage, and may participate in neural repair (Ohab and Carmichael 2008; Kahle and Bix 2013). Blocking free radical generation, such as with an antioxidant, reduces poststroke neurogenesis (Hu et al. 2014). It appears that free radical generation in peri-infarct tissue, together with other cytokines and growth factors, signals to multipotent neural stem cells and their progeny in the subventricular zone (Ohab and Carmichael, 2008; Ekdahl et al. 2009; Vieira et al. 2011; Kahle and Bix, 2013).

In the early stages after stroke, peri-infarct tissue exhibits synchronized electrical activity. Within the first several days after stroke, peri-infarct spreading depressions sweep through cortex. These are synchronized and long-duration neuronal depolarizations that produce a substantial metabolic demand on peri-infarct tissue. This demand may be so great that the tissue may die—in other words, peri-infarct spreading depressions may convert a region of penumbra into core (Heiss 2012; Pietrobon and Moskowitz 2014). However, peri-infarct spreading depressions also occur in tissue that will survive (von Bornstädt et al. 2015). They are followed in the first several days after stroke by synchronized low-frequency neuronal discharges across peri-infarct cortex, which occur at < 0.1 Hz (Carmichael and Chesselet 2002; Gulati et al. 2015). These two events are transitions from acute stroke injury to longer-term tissue reorganization. In serving as a trigger, this synchronized neuronal activity in peri-infarct cortex after stroke resembles similar activity patterns in the formation of brain connections during development in the retina, hippocampus, and cortex (Katz and Shatz 1996; Stellwagen and Shatz 2002; Egorov and Draguhn 2013). Synchronized neuronal activity may activate a downstream molecular program for neuronal growth or may stimulate the coactivation of many synapses in a particular region of neurons, stimulating Hebbian plasticity and synaptic sprouting after stroke, as it does in the developing brain.

4.4   Excitatory/Inhibitory Balance in Peri-Infarct Cortex

During days to weeks after stroke, the peri-infarct cortex loses its initial response to afferent inputs and then gradually regains this response, a process that occurs with widespread changes in the location and temporal flow of neuronal excitability, visualized with voltage-sensitive dye responses with optical imaging (Brown et al. 2009; Lim et al. 2014). These phenomena manifest at the macro level as changes in cortical maps. The initial loss of brain excitability in peri-infarct cortex and associated alterations in cortical motor and sensory maps corresponds to the period of maximal deficit in rodent stroke models. However, reversal of hypoexcitability and restoration of motor or sensory maps both occur on different timelines to behavioral recovery itself (see chapter 3) and so cortical map reorganization remains of indeterminate relevance in terms of repair.

Stroke causes an increase in GABAergic tonic inhibition in peri-infarct neurons (Clarkson et al. 2010). Tonic GABA inhibition is distinct from phasic or synaptic GABA inhibition. Phasic or synaptic GABA action is a mechanism of intracortical inhibition and can be measured with transcranial magnetic stimulation (TMS) and electrophysiology. Phasic GABA inhibition is fast and communicated by direct synaptic contact in a precise, circuit-specific way. In other words, phasic inhibition is what most of us mean when we discuss an inhibitory connection or "brain inhibition": one cell synapsing on another cell and inhibiting that cell every time it fires.

By contrast, tonic GABA inhibition is a slow inhibitory current that is mediated in part by levels of extracellular GABA in the milieu surrounding the neuron. Extrasynaptic GABA receptors bind this more "ambient" GABA and desensitize more slowly than synaptic GABA receptors. This means that extrasynaptic GABA receptors mediate an inward inhibitory chloride (Cl⁻) current that is potentially always present at some level. Because this current mediates a background level of inward negative current, an excitatory neuron must be depolarized even further to fire an action potential. The tonic GABA current thus establishes the threshold for action potential firing of a neuron (Glykys and Mody 2007). After stroke, there is downregulation of GABA uptake mechanisms in reactive astrocytes, leading to increased extracellular GABA levels. This increases the baseline inhibitory current in pyramidal/excitatory neurons. This process of increased tonic inhibition is initially neuroprotective. Blocking tonic GABA inhibition in the first several days after stroke causes an increase in ischemic cell death (Clarkson et al. 2010). However, the increase in tonic GABA inhibition persists long after the threat of ischemic cell death (Clarkson et al 2010). By persisting into the period of recovery, the increased tonic GABA current on neurons in peri-infarct tissue contributes to impaired circuit function in this area: these neurons are hypoexcitable. Blocking tonic GABA inhibition after the first several days from the stroke promotes motor recovery (Clarkson et al. 2010). Thus, increased tonic GABA signaling from diminished astrocyte GABA uptake does indeed inhibit recovery. Blockade of tonic GABA signaling can be termed a "tissue repair" approach as it has been directly linked to recovery. This is a tractable pharmacological therapy for stroke recovery and has been reported by several laboratories (Clarkson et al. 2010; Lake et al. 2015). In the mouse, where these data were obtained, the increase in tonic GABA inhibition after stroke occurs in peri-infarct tissue that is further away from the stroke than the scar.

Stroke also alters excitatory neuronal signaling in peri-infarct cortex. Neuronal firing frequencies are elevated in peri-infarct cortex for several months after focal stroke, with a maximal increase in firing rate in the first week (Schiene et al. 1996). Electrically shocking cortical slices prepared from the poststroke brain evokes longer depolarizations in nearby neurons in peri-infarct cortex compared to neurons from nonstroke cortex (Neumann-Haefelin et al. 1995). Also, in brain slice preparations, long-term potentiation is enhanced in peri-infarct cortex seven days after stroke when the subcortical tissue is electrically stimulated (Hagemann et al. 1998). In the basal ganglia adjacent to the stroke, there is increased excitatory neurotransmission, seen as increased spontaneous excitatory postsynaptic potentials (Centonze et al. 2007). Human studies with TMS have also identified reduced paired-pulse inhibition that is suggestive of hyperexcitability in ip-silesional motor cortex after stroke (Cicinelli et al. 2003), although TMS findings of hyperexcitability in human stroke may vary depending on the technique used (Classen et al. 1997). These data indicate that measures of stimulated glutamatergic signaling and of local fast inhibitory signaling show a hyperexcitable state that peaks in the first days to weeks after stroke.

How do reports of hyperexcitability square with the finding of increased tonic GABA signaling in stroke in this same brain region of peri-infarct cortex? Most of the studies that have reported "hyperexcitability" have done so with artificially evoked activity—such as electrically stimulating the connections that lead up to cortex or stimulating the brain with TMS. However, when measuring pyramidal neuron excitability in the natural resting state, an increase in tonic inhibition is seen (Clarkson et al. 2010). Furthermore, the findings of increased neuronal firing rates and excitability are not consistent with newer optical imaging findings that show a reduced firing rate and pattern of evoked activity in cortex after stroke, with voltage-sensitive dyes that measure evoked and resting firing rates, and with evoked activity from forelimb stimulation (Dijkhuizen et al. 2003; Clarkson et al. 2013; Lim et al. 2014). Thus, the actual physiological responsiveness of peri-infarct cortex after stroke is reduced, at least for several weeks, even though artificially stimulating this brain region can evoke abnormal and hyperexcitable responses.

One of the major systems in neurons to communicate enhanced neuronal activity is the cAMP response element-binding protein (CREB) transcription factor. When neurons are stimulated to a greater degree, CREB is induced, binds to specific sites in DNA, and activates a set of genes that set in place long-term changes in the neuron. These changes include enhanced excitability (Dong et al. 2006; Han et al. 2006) and enhanced neuronal responsiveness to inputs, such as induction of long-term potentiation (Wu et al. 2007). CREB activation mediates many aspects of memory in experimental animals, such as spatial memory and fear conditioning. CREB activation is important in cortical map plasticity in the adult after peripheral sensory loss (Glazewski et al. 1999; Barth et al. 2000) and in the formation of brain maps during development (Mower et al. 2002). With roles in cortical map development, map plasticity, and in the excitability changes that underlie learning and memory, it is perhaps not a surprise that CREB is fundamentally involved in motor recovery after stroke, as assessed in experimental models. This will be discussed in detail in the next chapter. The interesting observation in these studies of CREB in recovery is that the induction of this transcription factor after stroke is so powerful. In experimental animals, the induction occurs in even a small set of motor and premotor neurons adjacent to a stroke and can mediate early and complete motor recovery (Caracciolo et al. 2012). These findings indicate that specific motor and premotor circuits can mediate recovery of function after stroke—this process does not have to occur in widespread brain regions. These findings further indicate that enhancing excitable signaling in ipsilesional motor cortex adjacent to stroke sites enhances recovery.

How might this translate to the clinic? As will be discussed in later chapters, external measures to boost brain excitability, such as anodal transcranial direct-current stimulation (tDCS), enhance motor learning. Current trials are under way to determine if combining TMS or tDCS can be paired with neurorehabilitative training to enhance recovery after stroke. However, the initial clinical trials of TMS and tDCS in stroke recovery have not been positive (Plow et al. 2015; Grimaldi et al. 2016; Vallence and Ridding 2014). tDCS activates neurons and induces CREB and its downstream genes (Podda et al. 2016). It is possible that a pharmacological approach that induces CREB signaling may be able to replace the external electrical or magnetic stimulation paradigms, such an "excitability drug" could be paired with neurorehabilitation to promote recovery.

4.5 Dendritic Spine Alterations in Peri-Infarct Cortex

Stroke induces a change in neuronal morphology in the same regions as altered neuron electrophysiological responses and cortical map plasticity. Dendritic spines form the synaptic contacts of cortical pyramidal neurons. These are mostly stable in the adult cortex (Grutzendler et al. 2002; Trachtenberg et al. 2002) although they remodel in response to loss of afferent inputs or learning (Cheetham et al. 2008; Jasinska et al. 2010; Fu and Zuo 2011). There is initially a net loss of dendritic spines in peri-infarct cortex in the first week after stroke (Brown et al. 2008; Mostany et al. 2010). This occurs in regions with normal as well as abnormal blood flow (Mostany et al. 2010), indicating that this reduction is at least in part due to neuronal network damage from loss of axonal connections rather than partial ischemia in peri-infarct tissue. After this loss of connections, peri-infarct cortex within 1 mm of the infarct in the mouse recovers synaptic connections back to premorbid levels (Brown et al. 2008; Mostany et al. 2010), whereas neurons in regions 1 to 2 mm away from the infarct gain more synaptic connections compared to control (Mostany et al. 2010). This distance is well away from the border zone of scar around the stroke and includes peri-infarct regions adjacent to the stroke. These differences in dendritic spine responses in regions removed from the infarct core and the scar zone suggest that there are further distinctions in the type of tissue responses in stroke even in regions that are relatively distant from the infarct, at least in the mouse. The peri-infarct cortex, extending away from the stroke, can exhibit dendritic spine and synaptic loss and then gain. The actual branches of dendrites also remodel after stroke, with retraction and growth that is maximal two weeks after the infarct and occurs most prominently within 200 μm of the infarct core (Brown et al. 2010). This dendritic branch remodeling thus occurs in a region within the scar area of cortex, directly adjacent to the stroke.

Can we change the process of dendritic remodeling after stroke to promote recovery? This question is relevant because the dendritic spine is the structural component of the excitatory synapse. Changes in synaptic structure after stroke mean gains or losses in neuronal connections. If a candidate therapy preserves dendritic spines that might be lost in cortex adjacent to stroke or facilitates the formation of dendritic spines during task-specific activity, then this therapy might stimulate recovery. One way to do this is to manipulate the chemokine signaling system C-C chemokine receptor type 5 (CCR5). This receptor normally signals within immune cells in the inflammatory response to injury and infection. However, stroke induces CCR5 in neurons adjacent to the stroke site. Blocking CCR5 after stroke prevents the initial loss of dendritic spines that stroke normally causes in these neurons and promotes motor recovery (Joy et al. 2016). CCR5 acts in part by inducing CREB (Zhou et al. 2016). The CCR5 antagonist, maraviroc, has been approved for use in humans for treatment of HIV infection. Maraviroc promotes recovery of function after stroke and traumatic brain injury in animal models (Joy et al. 2015). These data position CCR5 as a promising molecular target for drugs such as maraviroc for stroke recovery.

In addition to dendritic spine plasticity, stroke also induces new axonal connections from neurons in peri-infarct cortex and in adjacent brain regions of premotor and somatosensory cortex, which are initiated in the first week after stroke and are reliably present one month after stroke (Carmichael et al. 2001; Li et al. 2010; Overman et al. 2012), particularly in the superficial cortical layers. In humans, it is of course not possible to label brain connections and track them precisely so as to determine if new connections form after stroke. However, neurons adjacent to stroke in peri-infarct cortex induce genes, such as *GAP43*, that are strongly linked to axonal sprouting (Ng et al. 1988). These data indicate that stroke, as the result of the loss of connections with the infarct core and damage signals within peri-infarct cortex, causes alterations in neuron morphology and in the pattern of connections in distinct spatial zones over specific time scales.

4.6 Reactive Astrocytes and Extracellular Matrix Changes in Peri-Infarct Cortex

Within peri-infarct tissue, additional anatomical zones are formed by reactive astrocytes. Reactive astrocytes are not a homogeneous cell type but instead have distinct morphological and functional properties in zones that extend away from the infarct or CNS injury site. This has been best detailed in the injured spinal cord and in normal brain, where distinct morphologies of reactive astrocytes likely demarcate different zones of tissue repair and recovery (Wanner et al. 2013) and/or of distinct circuit function (Khakh and Sofroniew 2015). Similarly, after stroke, distinct types of reactive astrocytes are also present. The scar region adjacent to stroke is formed by reactive astrocytes that have evolved into a bipolar morphology from their normal multipolar state and secrete large amounts of CSPGs (Gleichman and Carmichael 2014). The scar appears to be the only region where there is true astrocyte proliferation after stroke (Barreto et al. 2011; Shimada et al. 2011). Ablation of these border astrocytes after stroke and other injuries results in increased inflammation and increasing lesion size (Sofroniew 2012). This protective function of reactive astrocytes fits well with the transcriptional profile of reactive astrocytes after stroke compared to reactive astrocytes after an inflammatory stimulus. Reactive astrocytes in a middle cerebral artery occlusion model of stroke show a gene expression profile that is neuroprotective, with neurotrophic factors and cytokines, such as BDNF and connective tissue growth factor (CTGF) (Schwab et al. 2000), whereas in the case of inflammation, reactive astrocytes activate complement, interferon, and other inflammatory and synapse destruction pathways (Zamanian et al. 2012). At distances more removed from the stroke in peri-infarct cortex, astrocytes still "react" to the stroke and display increased glial fibrillary acidic protein (GFAP) expression but retain their normal multipolar morphology (Gleichman and Carmichael 2014; López-Valdés et al. 2014). Astrocytes show distinct molecular properties by brain region with respect to sonic hedgehog signaling, glutamate uptake, and extracellular recognition and patterning molecules (Garcia et al. 2010; Chaboub and Deneen 2012; Molofsky et al. 2012; Khakh and Sofroniew 2015). It is likely that differences in astrocyte subtypes persist in the reactive astrocytosis after stroke and lead to zones in which scar-forming molecules and other axonal growth inhibitors are differentially expressed. These distinct astrocyte zones will mean distinct regions of tissue repair and recovery and are likely to be an important area of future study in the field.

Stroke acts radially on peri-infarct tissue to alter not just the cells and signaling within this brain region but also the extracellular environment, or the extracellular matrix. Reactive astrocytes produce large molecules and deposit these into the extracellular matrix (Gleichman and Carmichael 2014; Khakh and Sofroniew 2015). These include hyaluronan and CSPGs. The upregulated CSPGs after stroke include neurocan, brevican, phosphacan versican, and aggrecan (Carmichael et al. 2005; Deguchi et al. 2005). These molecules serve to reinforce the tensile strength of the scar and to form molecular barriers. CSPGs form a barrier to cell migration that helps limit the spread of inflammatory cells from the infarct core and confine the spread of the initial stroke damage (Rolls et al. 2009). Hyaluronan serves to cross-link many matrix molecules to reinforce these barriers. In addition to astrocytes, neurons also produce CPSGs (Lander et al. 1998). CSPGs block axonal sprouting by causing growth cone collapse and have served as a target for neural repair therapies in stroke and spinal cord injury. In stroke, locally digesting CSPGs enhance recovery (Hill et al. 2012). However, most preclinical strategies to block or reduce CSPG signaling do not have clinical relevance, as they use large molecules (such as chondroitinase ABC) that are difficult to deliver and provoke an inflammatory reaction (Wilems and Sakiyama-Elbert 2015). The recent characterization of a specific receptor family for CSPGs, the leukocyte common antigen-related (LAR) family (Xu et al. 2015), has been successfully exploited in spinal cord injury to promote recovery (Lang, Cregg, et al. 2015) and may have applicability in stroke.

When brain tissue is stained for CSPGs after stroke, the upregulation of CSPGs appears diffuse in the extracellular matrix in peri-infarct tissue but heavily deposited in the scar immediately adjacent to the infarct. A different pattern of CSPG deposition is the perineuronal net. This is a dense organization of CSPGs around mostly inhibitory neurons in the cortex (Celio et al. 1998). Perineuronal nets are formed in the developing brain and serve to alter ion distribution and control inhibitory neuronal signaling (Takesian and Hensch 2013). When perineuronal nets form in the developing brain, they close off the most plastic period of connection formation: the critical period (Takesian and Hensch 2013). As discussed in a later chapter, the critical period is a time frame in the developing cortex in which it is uniquely sensitive to alterations in sensory inputs, such as visual or somato-sensory patterns. Perineuronal nets form as the cortex develops and progressively dimin-ish its plasticity (Wang and Fawcett 2012; Takesian and Hensch 2013). Artificially digesting perineuronal nets with an enzyme injected into the brain opens up the critical period (Piz-zorusso et al. 2002). Stroke triggers digestion of perineuronal nets in a radial distance from the infarct core (Carmichael et al. 2005). It is possible that this process also contrib-utes to transient plasticity and recovery in the peri-infarct cortex, in analogy with the criti-cal period in the developing brain. A way to prove this would be to block perineuronal net digestion after stroke and determine if recovery is reduced. This has not been done, and the role of the perineuronal nets in stroke recovery remains unexplored. The function of perineuronal nets in possibly reopening a sensitive period after stroke is discussed in the final chapter.

4.7   The Effects of Ischemia on Distant Connected Brain Regions

The effects of focal stroke are relayed to distant (non-adjacent) brain sites through loss of connections, secondary axonal degeneration, and microglial and astrocyte activation (fig-ure 4.3). This is quite distinct from the events described above for periinfarct cortex because the effects of stroke on connected brain regions are related to the premorbid net-work and not just to degree of adjacency.

Stroke first transmits electrical impulses that relay a damage signal to distant, connected brain regions. This produces an upregulation of immediate early genes within hours of the initial stroke in areas that are not exposed to the ischemia (Block et al. 2005). Stroke in one site causes the axons projecting from the now dead neurons to lose their synap-tic connections to distant sites. The axons from neurons that were in the stroke core will die. Dying axons from a loss of neuronal cell bodies is termed *Wallerian degeneration.* Wallerian degeneration can be seen clinically in MRI in long axonal tracts as an early di-minished diffusion of water and later as hyperintensities in T2 or fluid-attenuated inver-sion recovery (FLAIR) sequences, corresponding to periods of initial axon damage and then degeneration (Matsusue et al. 2007; DeVetten et al. 2010; Zhang et al. 2012). Dying axons release signals that cause reactive astrocytosis and microglial responses in con-nected regions within days to weeks after the stroke (Block et al. 2005; Zhang et al. 2012). Such regions include ipsilesional motor cortical areas, contralesional homologs, the basal ganglia, brainstem, and spinal cord, and lead to activation of the cytokines TNFα and interleukin 6 (IL-6) (Block et al. 2005). In the rodent spinal cord, TNFα and IL-6 are in-duced within the first two weeks after stroke, as are the growth factors BDNF and NT-3

(Sist et al. 2014). These distant sites are clearly not in the initial area of ischemia, and reflect the relayed nature of stroke.

The relayed nature of stroke also causes cell death in connected areas. Delayed apoptotic cell death occurs in the thalamus in experimental stroke models after cortical stroke (Wei et al. 2004) and similarly in the substantia nigra after striatal stroke (Zhang et al. 2012). After white matter stroke, connected cortical areas suffer delayed damage that is seen as cortical thinning on MRI (Duering et al. 2012) and as overall brain atrophy (Kloppenborg et al. 2012). In the aged brain with white matter stroke, there is increased microglial activation and inflammatory cytokine release in connected cortical areas compared to the young adult (Rosenzweig and Carmichael 2013). In summary, stroke can relay its deleterious effects along its network of connections, first through electrical and molecular signals early after injury and then through delayed signals from degenerating axons. These relayed signals, however, also induce tissue regeneration, as discussed in the next section.

4.8 Axonal Sprouting

Stroke induces a limited process of neural repair and associated behavioral recovery. This is seen in axonal sprouting and the formation of new connections, angiogenesis, vascular remodeling, and in glial and neuronal progenitor responses. Progenitor responses occur in local peri-infarct tissue (Ohab and Carmichael 2008; Kernie and Parent 2010). Axonal sprouting occurs in peri-infarct tissue and in premorbidly connected regions in ipsilesional cortex and the spinal cord (Murphy and Corbett 2009; Benowitz and Carmichael 2010; Overman and Carmichael 2014; Starkey and Schwab 2014).

Axonal sprouting occurs from neurons in both the ipsilesional and contralesional cortex. The reason that axonal sprouting is a substantial focus in stroke is due to its potential to form systems of new connections in the adult brain and construct circuits that may lead to either new functions, recovery of lost functions, or compensation of lost functions. The location of axonal sprouting after stroke and its contribution to functional recovery likely depend on the stroke size. Small- to medium-sized strokes leave a substantial amount of remaining neuronal tissue adjacent to the stroke core. In experimental studies in rodents and nonhuman primates with these types of strokes, remaining neurons form robust new connections in peri-infarct cortex, such as from motor cortex to somatosensory, premotor, prefrontal, and association areas (Dancause et al. 2005; Brown et al. 2009; Li et al. 2010; Overman et al. 2012; Clarkson et al. 2013), and mediate motor recovery (Overman et al. 2012). Most human strokes are small- to medium-sized in terms of proportion of the ipsilesional hemisphere involved (Fiehler et al. 2006; Ay et al. 2008; Whitehead et al. 2009) and match the size of stroke in the rodent brain in which peri-infarct axonal sprouting is observed (Carmichael et al. 2005), so it is likely that axonal sprouting in the ipsilesional hemisphere has relevance to the process of tissue repair in

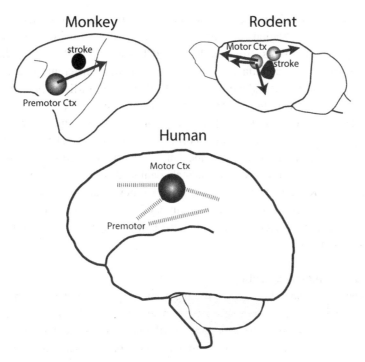

Figure 4.4
Axonal sprouting across species in stroke. Evidence from direct tracing of axonal connections in mice, rats, and squirrel monkeys indicates that stroke induces axonal sprouting in somatosensory, motor, and premotor connections. Growth-associated genes are induced in human tissue adjacent to the stroke site, suggesting that this process may occur also in patients after stroke. Modified from Carmichael (2014).

the human (figure 4.4). In large strokes in rodent models, there is a reduced amount of cortical tissue preserved in the ipsilesional hemisphere with the cortex contralesional to the stroke exhibiting axonal sprouting (Wahl et al. 2014). This axonal sprouting occurs from the local axons in the spinal cord that project from the contralesional cortex. These distant axons grow into the regions of spinal cord and brainstem that have lost their projections from the stroke site (Zai et al. 2009; Benowitz and Carmichael 2010; Wahl et al. 2014) (figure 4.4). These connections also mediate a degree of motor recovery after stroke (Wahl et al. 2014). As discussed at the end of chapter 3, however, evidence for a role of the contralesional cortex in stroke recovery in primates is largely lacking.

4.8.1   Reactive Axonal Sprouting
Axonal sprouting in peri-infarct cortex forms three patterns of connections that can be termed *reactive axonal sprouting, reparative axonal sprouting,* and *unbounded axonal sprouting.* Reactive axonal sprouting is local to the infarct and appears to largely be within

peri-infarct cortex that is just outside of and bordering the scar regions: new connections are seen in cortex in the regions of dense astrocytosis that form the glial scar and in the tissue directly adjacent to this stroke border zone (Clarkson et al. 2013). This kind of very local axonal sprouting in peri-infarct cortex is evident in many different stroke models and across species, including rat and mouse (figure 4.5). As noted earlier, a marker for axonal sprouting is also present in human stroke adjacent to the stroke site: the gene *GAP43* is induced in this region (Ng et al. 1988). This type of axonal sprouting is termed *reactive* because it appears as a component of the normal tissue turnover that forms the scar, just like reactive astrocytosis. This pattern of local or reactive axonal sprouting is seen in other types of brain lesions in which axons are acutely damaged, such as in the hippocampus or entorhinal cortex (Kelley and Steward 1997; McKinney et al. 1997; Dudek and Sutula 2007). Indeed, the general axonal reaction to injury is one of activation of a local growth program and local sprouting (Dickson et al. 2007; Lang et al. 2012; Mascaro et al. 2013) in many if not most neurons.

4.8.2   Reparative Axonal Sprouting

A second process of poststroke axonal sprouting occurs in peri-infarct cortex and connected areas and is associated with enhanced functional recovery. When specific molecular growth programs are stimulated after stroke, or glial growth inhibitory molecules are blocked, axonal sprouting after stroke occurs within motor, premotor, and somatosensory systems. This axonal sprouting occurs over a longer distance than reactive axonal sprouting. In the mouse, this pattern of axonal sprouting occurs over several millimeters (Li et al. 2010; Li et al. 2015; Overman et al. 2012), a fairly long distance of the mouse CNS. It links brain areas within a functional domain: sensorimotor/premotor areas. A limited process of reparative axonal sprouting occurs spontaneously in the nonhuman primate in this same functional domain, linking somatosensory and premotor areas (Dancause et al. 2005). Importantly, this type of reparative axonal sprouting is not a function of just distance from the infarct core—it is not the sole result of some simple diffusible signal from the infarct core. In rodents, axonal sprouting after motor cortex stroke occurs within premotor and somatosensory areas, but not to more closely adjacent lateral prefrontal areas (Li et al. 2010; Li et al. 2015; Overman et al. 2012; Omura et al. 2015). In squirrel monkeys, axonal sprouting after stroke can occur over a very long distance and again, not simply to the next adjacent cortical area, but within functionally linked premotor and somatosensory areas (Dancause et al. 2005). The signal for axonal sprouting is likely present in peri-infarct cortex, such as the induction of the growth factor GDF10 (Li et al. 2015), but the connections formed by axonal sprouting are directed to functionally related areas, some of which may be distant, and not just to neighboring areas. When reparative axonal sprouting is stimulated in the mouse, enhanced motor-to-premotor connections are formed and motor recovery is enhanced (Overman et al. 2012). When reparative axonal sprouting is blocked, such as by knocking down a molecular growth system (GDF10) (Li et al. 2015), or by

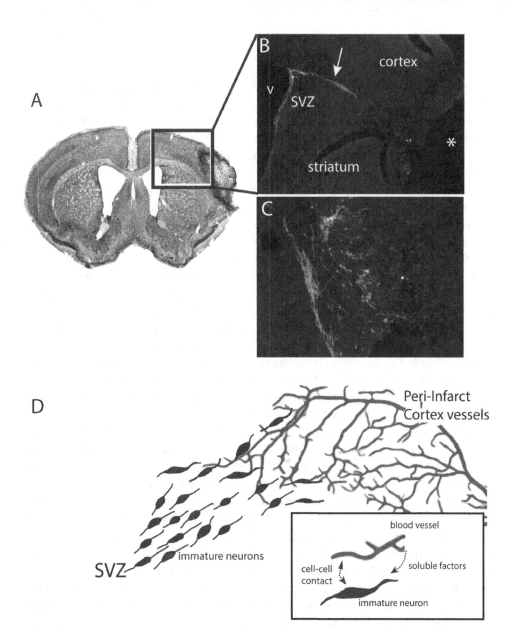

Figure 4.5
Neurogenesis and angiogenesis after stroke. (A) A coronal section through the frontal cortex of the mouse, stained for cell bodies. A stroke, produced by occlusion of the distal middle cerebral artery, involves somatosensory cortex and a portion of the underlying white matter. The boxed area is enlarged in **B**. (B) An image from a mouse in which a fluorescent protein is expressed in immature neurons (Cook et al. 2016). Immature neurons stream out from their normal location (arrow), the subventricular zone (SVZ), along the striatum toward the infarct (asterisk). An enlargement of the peri-infarct area (C) shows that these immature neurons extend local processes into the tissue. (D) A schematic of the migration of the immature neurons into the peri-infarct area and their association with blood vessels.

enhancing a glial growth inhibitor (ephrin A5) (Overman et al. 2012), recovery is reduced or blocked. These studies show that axonal sprouting from motor-to-premotor regions is causally mediating recovery, hence the term *reparative axonal sprouting*. In chapter 3, we provided further evidence that premotor regions can mediate recovery of upper limb movements in rodents and primates after stroke. Reparative axonal sprouting establishes new connections in brain regions linked to functional recovery in rodents (premotor cortex) and in humans, and does not occur automatically with all stroke lesions (Li et al. 2010; Li et al. 2015; Overman et al. 2012; Clarkson et al. 2013), unlike reactive axonal sprouting. Reparative axonal sprouting is associated with new patterns of information flow in the reorganized cortex (Brown et al. 2009). These properties indicate that this form of neuroplasticity after stroke is indeed "reparative."

A second example of reparative axonal sprouting in the rodent occurs in the sprouting of corticospinal axonal projections from the cortex contralateral to the stroke hemisphere into the cervical spinal cord. In normal stroke, unstimulated by a growth factor or a blocker of a glial growth inhibitor, there is a significant increase in these axonal projections from the unaffected hemisphere into the region of the cervical spinal cord that has been deafferented by the stroke (Benowitz and Carmichael 2010). This increase in normal stroke, i.e., unstimulated, is indeed significant but numerically very slight in comparison to the number of axons that innervate this territory of the cervical spinal cord from the ipsilateral hemisphere in the nonstroke condition. After stroke, when axonal sprouting is stimulated, there is a substantial increase in this corticospinal axonal growth. This is seen when axonal growth programs are stimulated, such as by inosine (Chen et al. 2002; Zai et al. 2009), or when glial growth inhibitors are blocked, such as with Nogo antagonists (Lindau et al. 2014), or when both inosine and Nogo blockers are delivered (Zai et al. 2011). This increased axonal sprouting response from the unaffected sensorimotor cortex into the ipsilateral cervical spinal cord is associated with increased behavioral recovery (Wahl et al, 2014). The relevance of this mechanism to recovery in human and non-human primates is questionable, however, as there is scant evidence for a role of the ipsilateral corticospinal tract from the contralesional hemisphere to motor recovery after stroke (see chapter 3). That said, from primary principles, this axonal sprouting fits the core definitions of reparative axonal sprouting: can be induced, occurs within functionally related areas of the CNS, and promotes increased motor recovery.

4.8.3 Unbounded Axonal Sprouting

Reactive axonal sprouting and reparative axonal sprouting after stroke occur within circumscribed areas of the CNS: the immediate vicinity of the infarct (reactive) or premotor and sensorimotor areas of the ipsilateral hemisphere (reparative). This pattern suggests that molecular or cellular limits exist within the sprouting response: axonal sprouting does not occur in any random direction. If this were the case, then axonal sprouting in peri-infarct cortex might be expected to produce new projections in brain areas that are simply the closest to the stroke or the closest to the neuroanatomical tracer injection site that is used to label these projections. Instead, new projections occur in nonrandom and spatially distinct patterns. This is most striking in quantitative maps of peri-infarct cortex motor system connections. The forelimb motor cortex always projects to adjacent motor cortex, somatosensory cortex, and premotor cortex when axonal sprouting is stimulated by the growth factor GDF10 (Li et al. 2015), by blockage of glial growth inhibitor NgR1 (Li et al. 2010) or ephrin A5 (Overman et al. 2012), or when axonal sprouting is mapped in a genetic strain of mice with natural, robust axonal growth after injury (Omura et al. 2015).

What would happen if poststroke axonal sprouting did not occur in functionally related brain or spinal cord areas? This is the case with unbounded axonal sprouting after stroke, which occurs when two conditions combine: glial growth inhibitors are blocked at the same time that the activity of the injured motor connections is increased. To date, there are two demonstrations of this. With stroke in the motor cortex, when the glial growth inhibitor ephrin A5 is blocked in peri-infarct cortex and the animal is then forced to overuse its affected forelimb, the motor cortex sprouts new connections into virtually every region of the ipsilesional hemisphere (Overman et al. 2012). This means that the forelimb motor cortex now connects with prefrontal, orbitofrontal, insular, parietal, and temporal areas—a dramatically divergent pattern of connectivity. In fact, as an aside, when one of us (S.T.C.) first saw this pattern of connections, we did not believe that this was possible. Although all analyses in the lab are performed with the investigator blinded to the condition, of course, this result was so unexpected that a separate lab member working on a totally different project was asked to analyze the data completely fresh. This study indicates that simultaneous blockade of astrocyte growth inhibition and enhanced behavioral activity of the motor system will enable the motor system to connect in a pattern that is not bounded by functional domains—connections form far afield of the sensorimotor system or the premotor areas and extend to executive, attentional, memory, and higher-order associational areas of the brain.

Unbounded axonal sprouting after stroke occurs in similar conditions of blockage of a glial growth inhibitor and enhanced behavioral activity within the spinal cord. When Nogo signaling is blocked and the animal is put into intensive, daily skilled reach training, axonal sprouting and functional recovery are enhanced if the Nogo blockade and the reach training are done sequentially (blockade then training) (Wahl et al. 2014). When these two are done simultaneously, axonal sprouting is increased and occurs into aberrant and functionally unrelated parts of the cervical spinal cord, such as the far dorsal horn. Behavioral performance does not improve and in fact is worse than control stroke conditions (Wahl et al. 2014).

These studies in poststroke axonal sprouting open up a new chapter in our understanding of brain connections, neurorehabilitative therapy, and enhanced functional recovery. The adult brain has a very limited capacity to form new connections after stroke, and this is induced by local cues in the vicinity of the scar in a process that resembles the formation of the scar itself—reactive axonal sprouting. That is, ischemic injury causes neurons to reactively generate small, local projections within the densest areas of reactive gliosis. This limited capacity of axonal sprouting in the brain after stroke can be increased by blocking glial growth inhibitors or stimulating a neuronal growth program, and this enhances motor recovery by inducing new more distributed patterns of motor, somatosensory, and premotor connections in cortex or new motor corticospinal projections in the cervical spinal cord—reparative axonal sprouting. Manipulating behavioral activity, for example, with intensive skilled reach training, while blocking glial growth inhibitors, releases growing connections from the control of tissue boundaries, and these connections can extend into vastly divergent functional areas—unbounded axonal sprouting. This process may degrade behavioral recovery. Future studies in this field will need to establish behavior/circuit interactions to develop the optimal approach for novel axonal growth therapies and neurorehabilitation paradigms to promote recovery.

4.8.4   Molecular Control of Poststroke Axonal Sprouting

The molecular programs that underlie poststroke axonal sprouting are unique tissue regeneration events (Li et al. 2010; Li et al. 2015). Studies of the gene expression profile of sprouting neurons after stroke indicate that a molecular program is activated that has a coordinated pattern of signaling from the extracellular environment into the cell (Zai et al. 2009; Li et al. 2010; Li et al. 2015). This includes specific growth factors and cytokines, cell surface receptors, intermediate cytoplasmic cascades, and transcriptional and epigenetic control points. There is an early upregulation of a molecular induction program for axonal sprouting within the first week after stroke and then a later maintenance program in axonal sprouting at three weeks after stroke (Li et al. 2010). A trigger for poststroke axonal sprouting is present within the first week after stroke. This distinction between an induction and maintenance program can be seen in the fact that extracellular signaling molecules, transcription factors, and epigenetic control molecules are activated in the induction phase, whereas cytoskeletal and synaptic proteins are more associated with the later, maintenance phase of axonal sprouting (Li et al. 2010).

Several clinically relevant questions can be applied to our understanding of this poststroke axonal sprouting transcriptome. Does it differ in the aged brain? This is a key question for human stroke, as most strokes occur in aged individuals. Does the molecular program of axonal sprouting after stroke resemble that seen in the initial axonal sprouting that underlies neuronal development? The molecular control of axonal sprouting is different in aged versus young adults. When directly comparing the transcriptional profile of sprouting neurons after stroke during the induction period of this response, there is little overlap of the young adult and aged sprouting transcriptome. Specifically, after correcting these transcriptional data sets for statistical significance, 1,346 genes are differentially up- or downregulated in young adult sprouting neurons (neurons that form a new connection in two-month-old rats after stroke) during the inductive phase after stroke, and 671 genes are up- or downregulated in aged sprouting neurons (sprouting neurons in stroke in two-year-old rats). When these two data sets are directly compared, only seventy-nine genes are commonly regulated by stroke in both young adult and aged sprouting neurons (Li et al. 2010). This indicates that in terms of tissue regeneration, stroke in the young versus aged brain represents a quite distinct, coordinated biological event.

Axonal sprouting neurons in the aged brain paradoxically upregulate genes that block axonal growth. These include the genes *EphA4* and *Lingo-1* (Li et al. 2010). Neurons that are induced to form a new connection in the aged brain also carry the seeds of their own destruction—axonal sprouting is coactivated with axonal growth inhibition. This is not the case for sprouting neurons in the young adult and may account for differences in tissue plasticity and repair with age. EphA4 is a receptor for, among other proteins, ephrin A5 and ephrin B3 (Coulthard et al. 2012). Ephrin A5 is expressed on reactive astrocytes and is more induced in the aged brain after stroke then the young adult brain (Li and Carmichael 2006; Overman et al. 2012): stroke in the aged brain activates both a receptor and a ligand for axonal growth inhibition. Functional tests of axonal sprouting and recovery in stroke bear this out. Ephrin A5 expression in reactive astrocytes blocks axonal sprouting and functional recovery in several different experimental models of stroke (Overman et al. 2012). Genetically deleting the EphA4 receptor also improves recovery in stroke (Lemmens et al. 2013). EphA4 is a promiscuous receptor and also binds ephrin B3, which is expressed in myelin (Benson et al. 2005). Blocking EphA4 induces even greater axonal sprouting than blockade of ephrin A5—suggesting the EphA4 receptor is a funneling point in the molecular signaling for axonal growth inhibition for astrocyte and myelin inhibitors. Lingo-1 is part of the Nogo receptor 1 (NgR1) signaling complex, which includes NgR1/p75 or TROY/Lingo-1 (Giger et al. 2010). Blockade of NgR1 signaling either directly by genetic knockout or pharmacologically with a Lingo-1 antagonist also causes more robust axonal sprouting after stroke (Li et al. 2010). These data on peri-infarct cortex axonal sprouting, derived from studies of an axonal sprouting transcriptome in neurons in this brain region, follow the effect of NgR1 antagonists in promoting axonal sprouting after stroke in the spinal cord (Lee et al. 2004; Zai et al. 2011). In each of these cases of Lingo-1 or EphA4 blockade in peri-infarct cortex, axonal sprouting occurs within sensorimotor circuits and is associated with functional recovery and thus fits in the category of reparative axonal sprouting.

4.9   Does Axonal Regeneration Recapitulate Development?

An unexpected finding in the molecular control of axonal sprouting after stroke was the identification of a *dependency state* in neurons located in peri-infarct cortex, for example, with insulin-like growth factor 1 (IGF-1) (Li et al. 2010). A dependency state in this case means that a neuron is dependent on something for survival. In the developing nervous system, neurons are initially dependent on successfully forming a new connection, and if they cannot, they will die. This dependency is due to the fact that a young neuron in the developing nervous system needs to transmit back to the cell body a signal from its target, usually a growth factor. IGF-1 is involved in such a retrograde signal for developing neuronal survival. In the sprouting transcriptome after stroke, IGF-1 is uniquely induced in aged sprouting neurons compared to young adult sprouting neurons, along with downstream molecular pathways in IGF-1 signaling (Li et al. 2010). This sounds like it would be fairly straightforward: IGF-1 is induced by stroke in nearby neurons, and this causes axonal sprouting, except IGF-1 delivery does not cause axonal sprouting after stroke. Indeed, IGF-1 delivery does not alter cortical motor connections at all, but blocking the normal IGF-1 signaling that occurs after stroke causes neuronal death—even well after the stroke has occurred (seven days after stroke). Blocking IGF-1 signaling in the normal, nonstroke brain does not cause neuronal death (Li et al. 2010). These findings show that plasticity in the adult brain after stroke comes at a cost. Neurons are indeed placed into a growth state, but they also become growth factor dependent for survival. This dependency state for a neuron on a growth factor resembles that of the developing brain. Neurons in the developing brain compete to form new connections, and the winner, the neuron that has formed the appropriate connection, is then stabilized by retrograde delivery of a growth factor, such as nerve growth factor (NGF), from its synaptic partner. This is the classic neurotrophic hypothesis (Clarke 1985; Barde 1989). However, adult neurons are no longer dependent on growth factor delivery for survival—they age out of this stage (Clarke 1985; Barde 1989; Li et al. 2010; Burke 2008). It appears that stroke may induce a similar degree of plasticity to development in the ability to form new connections and a similar type of dependency on IGF-1 signaling to sustain survival.

The finding of a dependency state and the overall superficial similarity of axonal sprouting after stroke to that seen in neural development has led to an open question in the neural repair field: does regeneration recapitulate development? When the transcriptional profile of sprouting neurons after stroke is directly compared to that of neurons during development, there is a substantial statistical difference between these two molecular programs. In fact, if unsupervised genome-wide association analysis is applied to 180 different transcriptomes from the literature, from development, to learning and memory paradigms, to spinal cord trauma and other injuries, the greatest distinction is between neurons that have been induced into an axonal sprouting state after stroke and early postnatal neurons that are still forming new connections (Li et al. 2015). These studies indicate that neural repair does not strictly follow neural development on a molecular level.

The process of axonal sprouting after stroke is a profound biological event for the adult brain. In a brain region that normally does not form such substantial new connections, a process is triggered in which local and long-distance projections are formed. This implies that there is a molecular trigger for this event. Working with candidate signaling molecules that are present in the aged neuron poststroke sprouting transcriptome, such a trigger was recently identified. The TGF-β family member GDF10, which previously did not have a known role in brain function, was found to trigger axonal sprouting. GDF10 is induced in peri-infarct cortex in mouse, rat, and nonhuman primates. GDF10 promotes axonal outgrowth, in vitro in many types of neurons and in vivo after stroke, as well as enhances motor recovery. GDF10 is thus a novel brain growth factor and stroke-induced trigger for recovery (Li et al. 2015). Interestingly, when endogenous levels of GDF10 are reduced, axonal sprouting after stroke is prevented, and motor recovery is reduced (Li et al. 2015). GDF10 signals through TGF-βRI and II and SMAD 2/3 to activate PI3 kinase gene systems and to inhibit PTEN and SOCS3 signaling. These gene systems mediate axonal sprouting in other contexts in the adult, such as in optic nerve and spinal cord injury (Sun et al. 2011; Lu et al. 2014; Danilov and Steward 2015). These data indicate that GDF10 is one molecule in what might be a complex trigger after stroke and coordinately activates parallel growth promotion cascades.

4.10 Neuronal and Glial Progenitor Responses after Stroke (Neurogenesis and Gliogenesis)

Ischemia induces proliferation and migration of neural and glial progenitors. For neural progenitors, the largest germinal matrix in the brain is the subventricular or subependymal zone (SVZ). The SVZ lies close to the lateral ventricles. Multipotent neural progenitors (which can give rise to glia and neurons) divide to produce more rapidly proliferating cells (termed *transit amplifying cells*), which then give rise to immature neurons (termed *neuroblasts*). In the rodent brain, neuroblasts from the SVZ migrate to the olfactory bulb (Goldman and Chen 2011; Lim and Alvarez-Buylla 2014). In the human brain, this migration to the olfactory bulb may be reduced, and neuroblasts may migrate into the basal ganglia (Ernst et al. 2014). Stroke signals to the SVZ, principally to transit amplifying cells, which proliferate, and to neuroblasts, which migrate to areas of damage. This response can produce new neurons in areas of damage (Ohab and Carmichael 2008; Ekdahl et al. 2009; Kernie and Parent 2010) (figure 4.5). If the stroke occurs in close proximity to the SVZ, such as in the striatum/basal ganglia, neurogenesis is robust and large numbers of neuroblasts migrate to areas adjacent to the stroke site. Most of these die, but some do mature and form local and long-distance connections. If the stroke site is distant to the SVZ, such as in cortex, fewer neuroblasts migrate to areas of injury, and these survive in vanishingly small numbers (Ohab and Carmichael 2008). Poststroke neurogenesis is stimulated by cytokines in the peri-infarct tissue, particularly SDF-1 and angiopoietin 2 (Ang-2) (Ohab and Carmichael 2008; Kahle and Bix 2013). It has been difficult to definitively determine if poststroke neurogenesis has a causal role in functional recovery after stroke, and future work will need to determine if this process is truly a "neural repair" event or a process of tissue reorganization that is not associated with behavioral recovery.

The reorganization response after stroke also includes angiogenesis and the modification of the vascular tree. In human studies, the degree of angiogenesis was correlated with the degree of survival in an early study (Krupinsky et al. 1994) but angiogenesis has not been directly tested for its role in behavioral recovery. Angiogenesis involves a coordinated response to local growth factors and alterations in the extracellular matrix that would appear on face value to be associated with recovery (Ergul et al. 2012). Angiogenic vessels are the core tissue in the center of a neurovascular niche after stroke that establishes clear aspects of tissue reorganization. In the scientific literature, angiogenesis after stroke is stimulated by a very large range of manipulations in pre-clinical stroke models, from cell therapies to growth factors to exercise. However, a causal role for angiogenesis in behavioral recovery has not been established, and specific molecular inactivation of angiogenesis in at least one stroke model did not impact behavioral recovery (Young et al 2014). This is unfortunate because in an experimental sense it is not difficult to deliver a candidate repair therapy in an experimental model and block angiogenesis at the same time. Such an approach would allow definitive study of the causal role of angiogenesis in recovery, rather than just stopping the line of investigation with evidence for an association of angiogenesis with recovery. In this discussion, angiogenesis will be treated as a tissue-reorganization-process with an agnostic view on its role in tissue "repair."

Angiogenesis and vascular remodeling occur most prominently in peri-infarct tissue (Liu et al. 2014), although there are changes in vascular structure in cortex contralateral to stroke (Ergul et al. 2012). Angiogenesis initiates early after stroke, within days (Hayashi et al. 2003). It is important to note that animal models of stroke differ in their initiation of angiogenesis. The photothrombotic stroke model, which is used in many optogenetic studies of stroke neural repair, does not produce substantial angiogenesis (Mostany et al. 2010; Tennant and Brown 2013). However, other stroke models with reperfusion and the production of a peri-infarct region of partial damage do induce angiogenesis (Ohab et al. 2006; Ergul et al. 2012; Young et al. 2014). Angiogenesis serves to bring new blood flow to peri-infarct tissue, but this occurs too late for an actual survival effect in the evolution of stroke damage. Instead, angiogenic vessels secrete growth factors and cytokines that modulate the local tissue environment (Brumm and Carmichael 2012). Angiogenesis is causally linked to neurogenesis in peri-infarct tissue, and these vessels are the source for Ang-2 and SDF-1, which induce migration of neuroblasts after stroke (Ohab and Carmichael 2008). In spinal cord injury, angiogenic vessels are the source of eicosanoids that induce axonal sprouting (Brumm and Carmichael 2012). However, there has been no direct link between axonal sprouting and angiogenesis in stroke.

4.11   Reconciling Regeneration with (Behavioral) Recovery

In this chapter we maintain fairly strict standards for distinguishing tissue repair from tissue reorganization. There have been frankly amazing advances in the biology of tissue reorganization after stroke, showing remodeling of neuronal networks and of tissue architecture. The stroke field has historically under-recognized these processes of axonal sprouting, angiogenesis, neurogenesis, gliogenesis, and synaptic plasticity, focusing instead on cell death and neuroprotection. The explosion over the last fifteen years in the recognition of robust tissue reorganization after stroke has, however, perhaps led to the hasty conclusions that tissue reorganization implies tissue repair that causes behavioral recovery. But all that glitters is not gold. Though candidate molecular, cellular, or rehabilitative therapies may stimulate a tissue-reorganization-event, in few cases have these been shown to causally promote recovery—done by blocking the tissue-repair-event and noting reduced recovery, and then inducing the tissue-repair-event and noting promoted recovery. For example, if transplantation of cell line X promotes behavioral recovery and also induces angiogenesis, it is critical to give cell line X and prevent angiogenesis to be able to correctly state that angiogenesis mediates recovery. This is experimentally possible in all cases but has rarely been done to date. The tissue-reorganization-events that are causally associated with recovery in the rodent are axonal sprouting in peri-infarct tissue in specific circuits (motor-to-premotor circuits) and alterations in excitatory signaling in peri-infarct cortex.

Axonal sprouting is not a random process of radial tissue reorganization, spreading out from the infarct. Axonal sprouting to non-adjacent connected areas, is associated with recovery whereas reactive sprouting close to the infarct is not (Overman et al. 2012; Li et al. 2015). This circuit-specific nature of axonal sprouting in stroke implies that in recovery there is an interaction of post-stroke tissue reorganization with pre-existing connectivity architecture—axonal sprouting within brain areas that contain movement representations for the forelimb/upper extremity. This two-way association of a process of poststroke reorganization with specific neuronal circuits or substrate could be further modified by activity patterns in that circuit, such as with neurorehabilitative therapies. This idea is discussed at the end of chapter 3; a tripartite theory of motor recovery after stroke that combines neural substrate, post-ischemic tissue-reorganization-processes, and behavioral activity. In that section of chapter 3 we reconcile the nonhuman primate literature, which indicates that second lesions of peri-lesion cortex, the site of many tissue-reorganization-processes, after the first lesion does not reinstate a motor deficit, but lesioning of premotor cortex, distant from the peri-lesion cortex, does indeed reinstate an original motor deficit. These results in the complex and large brain of the nonhuman primate suggest that a distributed network of motor areas retains the capability for recovering motor control in an affected limb after stroke if tissue reorganization, such as axonal sprouting, occurs within that specific substrate. In other words, brain motor areas have a substrate or architecture for motor control, and recovery occurs if this architecture can be facilitated when primary motor circuits are damaged. The recruitment of distributed limb motor-related areas into active movement control of the paretic limb can occur through tissue reorganization or altered behavioral activity, such as some element of neurorehabilitative therapy, which selectively recruits these areas into a functionally coactivated network. This hypothesis is specifically testable: axonal sprouting or other tissue-reorganization-events (such as angiogenesis) that are induced outside of neuronal architectures for motor control of the limb should not induce behavioral recovery. Inducing angiogenesis in premotor cortex after a stroke that destroys primary motor circuits should induce recovery, but inducing angiogenesis in posterior parietal cortex should not, according to this theory. There are exceptions to this theory in the rodent as they relate to spinal cord axonal sprouting after stroke. Adjacent hindlimb cortex can participate in recovery after large lesions in motor and premotor forelimb cortex in the rat by sprouting a collateral into the cervical (upper limb) spinal enlargement from neurons that originally projected to the more caudal spinal cord (Starkey et al. 2012). The applicability of this mechanism is questionable in primates, however, as there is as yet no good evidence that non-upper limb cortical representations can take over function in a comparable manner.

4.12 Does Activity in the Injured Brain Make the Injury Worse?

The previous sections discussed the role of neurorehabilitative therapy in stroke recovery in the weeks after stroke. Whenever discussion turns to early intense rehabilitation after stroke, the objection of a possible adverse effect of this increase in activity is raised with respect to exacerbation of lesion volume and a worse behavioral outcome in the affected limb. This objection originates from a series of well-cited studies by Schallert and colleagues in the rat with stroke-like lesions (Humm et al. 1999). They reported that immobilization of the unaffected forelimb with a hard cast for fifteen days after lesion induction led to less use of the affected side once the cast was removed. This was particularly striking because after lesion induction, immobilization of the unaffected limb in these studies had a greater worsening effect on recovery than immobilizing the affected limb. Immobilization of the unaffected limb not only had an adverse effect on behavior but was also accompanied by expansion in lesion volume (Kozlowski et al. 1996; Humm et al. 1998). What is less well appreciated is that in these early studies, the lesions were electrolytic rather than ischemic, making their relevance to stroke questionable. Electrolytic lesions were produced in these early studies by running an electrode back and forth through brain tissue while passing current. Subsequently, however, the same group of investigators asked the same question for ischemic lesions using a middle cerebral artery occlusion (MCAO) model in the rat. Here the results are more equivocal. In the case when forty-five minutes of MCAO caused moderate cortical ischemia, ten days of casting of the unaffected limb did not lead to exaggeration of infarct volume but did lead to worse behavioral performance in the affected side (Bland et al. 2001). For more severe cortical ischemia, induced by ninety minutes of three-vessel occlusion, there was no deleterious effect on lesion volume or outcome. In a filament MCAO model that caused subcortical (striatal) infarction, forced nonuse but not overuse of the affected forelimb led to detrimental behavioral outcomes but without exaggeration of lesion size (Bland et al. 2001). The same investigators failed to show a behavioral consequence of casting the unaffected limb despite exaggerations of cortical lesion volume (Bland et al. 2000). Indeed, in this study, as in the earlier subcortical study, it was *disuse* of the affected forelimb that had detrimental effects. Importantly, in these later experiments, the cast was smaller and lighter and the rats were housed in larger cages with littermates. Carmichael and colleagues have revisited the effects of overuse. They induced overuse of the affected forelimb one day after the stroke by using Botox in the unaffected limb; there was no increase in infarct size with this approach (Overman et al. 2012), but the same authors have demonstrated that there is instability in cortical excitability for about three to five days poststroke (Clarkson et al. 2010; Clarkson et al. 2011). These studies point to an instability period in early stroke in which increases in neuronal activity or behavioral activity may lead to exacerbation of stroke damage. In the rodent, this is up to 5 days after infarct.

4.13 Conclusions

Ischemic stroke occurs as a continuum in time of biological responses in the brain from initial ischemic cell death to secondary damage to regeneration and partial repair. Each element in this continuum is triggered by processes in the preceding stage. Stroke also occurs as a progressive spatially evolving biological response in the brain, from the initial ischemic core to radial brain regions at successive distances to the core and then to brain areas that are distant but connected to the core. The adult brain forms new connections after stroke, and these are part of the scarring response and also mediate recovery. The future challenge is to relate these processes occurring at the cellular and molecular level to equally fine-grained analysis of the recovered behavior they mediate. For example, the distinction between reparative and unbounded axonal sprouting is likely a matter of degree. It will be necessary to deliver pharmacological modulators of neural repair within well defined rehabilitative training paradigms (type, timing, intensity) to promote optimal repair.

5

A Hierarchical Framework for Tissue Repair after Stroke

5.1   First- and Second-Order Principles of Repair

The basic or elemental properties of neural repair were the subject of the previous chapter. These include the events of axonal sprouting, neurogenesis, gliogenesis, and changes in neuronal excitability in peri-infarct tissue. Specific elements in each of these tissue responses to stroke affect recovery. As examples, reactive astrocytes downregulate the uptake of the inhibitory transmitter GABA, which causes an elevation in tonic GABA signaling and hypoexcitability of pyramidal neurons adjacent to stroke (Clarkson et al. 2010). In poststroke neurogenesis, immature neurons migrate to areas of brain injury and are associated with recovery (Ohab and Carmichael 2008; Hermann and Chopp 2012; Kahle and Bix 2013). Oligodendrocytes respond to the ischemic event by associating with blood vessels and proliferating (Miyamoto et al. 2013; Rosenzweig and Carmichael 2013), but are blocked from full differentiation into mature and myelinating oligodendrocytes and become reactive astrocytes (Sozmen et al. 2016). With axonal sprouting, stroke induces a molecular program in adjacent neurons for them to grow new axons, and such connections to motor and premotor areas promote recovery (Overman et al. 2012; Li et al. 2015). The elemental events of reactive astrocytosis, gliogenesis, neurogenesis, axonal sprouting, and angiogenesis can be considered first-order principles in the observed tissue responses to stroke and correspond to single classes of cells (neurons, astrocytes, endothelial cells, oligodendrocyte precursor cells).

Tissue repair and recovery in the brain after stroke, however, cannot be properly understood by considering single types of cell in isolation. The distinct cell types—neurons, astrocytes, endothelial cells—cluster together in multicellular environments in cortex, white matter, and basal ganglia. It is the sum of these cells' responses and their interactions that lead to tissue repair and, more importantly, recovery after stroke. The interactions between different cell types can be considered the sum total of the whole brain's response to injury. The description of the responses of individual cell types to stroke and during recovery is at the level of first-order principles. Second-order principles of brain repair emerge from the interactions of these single cell types with each other. In summary: first-order principles describe the properties of individual cell types; second-order principles describe the interactions among cell types. For example, immature neurons that migrate to areas of damage associate with angiogenic blood vessels. These two cells are each literally "regenerative cells" after stroke, as they are newly born and create a new, albeit transient, tissue: new neurons and new blood vessels. Neurogenesis and angiogenesis occur together and are co-regulated, forming a regenerative neurovascular niche. The vascular signals to migrating neuroblasts in poststroke neurogenesis are part of the larger process of constructing this niche for neural repair, in which the cellular interactions of individual cells take on properties that are distinct from the cells themselves (Brumm and Carmichael 2012). This niche and its governing rules are a second-order principle in stroke recovery: the interaction of two first-order concepts producing an emergent second-order principle. Similar interactive (second-order principles) govern not just cells and molecules in stroke but also patients. As we will describe below, stroke recovery is usually incomplete, a first-order principle. Patients after stroke are less active, another first-order principle. These two first-order principles interact in a second-order principle: incomplete recovery interacts with physical inactivity to cause chronic progression. This chapter draws from recent discussions of emergent properties of stroke neural repair (Carmichael 2016) to re-frame the concepts of tissue reorganization in stroke not as elemental and emergent properties but as first and second-order principles. At this time in the field, there are eight second-order principles in stroke (figure 5.1), each of which has a section in this chapter.

5.2   Stroke Is Not just an Acute Killer but a Chronic Disabling Disease

Stroke mortality is declining. Three years ago, stroke slipped from the third leading cause of death to the fourth (Lackland et al. 2014). Stroke mortality continues to decline from this level (Benjamin et al. 2017). This decline in mortality is something to be welcomed and stems from the hard work and implementation of stroke guidelines in emergency and acute hospital care. This progress is now further stimulated by the successful implementation of the new stent/retriever devices for clot removal (Pereira et al. 2015). Outside of mortality, predictions of overall stroke incidence are in flux, with earlier predictions of a dramatic rise in stroke incidence as the population ages (Broderick 2004) and more recent epidemiological studies showing a decline in stroke incidence (Koton et al. 2014). However, even with a decline in stroke incidence, mortality is declining faster than the reduction in occurrence (Koton et al. 2014). The prevalence of stroke is increasing as a result. Thus, stroke is ever more a chronically disabling disease. In other words, stroke victims survive their stroke but not their disability. With stroke disability, up to 80 percent of persons may ultimately recover the ability to walk short distances, but most do not achieve the ability for community ambulation (Bogey and Hornby 2007). Approximately 70 percent to 80 percent of people who sustain a stroke have upper extremity impairment

First Order Biological Principles   Second Order Biological Principles   Therapeutic Principles

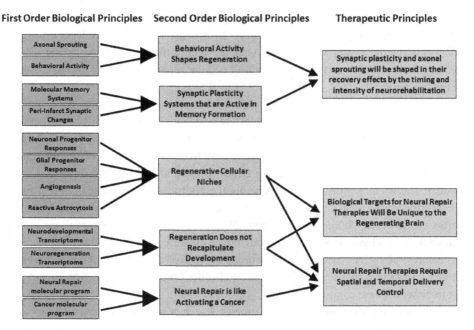

Figure 5.1
A neurophilosophy of neural repair in stroke. The elemental or basic principles of tissue repair have a more complex and emergent meaning when they are taken together, which then generates principles in therapeutic application of these principles.

(Parker et al. 1986; Nakayama et al. 1994). Many of these do not regain functional use of the paretic arm, and at six months after stroke, a substantial proportion (25 percent to 53 percent) of people remain dependent in at least one activity of daily living (ADL) task (Gresham et al. 1975; O'Mahony et al. 1999).

Stroke victims show behavioral declines over time after their initial gains from neurorehabilitation (Sonde et al. 2000; Kernan et al. 2005; Thorsén et al. 2006; Dhamoon et al. 2012). Most of these declines are due to inactivity and lack of use of their paretic limb in everyday activities (Dobkin 2003; Ivey et al. 2005; Everson-Hock et al. 2016). The interaction of increasing prevalence of stroke disability with increasing disability over time after the initial stroke means that stroke is a chronically progressive, disabling disease with no effective treatments at this time, which makes it an urgent research priority.

5.3 Behavioral Activity Shapes Tissue Regeneration

Stroke changes behavioral activity by directly altering neurological function, for example, hemiparesis or aphasia. Clinicians in turn alter behavioral activity after stroke through neurorehabilitation; regimens of repetitive and task-oriented training. Neuronal activity is the coin of the realm in neurological disease, distinguishing it from liver, bone mar-row, intestinal, or other systemic diseases. In neuroscience, activity in brain regions affects the cellular events within them. A second-order principle in stroke neural repair is that neural activity interacts with elements of cellular repair after stroke to create directed tissue regeneration.

Axonal sprouting and the formation of new connections after stroke occur in brain regions that are damaged or partially deafferented from the stroke. Stroke triggers new connections to form in ipsilesional motor, somatosensory, and premotor cortex and addi-tionally, in rodent models, projections are formed from contralesional cortex to striatum, midbrain, and spinal cord (Zai et al. 2009; Benowitz and Carmichael 2010; Wahl et al. 2014). In other studies in healthy animals, behavioral activity such as overuse of or learning in a forelimb induces local changes in synaptic connections within the corresponding motor cortex (Kleim et al. 2004; Overman et al. 2012). These two sets of observations indicate that stroke and behavioral activity can both induce the formation of new connections. When the behavioral activity patterns of a neurorehabilitative therapy are combined with the axonal sprouting response to ischemia, second-order properties appear that are not pres-ent in either of these conditions alone.

Limb overuse after stroke produces a greater degree of axonal sprouting than limb over-use in the normal brain (Overman et al. 2012), although the new projections that are formed are confined largely within the motor system (termed *reparative axonal sprouting* in the preceding chapter). In ipsilesional cortex, such reparative axonal sprouting occurs in motor-to-premotor connections. However, when a growth inhibitory protein such as ephrin A5 is blocked after stroke *and* there is concomitant overuse of the rodent's af-fected limb, there is a magnitude of axonal sprouting that is remarkable for an adult brain. New projections are formed from motor cortex to widespread ipsilesional areas in prefrontal, orbitofrontal, temporal, and parietal cortical areas (Overman et al. 2012). The magnitude and pattern of this axonal sprouting pattern would not be predicted from re-sponses to the two individual conditions alone: behavioral activity or stroke plus glial growth inhibition blockade. We called this type of axonal sprouting *unbounded axonal sprouting* in the preceding chapter, because the formation of new connections extends beyond the boundaries of one functional brain system, such as the sensorimotor system. This degree of axonal sprouting may not be beneficial and indicates that careful attention must be focused on the interaction between behavioral therapies and new drugs that block axonal growth inhibitors in stroke. However, this second-order principle also suggests that neurorehabilitative therapy will have great power to shape neuronal connections when these connections are released from the normal inhibitory conditions of the post-stroke brain.

5.4   The Suffered Is the Learned

The process of spontaneous recovery after stroke shares some similarities with memory-related changes at the molecular and cellular level. On a cellular level, processes of memory formation and neuronal network changes in the poststroke brain are both associated with long-term potentiation (LTP)-like phenomena and dendritic spine morphogenesis (Brown et al. 2007; Mostany et al. 2010; Bosch and Hayashi 2012). On a molecular level, learning and memory paradigms, such as in the hippocampus (Bosch and Hayashi 2012; Brown et al. 2007), are associated with expression changes in stathmin, RB3, GAP43, and the Nogo signaling system, and these same molecular pathways are involved in motor recovery after stroke (reviewed in Carmichael 2012). A second-order principle in neural repair after stroke would seem to be that, given the overlap between molecular and cellular aspects of recovery after stroke and normal learning and memory, that the synaptic plasticity systems that mediate memory formation are active after brain injury to mediate recovery of function. There has been recent experimental testing of this hypothesis. Two signaling systems that affect the excitability state and synaptic signaling systems that underlie learning and memory have been implicated in motor recovery after stroke. Tonic GABA signaling occurs through extrasynaptic $GABA_A$ receptors, particularly via receptors with $\alpha5$ and δ subunits. Tonic GABA receptors respond to extrasynaptic or ambient levels of GABA, are more sensitive than synaptic $GABA_A$ receptors, and desensitize more slowly. These inhibitory channels can thus mediate an inhibitory chloride current that controls the baseline firing threshold of pyramidal neurons (Glykys and Mody 2007). Phasic or extrasynaptic GABA signaling is a target in the learning and memory field, as blocking this current promotes neuronal excitability in the hippocampus and enhances learning and memory in many animal models (Martin et al. 2010; Rudolph and Möhler 2014). Tonic GABA signaling is elevated in other central nervous system (CNS) disease states and contributes to cognitive dysfunction, for example, in Down syndrome (Rudolph and Möhler 2014). In stroke, tonic GABA signaling is increased and produces a hypoexcitable state in pyramidal neurons in brain regions that normally mediate motor recovery. This can be pharmacologically reversed to enhance recovery in several rodent models at a considerable delay after stroke (Clarkson et al. 2010; Lake et al. 2015).

Is tonic GABA inhibition altered in human stroke? No technique in human studies (transcranial magnetic stimulation [TMS], magnetic resonance imaging [MRI], or positron emission tomography [PET]) can specifically measure tonic over phasic inhibition (Paulus et al. 2008). The closest approach to measure an element of GABA signaling that might reflect or relate to tonic GABA signaling would be to measure total extrasynaptic GABA levels, which are increased in the mouse because of reduced astrocyte uptake of GABA after stroke (Clarkson et al. 2010). However, the only ability to measure GABA in humans is with ^{18}F-flumazenil (which measures synaptic GABA receptor binding occupancy) and GABA magnetic resonance spectroscopy MRI (which measures total GABA levels in a 2-mm^3 block of brain tissue). Neither of these approaches is sensitive to phasic (synaptic) versus tonic (extrasynaptic) GABA levels or activity (Paulus et al. 2008; Stagg et al. 2011). Nonetheless, there are reports that good functional recovery is correlated with declining GABA levels after stroke (Kim et al. 2014; Blicher et al. 2015) and that motor learning is associated with reduced GABA levels in motor cortex (Sampaio-Baptista et al. 2015), which supports the idea that reducing the effects of GABA after stroke promotes recovery.

Synaptic activity mediated by the AMPA subtype of the glutamate receptor is also a key process in memory formation. Manipulations that enhance AMPA receptor signaling or increase the number of AMPA receptors in the postsynaptic membrane increase LTP and enhance performance in many models of learning and memory (Partin 2015). Positive allosteric modulators of the AMPA receptor, which enhance AMPA receptor signaling only when glutamate is bound to the receptor, were originally developed and tested as memory-enhancing drugs (Partin 2015). Called AMPAkines, these drugs stimulate learning and memory under both normal conditions and disease models (Lynch et al. 2014; Partin 2015). High-impact AMPAkines that induce downstream production of brain-derived neurotrophic factor (BDNF) promote motor recovery in rodents. This action is via BDNF production, rather than through other effects of glutamate receptor function. The action of AMPAkines after stroke occurs specifically in ipsilesional cortex, further indicating that stroke specifically alters neuronal networks in regions of recovery in a way that involved activating BDNF signaling (Clarkson et al. 2011). These data for tonic GABA and AMPA receptor signaling are supported by other studies in which drugs that affect molecular systems first described in learning and memory contexts also improve motor recovery after stroke, such as isoform-specific phosphodiesterase inhibitors (MacDonald et al. 2007). The ancient Greeks used the term *ta pathemeta mathmeta*: the suffered is the learned. A second-order principle in neural repair after stroke is that the damage suffered from the infarct sets in place mechanisms of heightened learning that can be manipulated to promote recovery.

The issue of brain excitability in stroke touches on many different past concepts and has been responsible for the promulgation of some nonscientific "lore" when it comes to understanding of inhibitory and excitatory signaling after stroke. Stroke clearly causes cell death through a process of excitotoxicity: a positive feedback loop in which hypoxia leads to neuronal depolarization, glutamate release, and further neuronal depolarization (discussed in the preceding chapter). Research in the 1990s and early 2000s indicated that the brain tissue that survives the stroke may show evidence of diminished inhibition and prolonged and abnormal excitatory signaling. These studies suggested a state of hyperexcitability in the brain after stroke: diminished paired-pulse inhibition, prolonged excitatory field potentials to stimulation, and enhanced LTP (Schiene et al. 1996; Buchkremer-Ratzmann and Witte 1997; Hagemann et al. 1998; Qü et al. 1998). GABA receptor binding sites and specific GABA receptor subunit expression are changed after stroke with overall decreases in many $GABA_A$ receptor subunits (Redecker et al. 2002; Kharlamov et al. 2008). More recent studies indicate that specific GABA receptor subunits are lost from the initial segment of the neuron—the generation zone of the action potential (Hinman et al. 2013). This loss may lead to hyperexcitability of the neuron. These data sets relate to phasic synaptic inhibition. However, as noted above, tonic GABA inhibition is actually increased in the cortex adjacent to the stroke site and impedes recovery in rodents, suggesting a more nuanced alteration in GABA inhibition after stroke: not all brain inhibitory systems are affected in the same way by stroke.

Recent evidence in rodents indicates further subtlety in the brain's inhibitory system response to stroke. Detailed mapping of GABA α1 receptor subunits in the cortex adjacent to stroke indicates that in layer V there is an increase in these receptors, which is associated with an increase in GABA synaptic/phasic inhibitory currents. Enhancing these phasic currents in a mouse model of stroke with subsedative levels of the GABA agonist zolpidem (Ambien®) enhanced behavioral recovery (Hiu et al. 2015). Thus, the brain's inhibitory systems do not respond uniformly to stroke, with differences with respect to distinct GABA signaling system (tonic vs. phasic), brain layer, and GABA receptor subunit. Future work should exploit these details to determine how to target selective inhibitory systems to promote recovery, as presently appears possible with tonic GABA antagonists/inverse agonists and zolpidem.

5.5 The Motor Recovery Engram and CREB

A main hub in the pathways within a neuron that control excitability is the transcription factor cyclic AMP response element binding protein, or CREB. CREB is activated by increased neuronal firing, by growth factors and cytokines that activate neurons, and by other neuronal stimuli that enhance neuronal signaling. Thus, many inputs to a cell exert their effect on a neuron via CREB. In turn, CREB's actions determine the threshold for a neuron firing an action potential by increasing intrinsic neuronal excitability and increasing dendritic spine morphogenesis (Dong et al. 2006; Kim et al. 2013). CREB facilitates LTP (Wu et al. 2007). LTP is altered in both experimental animals and in humans after stroke (Di Filippo et al. 2008), suggesting that it may be involved in the changes in synaptic signaling that underlie recovery. Transcranial direct current stimulation, which enhances cortical excitability and can enhance motor function transiently after stroke (Kang et al. 2016), induces CREB in neurons (Podda et al. 2016). Thus, CREB is a prime candidate molecule that may play a role in functional recovery in stroke. The preceding section described molecular systems that enhance neuronal excitability, initially identified in studies of learning and memory, which play a role in stroke recovery. The transcription factor CREB follows in this intellectual pathway.

In rodent stroke models, CREB directly mediates motor recovery via a small subset of motor cortical circuits. To test the role of CREB in motor recovery, this gene was induced

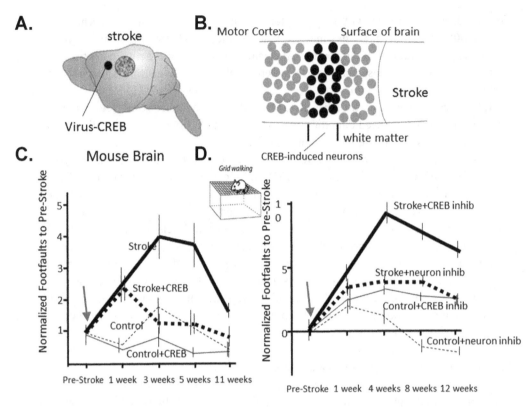

Figure 5.2
CREB in the control of motor recovery in stroke. (A) Schematic of approach to understand CREB in stroke. Stroke is produced in motor cortex of the mouse brain. A small injection of a virus that turns on CREB is placed in motor cortex just in front of the stroke site. (B) Schematic of the CREB-induced neurons. This schematic shows a section through the stroke and the site of CREB induction. The black neurons have had extra copies of CREB inserted. This is a fraction of neurons in front of the stroke site. (C) Performance of CREB-induced mice after stroke. The schematic at the top shows a mouse walking on a challenging grid, which even in nonstroke mice occasional foot faults are made. The graph shows the increase in foot faults that occurs after stroke. CREB induction in a small number of motor cortical neurons in front of the stroke induces a rapid and complete recovery of gait. There is no statistical difference in the control conditions. (D) Graph has same representation as (C). The top two graph lines show the motor control after stroke: the effect of inhibition of motor neuronal output after first induction of CREB (top, solid line) or inhibition of motor neuronal output without inhibition of CREB (middle, dotted line). The bottom two graph lines show the effect of inhibition of motor neuronal output with (thin solid line) or without (thin dotted line) first induction of CREB. Arrows in C and D show timing of stroke. Adapted from Caracciolo et al. (2014).

using a viral approach in a small number of neurons in the motor cortex anterior to the stroke: roughly 15 percent of all neurons in motor cortex (figure 5.2) (Caracciolo et al. 2012, 2014). This test of the role of CREB in motor recovery after stroke sets out a deliberately difficult test for a molecular system purportedly implicated in recovery because CREB is influenced in only a small portion of motor cortex near the stroke. Rather than turning this system on in the whole brain or in a general brain system (such as the entire motor system), this approach only activated CREB in a subset of excitatory/pyramidal neurons in motor cortex near the stroke site. Remarkably, this approach induced rapid and complete motor recovery in several tests of gait and of forelimb function. Activating CREB using the same approach but in neurons in somatosensory association areas did not promote functional recovery. Blocking CREB using an inducible approach to inhibit it prior to motor testing prevented recovery. These data indicate that CREB is essential for motor recovery as measured in this stroke model: gain of CREB function enhances motor recovery; loss of CREB function blocks motor recovery.

A significant test of the role of CREB in motor recovery after stroke would be if CREB could be activated in motor neurons in circuits near the stroke site and if these motor neurons could then be turned off (in essence a double-lesion approach as discussed in chapter 3). The prediction would be that activating CREB would enhance motor recovery, and subsequent turning off of these neurons would reinstate the deficit. This can be done using a specific inactivation approach. The above studies used a virus, injected into the motor cortex, to introduce extra copies of the *CREB* gene into motor cortex neurons. In these neuronal inactivation studies, the virus not only can introduce the gene for CREB but can also introduce a gene for artificial receptors that modulate neuronal activity. These receptors are called designer receptors exclusively activated by designer drugs (DREADD). One of these receptors, hM4Di, inhibits neurons only when mice are given the chemical clozapine N-oxide (CNO) (Roth 2016). In this experiment, neurons in the motor cortex after stroke will have CREB activated, and then these neurons can be reversibly turned off with CNO. In control conditions, mice will have just the hM4Di inhibitory receptor but no CREB. A prediction is that mice with CREB induction will recover more rapidly and completely, but that this recovery will not be present when they are tested after CNO delivery.

The results are actually more surprising than this, as evidenced by two observations. First, the prediction holds true. Activating CREB in neurons in peri-infarct motor cortex induces recovery. Turning these neurons off reverses the recovery. Then, when the drug CNO washes out, these mice return to a highly recovered motor state after stroke. This means that motor recovery after stroke can literally be turned on and off using this approach (figure 5.2). Second, when CREB is induced in neurons after stroke and these neurons are turned off, the deficit is actually much worse than the original stroke deficit. These mice are profoundly impaired—more impaired than with stroke alone (Caracciolo et al. 2015). This result means that inducing CREB in a subset of motor cortical neurons, and then turning these neurons off, has an effect on motor control that is far greater than if CREB were not induced in the first place.

What is happening here? The results suggest the presence of a "motor recovery engram" (figure 5.3). An engram is a network of neurons that are coactive in a cognitive state. This definition is best established in memory systems. Coactivation of a network of neurons in the lateral amygdala represents a fear memory. The neurons that are active in the storage of

Motor Cortex

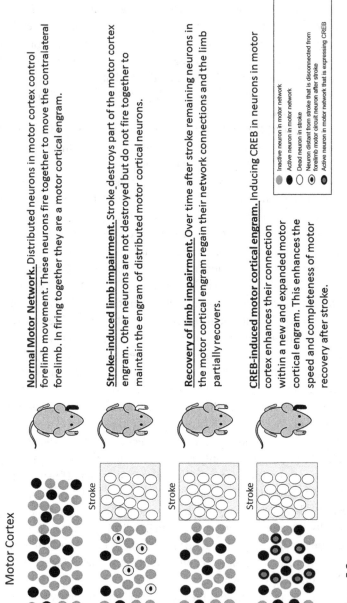

Normal Motor Network. Distributed neurons in motor cortex control forelimb movement. These neurons fire together to move the contralateral forelimb. In firing together they are a motor cortical engram.

Stroke-induced limb impairment. Stroke destroys part of the motor cortex engram. Other neurons are not destroyed but do not fire together to maintain the engram of distributed motor cortical neurons.

Recovery of limb impairment. Over time after stroke remaining neurons in the motor cortical engram regain their network connections and the limb partially recovers.

CREB-induced motor cortical engram. Inducing CREB in neurons in motor cortex enhances their connection within a new and expanded motor cortical engram. This enhances the speed and completeness of motor recovery after stroke.

- Inactive neuron in motor network
- Active neuron in motor network
- Dead neuron in stroke
- Neuron distant from stroke that is disconnected from forelimb motor circuit neuron after stroke
- Active neuron in motor network that is expressing CREB

Figure 5.3

Motor engram in stroke and CREB. A region of neurons in motor cortex is represented as circles in the panels at left. During a normal motor movement of the forelimb, a network of neurons is functionally active together to move that limb. After stroke, this network is partially degraded. Neurons are not dead, but they do not fire together in a coordinated manner to move the limb (see also figure 5.4). The loss of forelimb function that occurs is shown as the white limb in the model mouse. During recovery after stroke, there is partial restoration in the coordinated firing activity of motor neurons that code for limb movement, and the limb motor control is improved. With CREB induction, the neurons that have the induced CREB are more active and more likely to be recruited into a forelimb motor network. With the addition of CREB-induced neurons into this network, motor control of the forelimb is enhanced after stroke.

this fear memory can be identified with a molecule that signals neuronal activation, such as an immediate early gene. Elimination of the neurons in the fear memory engram by selective ablation of the cells, or through a similar approach as in the CREB stroke studies with DREADD, disrupts the memory. Similar data exist for learning and memory engrams in cortex and hippocampus (Han et al. 2007; Zhou et al. 2009; Sano et al. 2014; Josselyn et al. 2015; Rashid et al. 2016). These data suggest that CREB is regulating a motor engram after stroke: a network of coactivated neurons in motor cortex that control forelimb movement.

With the concept of a motor engram in mind, these CREB findings in stroke recovery suggest several things that lead to a new principle with regard to CREB function in the brain after stroke. CREB induces a plastic state, whereby neurons with extra copies of CREB are able to capture more motor cortical circuits and engage a greater motor control network, or motor engram, for recovery. An additional study tested this idea. Mice with CREB induced in motor cortex received stimulation of the CREB-activated neurons after stroke, and this stimulation was compared to the control state with neurons that did not have CREB induced. Stimulation was directed to the forelimb site in motor cortex. Stimulation of CREB-induced neurons caused movement in many body parts, such as trunk and face, compared to stimulation of normal (non-CREB-induced) neurons after stroke (Caracciolo et al. 2015), which caused mostly just forelimb movements. Such an enhanced ability of CREB-induced neurons to integrate into and control a brain circuit that is greater than the normal circuit in the brain has been described previously for the memory system. CREB-induced neurons in the amygdala, hippocampus, and cortex preferentially integrate into a memory circuit and control the function of that circuit, such that ablating these neurons or turning them off disables the memory (Han et al. 2007; Zhou et al. 2009; Sano et al. 2014). These effects mean that CREB controls neuronal "allocation": the integration or assignment of a neuron into a functionally active circuit (Han et al. 2007; Zhou et al. 2009; Sano et al. 2014; Josselyn et al. 2015; Rashid et al. 2016). CREB activation within neurons makes them more easily excited, facilitates their synaptic signaling (such as LTP), and enhances the formation of their connections (via new dendritic spines). In essence, these CREB-induced changes create a neuron that really cares what its neighbors are saying and connects with them more strongly when they are active than would a neuron without CREB induction. Stated again, neuronal allocation during a memory task establishes the engram for that task: the network of neurons that, when coactivated, mediate that memory (Josselyn et al. 2015; Rashid et al. 2016).

In summary, these studies with CREB suggest that there is a motor engram for recovery after stroke. A small number of neurons in the motor cortex adjacent to the stroke, after being induced to express CREB, capture more motor cortical circuits and enable motor recovery. When these circuits are inactivated, the motor system essentially collapses, presumably because these "hyperconnected" CREB motor neurons have taken up more of a role in motor function than noninduced CREB neurons. However, the effect of inactivation of CREB-induced neurons on motor control is much greater after stroke than in the control, nonstroke brain. Thus, there appear to be two levels of plasticity in the motor system after stroke: an activity-dependent plasticity that signals through CREB and another type of plasticity that is induced by the stroke itself. The cortex after stroke is unstable territory, and the molecular manipulations that enhance recovery appear to do so by enabling allocation of neurons into larger motor recovery circuits, or engrams. The flip side of this process is that these circuits, if negatively perturbed, will result in a substantial loss of function. This overintegration or overcommitment of surviving motor circuits in the recovery of movement after stroke may account for the phenomenon of recrudescence: these circuits are more sensitive to perturbations that would not cause loss of function in non-recovered brain circuits.

5.6   Plasticity Is a Risk for Cell Death: Timing for a Neural Repair Therapy after Stroke

Changes in molecular systems that increase cellular excitability after stroke enhance synaptic plasticity by diminishing tonic GABA inhibition, enhancing glutamate receptor signaling, or increasing the transcription factor CREB. However, enhancing synaptic plasticity may also destabilize a brain's ability to deal with ischemia. Put more simply, enhancing neuroplasticity at times in which the acute insult of the stroke is still present will make the stroke worse. Blocking tonic GABA signaling, or enhancing signaling through the AMPA glutamate receptor, concomitant with initiation of therapy within three to five days after stroke, make the stroke worse by increasing infarct size. Conversely, when these therapies were initiated beyond five days, they did not change infarct size and instead enhanced motor recovery (Clarkson et al. 2010; Clarkson et al. 2011). Many potential neural repair therapies enhance endogenous neuronal plasticity in ways that activate neuronal excitability, such as enhancing signaling of the transcription factor CREB (MacDonald et al. 2007), and might be expected to also follow this timeline. This general principle of a trade-off between enhancing the plasticity of a cell versus ensuring its ability to withstand stress is seen in other systems. Manipulations of the growth cone protein GAP43 enhance the ability of a neuron to grow a new axon after injury but also increase the level of cell death from that very same injury (Harding et al. 1999; Huang et al. 2013). Neuronal cell types with the greatest resistance to cell death after injury usually also have the lowest regenerative capacity, such as Purkinje cells, whereas neuronal cell types with the greatest regenerative capacity after injury also suffer the most cell death when injured, such as inferior olivary neurons (Dusart et al. 2005). In the optic nerve, lesions closest to the retina ganglion cell bodies provoke the greatest amount of cell death but also induce a greater regenerative response—compared to lesions distant from the cell body in which there is little cell death and also little regeneration (Dusart et al. 2005).Thus animal models suggest that, increasing the excitability of the motor system very early after stroke may exacerbate the stroke insult even though this same activity will lead to improved recovery at a later period after stroke (Bland et al. 2000). This concept is discussed in more detail in chapter 3 as it relates to studies of early neurorehabilitative interventions in humans.

Based on this transition from stroke damage to stroke repair, the correct timing for administration of an activity or plasticity-enhancing therapy in human stroke, based on animal model data, is not exactly clear. The early events of inflammatory cell infiltration and astrocytosis appear to follow similar time courses in rodents and humans. The recovery curve for rodents also has a similar shape and ceiling to that of humans, except it is constricted to the first month after stroke in rodents and extended to the first three months after stroke in humans (Krakauer et al. 2012; Dobkin and Carmichael 2016). The animal model literature suggests that a neuroplasticity therapy started earlier than the first week after stroke in humans positions this therapy within a window of risk for exacerbation of the initial stroke damage.

The opposite applies in regard to this second-order principle: treatments that protect the brain in stroke will worsen behavioral performance if given during the recovery phase (Clarkson et al. 2011). Glutamate receptor antagonists and GABA receptor agonists reduce neuronal excitability and stroke size when given early after stroke but degrade behavioral performance and recovery when given later after stroke (Lazar et al. 2010; Clarkson et al. 2011). The upshot of this second-order principle in stroke neural repair is that treatments that promote plasticity and recovery must clearly be distinguished from treatments that promote stability and protection, and their optimal windows of application determined.

5.7   The Brain Forms Regenerative Cellular Niches during Repair and Recovery

In poststroke neurogenesis, multipotent neural stem cells in the subventricular zone respond to stroke, even when distant from this site, and proliferate. Immature neurons migrate to areas of injury and can differentiate into mature neurons with local synaptic connections and long-distance projections (Ohab and Carmichael 2008; Hermann and Chopp 2012; Kahle and Bix 2013). Ablation of newly derived immature neurons after stroke causes reduced recovery in rodents (Jin et al. 2010). Immature neurons localize to angiogenic blood vessels in damaged tissue and are stimulated to migrate by growth factors or cytokines released by these vessels (Ohab et al. 2006; Kahle and Bix 2013). However, despite a robust initial neurogenic response, most of these immature neurons die. Poststroke neurogenesis has been reported in human stroke, by using tissue staining for protein markers of immature neurons in autopsy material (Jin et al. 2006; Macas et al. 2006; Ekonomou et al. 2012). However, a lack of poststroke neurogenesis has been reported in human cortical stroke, using ^{14}C labeling of newly born cells (Huttner et al. 2014). Both techniques have limitations in specificity and sensitivity, and they may also miss a transient neurogenic response after stroke that is limited in size and then stops (Boekhoorn et al. 2006; Macklis 2012). A neurovascular niche for neurogenesis does exist in the nonhuman primate brain (Tonchev et al. 2007) but has not been determined in humans (Conover and Todd 2016). Because of the nature of human studies, a definitive finding of clinical poststroke neurogenesis remains lacking.

Strokes in cortex, striatum, and white matter stimulate oligodendrocyte progenitor cells (OPCs) to divide and migrate to the lesion in animal models (Sozmen et al. 2009; Sozmen et al. 2016), and in humans (Sanin et al. 2013). Further white matter remodeling occurs in humans with stroke and with therapies that promote recovery in human stroke (Wan et al. 2014; Zheng and Schlaug 2015). OPCs carry the capacity to differentiate into mature oligodendrocytes and are in a position to mediate neural repair, as occurs in the initial stages of multiple sclerosis. However, OPCs after stroke do not appear to differentiate into oligodendrocytes as they do in multiple sclerosis (Sozmen et al. 2016). White matter stroke, damaging myelinated fiber tracts, is even worse in aged animals. This age effect on OPCs and white matter responses appears to be mediated by greater local inflammation in the aged brain (Rosenzweig and Carmichael 2013) and by intrinsic differences in aged OPCs (Miyamoto et al. 2013).

Neurogenesis and gliogenesis involve cellular programs of tropism, migration, and stimulation (in neurogenesis) or inhibition (in gliogenesis) due to inflammation. These properties of immature neurons and glial cells are first-order principles: rules that govern behavior of each cell type. However, these elemental cellular responses occur within the larger context of multicellular niches. Poststroke neurogenesis occurs within a neurovascular niche in which angiogenesis and neurogenesis are causally interconnected (Ohab et al. 2006). Gliogenesis occurs in a zone of reactive astrocytes and damaged axons and appears limited by cues from these cellular or subcellular compartments (Sozmen et al. 2009; Sozmen et al. 2012; Sozmen et al. 2016). The second-order property of tissue regeneration relates to the communication among "repairing" cells in the brain after stroke. After stroke, neurogenesis and angiogenesis occur in unique regenerative cellular niches within damaged tissues that are transient and unique to the injured brain. These cell types signal to each other using diffusible cytokines and growth factors, such as stromal cell-derived factor 1 (SDF-1) and angiopoietin 1, and likely cell-cell contact signals.

The concept of a progenitor niche in neural repair after stroke is informed by the concept of the stem cell niche. The original description of the stem cell niche explained the maintenance of bone marrow stem cells in aging and in response to chemotherapy (Schofield 1977). Scadden (2006) characterized this environment: "Stem-cell populations are established in 'niches'—specific anatomic locations that regulate how they participate in tissue generation, maintenance and repair. . . . It constitutes a basic unit of tissue physiology, integrating signals that mediate the balanced response of stem cells to the needs of organisms." Progenitor cells in stroke engage in tissue repair within transitory regenerative niches. These niches derive from the constituent cells and induce the initial progenitor response, mediate cues from the surrounding environment, communicate cellular or tissue signals that reflect age and likely comorbid diseases, and ultimately define the outcome of recovery. A promising area of future research will be to define the molecular signaling systems within these niches to enhance progenitor responses, maturation, and repair.

5.8 Engaging CNS Tissue Regeneration Is Like Activating a Cancer

Neural repair means activating a growth program in an adult neuron to form new connections (Zai et al. 2009; Li et al. 2010; Li et al. 2015). Tumor suppressor proteins, such as PTEN and SOCS3, control key nodes in the molecular machinery of this regenerative response of poststroke axonal sprouting (Park et al. 2008; Sun et al. 2011). PTEN is also mutated in up to 70 percent of prostate cancers (Chen et al. 2005) and 40 percent of glioblastomas (Lino and Merlo 2011). After stroke, progenitor cells are induced into a growth program that involves cell division, migration to a tropic cue, and association with angiogenic vessels (Ohab and Carmichael 2008; Hermann and Chopp 2012; Kahle and Bix 2013). This is a similar cellular response as tumor metastasis, in which primary tumor cells invade adjacent tissue, circulate, and localize to angiogenic vessels (Ghajar et al. 2013; Wan et al. 2013). Like metastatic tumor cells (Wan et al. 2013), migrating progenitors after stroke secrete matrix metalloproteinases (MMPs) to localize to the target tissue (Lee et al. 2006). Molecular receptors systems, such as transforming growth factor beta (TGF-β) receptors, induce a growth state and axonal sprouting in stroke (Li et al. 2015) and also are key molecules in metastatic transformation in cancer, such as the epithelial-mesenchymal transition (Lamouille et al. 2014).

In this section, a second-order principle means that a principle originally developed to describe one phenomenon has meaning in a second, previously unrelated phenomenon, and when the two are considered together a new principle emerges. In this case, tissue repair after stroke (first-order principle) is similar to oncogenesis and metastasis (another first-order principle) and this means that a second-order principle in stroke recovery is that treating for 'brain repair' may induce cancer. In addition to tissue regeneration in the CNS, similar pathways regulate tissue regeneration in other organ systems within molecular pathways that are also critical in cancer development (Johnson and Halder 2014). This second-order principle of tissue regeneration is obviously problematic for the design of neural repair therapies.

5.9 Neural Repair Therapies Must be Focused in Time and Space

Therapies that stimulate neural repair after stroke will need to be controlled in their effect in both time and space. Temporal and spatial control is necessary because of the overlap in molecular programs in oncogenesis and neural repair. Also, outside of this cancer overlap, many of the growth factors or cytokines that stimulate neural repair in preclinical models have widespread effects in other body tissues. Examples of these are erythropoietin, fibroblast growth factors, and granulocyte colony stimulating factor (G-CSF). In preclinical studies, these molecules stimulated axonal sprouting, neurogenesis, and other aspects of neural repair. However, in clinical trials, their off-CNS effect limited their use, with renal, hemodynamic, bone marrow, and thrombogenic complications (Ehrenreich et al. 2009; Ringelstein et al. 2013). In addition to these past examples, recent discoveries in the neural repair field further make this point. The TGF-β family member GDF10 serves as a signal after stroke that activates a gene expression program in peri-infarct tissue, which produces axonal sprouting and motor recovery (Li et al. 2015). However, this molecule signals through TGF-βRI and II, and this system also plays a role in tumor metastasis (Lamouille et al. 2014). A GDF10 therapy would need to be delivered locally to peri-infarct tissue, in the subacute phase after stroke when axonal sprouting is active—so as to minimize a potential oncogenic or non-CNS complication. Similar therapeutic control has been proposed for PTEN inhibition (Naguib and Trotman 2013), which would be another mechanism to enhance axonal sprouting and recovery. One mechanism for spatial and temporal control of delivery of a neural repair drug is to use tissue bioengineering, with hydrogels that self-assemble in brain and locally release a small molecule or biologic to the peri-infarct tissue (Ma et al. 2007; Li et al. 2010; Overman et al. 2012; Li et al. 2015).

5.10   Regeneration Does Not Recapitulate Development: The Meaning of "Phenotype"

Many cellular aspects of tissue regeneration resemble the initial process of organ formation in development. In the cortex adjacent to the stroke core, neurons lose their perineuronal net, show altered intracortical inhibition, and become growth factor dependent (Clarkson et al. 2010; Witte and Stoll 1997; Carmichael et al. 2005; Hobohm et al. 2005; Li et al. 2010). These are hallmarks of neurons in the critical period of neuronal development, when cortex is uniquely plastic to environmental alterations in physiology and structure (Levelt and Hübener 2012). In poststroke neurogenesis, multipotent neural stem cells give rise to immature neurons, which migrate long distances and mature in small numbers into neurons with synaptic connections and long-distance projections (Ohab and Carmichael 2008; Hermann and Chopp 2012; Kahle and Bix 2013). This resembles both pyramidal neuron development in cortex and inhibitory neuron development in the forebrain (Southwell et al. 2014). Such similarity in tissue repair in stroke to neurodevelopment has prompted suggestions that brain regeneration recapitulates development (Cramer and Chopp 2000). Similar comparisons of regeneration to development have been made in other systems, such as kidney, bone, liver, and skin (Kelley-Loughnane et al. 2002; Kawakami et al. 2013; Hadjiargyrou and O'Keefe 2014).

The proposed similarities in mechanism between regeneration and development derive from similarities in phenotype: axonal extension and synaptogenesis, perineuronal net condition, growth factor dependency, and neuroblast migration. Phenotypes can be similar across many conditions in medicine, but what is critical for understanding causality and then moving a field forward in therapeutics is whether there is a true pathophysiological overlap—which is established not just be similarities in cellular events but overlap in the causative molecular biological events. For example, in peripheral axonal neuropathies, there is overlap in many diseases at the cellular level: axonal damage, length-dependent effects, distal accumulation of neurofilaments, mitochondria and other organelles, a central cell body reaction and occasional demyelination. However, axonal peripheral neuropathies can be caused by metabolic disease (diabetes), toxins (alcohol), and immune attack (IgM deposition). The causality in peripheral nerve injury of these different diseases or agents is distinct: their molecular effects on the nerve are different (though their microscopic appearance is the same) and their treatment is very different. Phenotypic similarity is misleading in the science that goes into an understanding in depth of the disease. What is critical for the biology of regeneration and for a possible regeneration therapy is whether the underlying molecular profile of CNS regeneration recapitulates the molecular control of neurodevelopment. Initial transcriptional profiling of neurons that form new connections after stroke, retinal ganglion cells that regenerate an injured axon in the optic nerve, or dorsal root ganglion cells that regrow a connection after peripheral nerve injury, all suggest some overlap to genes that are active in the developing nervous system, but overall a distinct set of genes is regulated in these injury responses (Bosse et al. 2006; Li et al. 2010). Direct comparison of the transcriptome of neurons exposed to a regenerating stimulus after stroke and the transcriptomes of neurons at several stages of neuronal development from many different labs clearly indicates that the molecular expression profile of regeneration is statistically distinct from the developmental transcriptome (Li et al. 2015). Thus, on a molecular level, regeneration does not recapitulate development in the brain after stroke. Such a lack of correspondence in the molecular control of regeneration to that of development was also shown in muscle repair (Wang and Conboy 2010). Indeed, even in highly regenerative animals, like the newt or salamander in which a whole limb develops after injury, the molecular and cellular processes of regeneration is distinct from development (Nacu and Tanaka 2011).

5.11 Neuronal Networks in Motor Recovery: Second-Order Principles Interact

In chapter 4, we described the focal core of stroke damage and the radial zones of partial damage outside of this core. In both animal model studies of stroke and in the few studies done in human stroke, the brain areas adjacent or connected to the stroke site do not lose neurons but their physiology is disrupted. In optogenetic studies of neuronal activity in spared tissue, stroke causes a loss of responsiveness to sensory inputs or local cortical excitation, which is then regained over time. This initial loss of responsiveness is greatest in areas that are (or were) connected to the stroke core and least in areas that were less connected to the stroke (Lim et al. 2014). It appears that stroke can physiologically disrupt neuronal networks at a distance, although the behavioral relevance of this disruption is harder to determine.

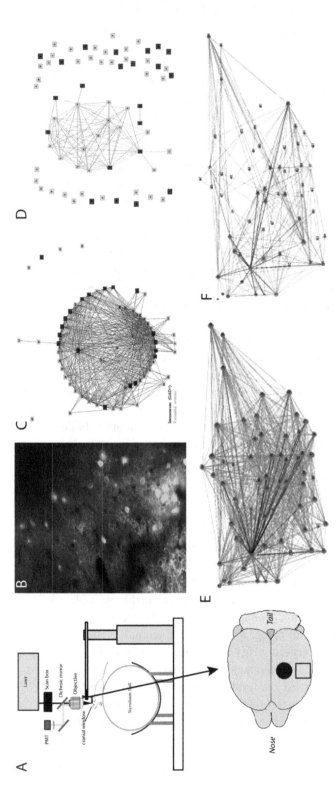

Figure 5.4

Network connectivity in stroke. To study the connections of neuronal networks in stroke in real time during the course of recovery, neuronal calcium activity was imaged in the living mouse. Cortical Ca²⁺ somatic transients were imaged with two-photon calcium imaging in mice labeled with the calcium indicator protein, GCaMP6s. (A) The imaging setup in which cortex is imaged in a living mouse as it moves on a floating Styrofoam ball. Selected fields of view (size, 500×500 μm) were recorded from somastosensory cortex one day before and then twenty-five days after stroke in the adjacent motor cortex. The arrow shows a schematic of the top of the mouse brain with the stroke (dark circle) and the area of imaging (black box). (B) A representative field of view with neurons that are expressing GCaMPs seen as bright. Regions of interest (ROIs) corresponding to identifiable cell bodies were selected using a semiautomated algorithm. ΔF/F0 in change in fluorescence during limb movement was calculated as (F—F0)/F0, and then a deconvolution algorithm was applied to remove the nonphysiological slow decay in the GCaMP6s signal and sharpen Ca²⁺ transients. To determine significant correlation value, Monte Carlo simulation was applied. To visualize changes in microcircuit activity, a connectome procedure for displaying neuronal functional connectivity based on graph theory elements was used. Connections between neurons were determined based on the significant correlations above the threshold cutoff (p < 0.05 from the Monte Carlo simulation)

as an indicator of functional connectivity. If a neuron is significantly correlated to 95 percent of all other neurons in the field, the neuron is considered functionally connected to these neurons in the field of view and the edges should be drowned for them. (C) The connectome in somatosensory cortex before stroke. (D) The connectome twenty-five days after stroke. Many neurons in somatosensory cortex are not functionally connected to their neighbors, although they are firing in normal patterns. These are "orphan neurons" in that they are not dead and are firing normally but not in a coordinated way with their neighbors as before stroke. (E) Functional connectivity data as in **C** and **D** but shown as a function of location of cells in the brain. The cells in **C** and **D** are arranged to place them close to their most highly connected partners, irrespective of location. In **E** (before stroke) and **F** (twenty-five days after stroke), the location of the neurons is according to their actual location in the cortex in the imaging window. Edges in these data indicate a significant relationship of calcium firing between neurons: they are firing together. This is generated in the same way as the data as in **C** and **D**. Comparing **E** and **F**, it can be seen that what is primarily lost in correlated firing is the near connections. Long-distance connections are preserved. Using graph theory terminology, the small worldness that is present in normal neuronal functional networks is lost after stroke. Adapted from Latifi et al. (2016).

To observe physiological disruption at a distance, Latifi et al. characterized neuronal activity in ipsilesional somatosensory cortex after focal motor cortical stroke (figure 5.4, adapted from Latifi et al. 2016). In this approach, neurons express the calcium indicator dye GCaMP6s. When neurons are active, they fire calcium spikes that can be recorded with an optical microscope over the surface of the brain. Neuronal firing activity is then correlated within all of the neurons in that area. In this case, somatosensory cortical neuronal activity was recorded before and after stroke. Using network analysis approaches (figure 5.4), the activity in neurons that is correlated—the firing of neurons together—is then analyzed and neurons that fire together are studied in a network. This is like the social network analysis of a "friends group" in online social systems, a sort of Facebook for neurons. Prior to stroke, the neurons in somatosensory cortex are intensely firing in close association together. These neurons form a tight and correlated network. In the metaphor, most of these neurons are friends. After stroke, there is a loss of correlated network activity. Neurons are still alive and are still firing (these data not shown), but they lose a correlated firing pattern. Neurons are not firing together as a network. There has been some unfriending. In a term, there are many "orphan" neurons that are not functionally connected to their neighbors—despite that fact that these neurons are healthy and have normal firing patterns, they are just not firing within a relationship or functional network to their neighbors.

One way to analyze this is in the way that an online social networking program would (figure 5.4C,D), with each neuron as a node and the connections between neurons as edges, and a representation of the neurons as close or far away from each other based on the strength of their connection. Another way to represent this network is to keep the actual physical location of the neurons in the brain constant and then put in the significant firing correlations. In the normal brain, there are correlated firing patterns with neurons that are close and far, but the preponderance of connections are with close neurons (figure 5.4E). After stroke, there is a loss of functional connections, as we saw in the mapping of this relationship without physical location (figure 5.4E), but this loss is predominantly in the correlated firing between neurons that are close to each other, i.e., neurons after stroke have lost their small world network property. As stated above, the behavioral relevance of this loss of a network property has yet to be fully determined, but this approach provides elemental insights into the exact brain circuits that move a limb and how they respond to stroke and may be in a position to mediate recovery (or not).

5.12 Conclusions

Stroke is not only an acute killer but also a chronic and progressively disabling disease. During the limited recovery that occurs after stroke, the brain activates growth programs in surviving neurons, which can form new connections in ways that are potentiated by behavioral activity or neurorehabilitative training. These cellular responses are driven by underlying molecular programs that are unique to CNS regeneration, overlapping less with neurodevelopment than was previously assumed. Tissue repair after stroke occurs within transient regenerative cellular niches that communicate cues to the regenerating cells. Neurons respond to these cues by growing new connections, whereas neural progenitors repair damaged tissue; processes which share molecular features with malignant transformation of tissues and metastasis. These second-order interactive properties of neural repair present targets for future therapies.

6

Chronic Hemiparesis
Motor Learning, Compensation, and the Challenge of Reversing Impairment in Late Stroke

6.1 Introduction

Patients with chronic stroke (beyond six months) are far and away the main focus of research studies on stroke recovery largely because they are easier to schedule and recruit, but clinical trial results in this population have been almost uniformly disappointing. There is an irony to this intense research focus on chronic stroke because in the real world, therapy for these patients is close to nonexistent. Indeed, if aliens from outer space were to observe how stroke rehabilitation is given, at least in the United States, they would be struck by some puzzling contradictions. First, as we have seen in previous chapters, there is abundant evidence in humans and in animal models that almost all recovery from stroke at the level of impairment occurs in the first month after stroke and is mainly attributable to endogenous repair processes. Animal models also show that the effects of spontaneous recovery can be augmented by intense training regimens but only within a short time window. The aliens would therefore no doubt be surprised that acute rehabilitation units all deemphasize impairment and instead teach compensatory strategies in what is likely to be a precious time-limited window of opportunity. Even with respect to compensation, the upper limb is all but ignored; the average time spent per day on the upper limb is around thirty minutes (Lang et al. 2009), and this might not include the weekends or holidays. Second, the aliens would notice that clinical trials and research efforts that have directly addressed the need for therapy at higher doses and intensities have been conducted in the chronic stroke period. Third, as we have already mentioned, patients with chronic stroke rarely receive significant amounts of therapy. Indeed, in some trials in chronic stroke, a control group has not been included because standard of care is basically lack of care. When control groups are included, they receive the low amount of care that is usually given acutely. Overall, it would be hard for the aliens not to conclude that some perverse nonscientific law is in operation.

As we described in chapter 3, Thomas E. Twitchell provided a detailed description of the stereotypical stages of recovery of the upper limb after stroke. Signe Brunnstrom and others then devised treatment approaches with the goal of increasing the chances that patients would not get stuck at an early stage with weakness, spasticity, and synergistic movements (i.e., the classic patient with moderate to severe chronic hemiparesis). As we describe in chapter 1, at some point in the 1980s, emphasis switched to motor learning approaches, compensation, and activities of daily living (ADLs). The main impetus for this was the realization that facilitation of normal-looking movements and a focus on muscle tone and reflexes was not generalizing to ADLs (Carr 1987). In 1984, the Functional Independence Measure (FIM) was devised with the intent to be used as a universal assessment tool for rehabilitation (Keith et al. 1987). Unfortunately, however, the measurement became the treatment target. This is known as Goodhart's law, originally used in the context of economics, which states, "As soon as the government attempts to regulate any particular set of financial assets, these become unreliable as indicators of economic trends," more generally when a measure becomes a target, it ceases to be a good measure (Strathern 1997). This slide from measurement to treatment was unwittingly bolstered by new theories of motor control and motor learning, which moved away from reflex accounts of movement. To paraphrase James Gordon, the goal of rehabilitation should be to have patients actively solve a motor problem rather than passively relearn a normal movement pattern (Carr 1987; Gordon 1987). Although this argument is not without merit, it exemplifies the reasoning behind the move away from impairment and toward function and compensation.

6.2   Motor Learning Principles for Neurorehabilitation

Motor learning is the core idea that drives current task-based approaches to neurorehabilitation, as well as robotic and constraint-based therapies in chronic stroke. Statements to the effect that recovery after stroke is a form of learning or relearning are ubiquitous. The point we want to make here is that there is a fundamental difference between thinking that motor learning can reverse impairment versus thinking that it can be used to learn compensatory strategies. As we shall see, there is very little evidence for the former in chronic stroke. The relearning premise for neurorehabilitation rests on three other underexamined assumptions: first, that the target of the paretic deficit to be rehabilitated through learning has been properly identified; second, that a framework exists for choosing the proper learning mechanism in a systematic way; and third, that learning mechanisms are intact in patients after stroke despite impaired motor performance.

The fundamental problem for motor learning is to find the appropriate motor commands that will bring about a desired task outcome. Motor learning is a fuzzy category that encompasses action selection guided by instruction, reward, or error and subsequent practice-dependent improvement in the execution of the selected actions. *Skill* is a widely used term but is hard to define. Here, it will suffice to say that one is *skilled* at a task when practice has led to it being performed better than baseline because of selection of optimal mean actions that are then executed with precision at high speed and short latency.

6.2.1   Instruction and Imitation

Although underemphasized in the neuroscience literature, instruction and imitation are fundamental to the acquisition of motor skills, as attested to by the ubiquity of coaching and teaching in sport, music, and dance. Similarly, physical and occupational therapists have an essential instructional role in rehabilitation. We have recently suggested (Stanley and Krakauer 2013) that one reason why the crucial roles of knowledge, instruction, and imitation for motor learning have been underappreciated is because of an overemphasis on simple implicit adaptation tasks due to the classic result in the patient H.M., who retained memory of mirror-drawing ability across days despite no explicit memory of ever having performed the task (Milner 1962). Unfortunately, this led to overgeneralization of the notion of procedural learning/memory from this simple task to all motor skills. With just a moment's thought, however, it is obvious that everyday motor skills such as cooking or driving cannot be extrapolated from motor adaptation tasks and could not be learned without knowledge and instruction. A stark demonstration of this was made in a recent paper; subjects could not learn a motor task with even minimal redundant structure without explicit awareness of this structure (Manley et al. 2014). We, and others, have recently shown that even adaptation tasks have a crucial explicit component (Taylor et al. 2014) and that knowledge enhances skill (Wong et al. 2015). Thus, it is very important not to make the mistake of considering motor learning synonymous with low-level, automatic implicit processes.

6.2.2   Reinforcement Learning

Actions will be selected with increased or decreased frequency depending on the schedule of rewards and punishments, respectively. Reward can be intrinsic, based on self-perceived success or failure, or it can be based on extrinsically provided loss or gain in points or praise. Rewards can be short term or long term, and the balance between present and future rewards is of central importance to computational theories in the field of reinforcement learning. A local action solution that is "just good enough" may be exploited and then become habitual, even though with more time and exploration, a more optimal action could have been found. For example, if a person is given a tennis racket and told to hit a backhand, he or she may well find a way to do so on his or her own but is very unlikely to discover the best technique, which would require coaching and more extended practice. Later in the chapter, we will argue that compensatory strategies after stroke often represent precisely this kind of premature adoption of habitual "just good enough" actions. Constraint-induced movement therapy (CIMT) is based on ideas from reinforcement learning. The good arm is constrained in an attempt to prevent adoption of the bad habit of choosing the unaffected arm to perform tasks rather than doing the harder work of improving the affected side. The affected side is trained with shaping, which is a form of titrated operant conditioning (Taub et al. 1994). Operant approaches are also used in an attempt to train out of synergy movements in patients with arm paresis. Zachary Wright,

Zev Rymer, and Mark Slutzky used a myoelectric-computer interface to operantly decouple anterior deltoid and bicep coactivation (part of the flexor synergy after stroke) in patients with chronic stroke. The basic approach was to have different combinations of muscle activation steer a cursor on a screen. The mapping of electromyography (EMG) to two-dimensional (2D) cursor space was such that it encouraged decoupling. The results were encouraging, in that the patients showed small gains in the upper extremity Fugl-Meyer Motor Assessment (FMA) (Wright et al. 2014). The question is whether this kind of approach can ever be expected to have large effect sizes.

6.2.3 Sensorimotor Adaptation

Adaptation refers to reduction of errors in response to a perturbation. Sensorimotor adaptation tasks have been extensively studied experimentally and modeled computationally (Shadmehr and Mussa-Ivaldi 1994; Krakauer et al. 2000; Thoroughman and Shadmehr 2000; Smith et al. 2006). The prevailing idea is that adaptation occurs through cerebellar-dependent reduction of errors through updating of a forward model via sensory prediction errors (Shadmehr et al. 2010). This is a form of supervised learning with the goal of minimizing the error between predictions and observations. The learning rule in many supervised learning algorithms, such as gradient descent, depends only on the size and direction of the error and hence is often referred to as error-driven learning. The relevance of sensorimotor adaptation to rehabilitation protocols remains unclear, however, because although imposed errors can lead to fast and large changes in behavior, these changes do not seem to last once the perturbation is removed. For example, the paretic arm can be made to adapt to a viscous force field set to amplify baseline directional reaching biases. When the force field is switched off, aftereffects are now in a direction that negates the biases—an approach called "error augmentation" (Patton et al. 2006; Abdollahi et al. 2014). A similar approach has been used using a split-belt treadmill to reduce step asymmetry in hemiparetic gait (Reisman et al. 2007). In both cases, however, the desirable aftereffects are very short lived. In the case of force-field adaptation of the arm, aftereffects lasted for only 30 to 60 movements after 600 training movements (Patton et al. 2006). More recently in the case of gait, it has been shown that repeated exposure over multiple sessions prolongs split-belt treadmill aftereffects in patients with stroke (Reisman et al. 2013). Interestingly, repeated exposure is also required for prism adaptation in the treatment of neglect after stroke (Rossetti et al. 1998; Newport and Schenk 2012). One explanation for the short-lived nature of adaptation is that newly adapted behaviors are out-competed by baseline behaviors that have been reinforced over much longer periods of time and have become habits. In support of this idea is the recent finding that if a newly adapted behavior, once it has reached asymptote, is reinforced by switching from continuous vector error to binary feedback, the adapted behavior is retained for longer (Shmuelof et al. 2012; Galea et al. 2015; Therrien et al. 2016). Thus, if adaptation paradigms are going to be used to have patients quickly converge on desired behaviors, then error-based and reinforcement-based learning mechanisms will likely need to be combined. A potential way to do this would be to adapt a patient first and then reinforce the aftereffect.

6.2.4 Use-Dependent Plasticity

Use-dependent plasticity (UDP) refers to the neural changes induced by repetition of movements in the absence of errors. The behavioral consequences of UDP are referred to as use-dependent learning. It will be argued here that the assumption that UDP leads to a form of motor learning or motor memory relevant to neurorehabilitation is likely incorrect. Motor learning through practice to improve task performance is not synonymous with movement repetition. This erroneous conflation has arisen because of the tendency to blur the distinction between plasticity and learning. Plasticity refers to the capacity of the nervous system to change its input–output characteristics with various forms of training. These input–output relationships can be assayed in a variety of ways, which include single-unit recording in animal models and noninvasive brain stimulation in humans. Learning does imply that a plastic change has occurred, but a plastic change does not imply that behavioral learning has occurred. Thinking otherwise is to commit the classic logical fallacy called "affirming the consequent": (1) If P, then Q. (2) Q. (3) Therefore, P. Unfortunately, a sizable literature appears to consider UDP important to neurorehabilitation based largely on this logical fallacy.

To appreciate the misunderstanding about UDP, consider the classic paper in this area by Joseph Classen and colleagues (1998). Transcranial magnetic stimulation (TMS) of the motor cortex was used to evoke isolated and directionally consistent thumb movements through activation of the abductor pollicis brevis muscle. Subjects were then required to practice thumb movements for thirty minutes in the direction approximately opposite to that elicited by TMS. The critical finding was that subsequent TMS was found to evoke movements in or near the direction practiced rather than in the pretraining baseline direction. This is a very interesting result with regard to how movement repetition (it is not really training insomuch as the goal is not to improve performance in any way) can lead to changes in cortical representation. Indeed, a very similar mechanism is likely at play in the series of controversial papers published by Michael Graziano and colleagues showing that long duration trains of intracortical microstimulation of monkey motor cortical areas elicit movements that look like natural movements performed at high frequency in everyday life (Graziano et al. 2002). More recently, it has been shown that TMS in piano players elicits piano-relevant finger postures not seen in non–piano players (Gentner et al. 2010).

The crucial point when considering UDP-like results related to movements elicited by stimulation of the brain is that it is not at all clear how they relate to *voluntary* movements. To appreciate this objection, consider the thumb experiment; although TMS after training causes the thumb to move in a direction roughly similar to the one practiced, if subjects are asked to move their thumb in the original pretrained direction, they do not suddenly find themselves going in reverse! That is to say, the plastic changes assayed with TMS have not changed voluntary behavior to any appreciable degree. It is as if the dramatic TMS-induced near-180° flip falsely inflated expectations about the potential efficacy of repetition for rehabilitation. The effects of repetition on voluntary behavior are much more modest and short-lived, with small reductions in variance in the repeated direction and small biases in other directions (Diedrichsen et al. 2010; Huang et al. 2011; Verstynen and Sabes 2011). In addition, the therapeutic benefit of movement repetition is cast in doubt with the demonstrable failure of haptic and robotic guidance approaches to motor training (Winstein et al. 1994; Liu et al. 2006; Reinkensmeyer and Patton 2009). Thus, at the current time, experiments that induce UDP are informative about how the brain changes with repetition, but repetition is not practice and has only transient effects on behavior (i.e., repetition induces plasticity but plasticity does not equal learning). It appears that the interest in UDP is out of proportion to its practical usefulness for neurorehabilitation.

6.2.5   Sequence Learning

We shall only touch on this briefly, which may come as something as a surprise as the motor learning field is probably even more fixated on sequence learning than it is on sensorimotor adaptation. Every action requires a sequence of muscle activations: the triphasic EMG response, which means that in essence, every single movement is a sequence task. The questions one then has to ask are as follows: why do we need to study discrete sequence learning tasks (the most studied kind) at all, and do they provide insight into continuous movements like a reach or a tennis serve? Overall, we think the answer to both questions is no. This book is not the place to go into an in-depth critique of the pros and cons of sequence learning as a model for motor skill learning. Suffice to say that the most popular sequence task in general and the one most studied in patients after stroke, the serial reaction time task (SRTT) (Nissen and Bullemer 1987; Robertson 2007; Orrel et al. 2007), is in our view not a motor learning task at all and is flawed even on its own terms, which therefore makes it unsuitable for the study of motor learning in stroke. There are two main reasons for our view. The first is that the SRTT is considered a motor learning task only because it supposedly probes implicit learning. As we have discussed above and written about elsewhere, equating implicit learning with motor learning is incorrect (Ghilardi et al. 2009; Stanley and Krakauer 2013; Taylor et al. 2014; Huberdeau et al. 2015). The second reason is that the performance metric in the SRTT, the response time, is the sum of the reaction time and the movement time, two variables that measure distinct aspects of planning and movement and are therefore subject to entirely different influences. To add these two variables together is to create a new arbitrary and artificial variable that can only mislead when it comes to sequence learning in particular and motor learning in general (Krakauer and Shadmehr 2006; Moisello et al. 2009; Wong et al. 2015).

6.2.6 Motor Acuity and Motor Skill

We have recently introduced the term *motor acuity*, drawing a direct parallel with perceptual acuity, to describe motor skill captured by shifts in the speed-accuracy trade-off function for a task, which is accomplished principally through practice-induced reductions in movement variability (Shmuelof et al. 2012). Practice also leads to smoother and more stereotyped movements. This kind of skill learning is probably what dominates in the prehension tasks studied in animal models and discussed in detail in chapter 3. Surprisingly, this kind of learning is not studied very much in healthy humans, where the emphasis is primarily on sensorimotor adaptation and sequence learning. To address this gap, Lior Shmuelof, John Krakauer, and Pietro Mazzoni introduced the arc-pointing task, in which subjects have to make visually guided movements about the wrist to steer a cursor though a U-shaped tube as fast as possible without hitting the walls (Shmuelof et al. 2012). The kind of motor skill required to accomplish the arc-pointing task is directly comparable to skilled reaching in rodents. Prehension itself is not a good learning model in humans because we are so good at baseline that it is hard to obtain a good dynamic range over which to study the effects of practice.

It is important to point out that the motor learning field does not yet possess an adequate computational model for practice-induced increases in motor skill. We do not know precisely which neural changes are responsible for increased skill, either structurally or physiologically. As we have already discussed in chapter 3, expansions in cortical maps do not appear to be necessary for motor skill expression. Just finding correlations between any kind of structural change in neural substrate and improved behavioral performance does not constitute understanding (Krakauer et al. 2017). We also do not know what the learning algorithm is that leads to improved motor performance trial by trial. Studies of practice-induced improvements in reaching and prehension in humans, nonhuman primates, and rodents all suggest a critical causal role for the contralateral primary motor and premotor cortices (Lemon 2008; Shmuelof and Krakauer 2011; Alstermark and Isa 2012). Thus, improvements in motor acuity that occur with practice may be driven by increased signal-to-noise in motor cortical representations and better feedback control policies. These cortical changes are perhaps aided by improved state estimation by forward models, possibly in the cerebellum (Shadmehr and Krakauer 2008). It has been proposed that unsupervised or statistical learning is the algorithm used by cortex (Doya 1999). At this point, however, these suggestions do not go much beyond informed conjecture.

In summary, instruction, imitation, supervised learning, and reinforcement learning apply better to the early stages of learning when the desired mean movement has to be selected (i.e., they explain how the action that requires practice is converged upon but not why subsequent practice will lead to improved performance). For example, a patient can be encouraged to reach by extending the elbow and avoiding trunk rotation either through instruction or through an operant process called shaping. Once the average movement is properly selected, however, subsequent practice is required, and as we discuss above, we do not know how practice works. Caution and skepticism are therefore required in response to statements to the effect that task-oriented training, the dominant learning-based approach in neurorehabilitation, is based on motor learning principles because we do not yet know what these principles are. All we have at the current time is a menu of empirical approaches derived from the motor learning literature (e.g., spaced vs. massed practice) (Schmidt and Lee 1988). It would be better for the field to admit we are closer in what we "know" to sports coaches, music teachers, and choreographers, a form of tacit knowledge (Polanyi 1966).

6.2.7   Retention and Generalization

Skill acquisition is not of great use if what is learned is not retained across sessions or generalizable across tasks. Indeed, one could argue that a gain is attributable to motor learning rather than repair if it is partially forgotten without continued use because skills need ongoing practice to be fully retained. One can in fact forget how to ride a bike insomuch that one has to practice to get full proficiency back after years of nonriding, hence the origin of the phrase "I'm rusty." Any form of forgetting is presumably less true for spontaneous recovery and repair processes—once it is fixed, it is fixed. This difference can be understood with the example of handwriting with the nondominant arm—with practice, we can all get better at it (Lindemann and Wright 1998). If we stop practicing, we will start to write poorly again, but this does not mean that our basic ability to move our hand has in any way worsened. Similarly, a patient who has recovered to make out-of-synergy movements within the first three months after stroke will not revert to a prior flexor synergy stage because this reflects repair, not learning. In addition to retention, training the limb on a task in the rehabilitation clinic needs to generalize to other activities of daily living. It is surprising how little investigation there has been of retention and generalization of motor learning in the context of neurorehabilitation (although there are notable exceptions; Reisman et al. 2013; Schaefer et al. 2013). Limited generalization is attributable to the fact that rehabilitation of chronic stroke depends on motor learning and therefore suffers from the same "curse of task specificity" as normal motor learning (Bavelier et al. 2012). We shall discuss this later in the chapter. To summarize: when skill at a particular task is acquired through learning, continued practice is required to maintain full retention, and generalization beyond the trained task is usually limited.

6.2.8   Motor Learning after Stroke

All the kinds of motor learning described above for healthy subjects are predicated on the existence of normal neural substrate for the expression of that learning, that is, that the motor system is able to send and execute the commanded movement, whatever it may be. It should be immediately apparent that if the neural substrate that transmits the motor commands is damaged (e.g., the corticospinal tract [CST] after a capsular infarct), then learning might not be expressible, even if normal. More fundamentally, motor learning is not going to reverse the chronic CST lesion; as we have pointed out in chapter 3, learning is not repair. It will be argued here that motor learning in response to rehabilitative training in patients with chronic stroke can only operate within the residual performance envelope possessed by the remaining nervous system after spontaneous biological recovery is complete. The null position taken in this chapter is that motor learning in response to training in the period after spontaneous biological recovery is complete cannot reverse loss of motor control but can only help select and then allow for more skilled execution of compensatory strategies. The one caveat to this view, and it is potentially an important one, is that patients' residual capacity itself maybe latent (i.e., they have a larger performance envelope than they realize), and some forms of learning may be able to unmask and retrain this capacity.

Thus, there are three distinct ways to think about motor learning after stroke: (1) Selection, through self-exploration, instruction, or operant conditioning, of a new strategy to accomplish a task—for example, moving the trunk forward during a reach because of loss of the ability to extend the elbow (Roby-Brami et al. 2003). (2) Practicing to reach with this new strategy and becoming more skilled at it. Note that this can happen in the absence of any improvement in elbow extension. (3) Stopping a patient from using his or her new reach strategy and focusing on residual elbow extension ability and, if present, attempting to augment it through, for example, an operant process. These three ways of thinking about motor learning—selection of an alternative action (compensation), practice-induced increases in the skilled execution (acuity) of the compensatory action, and exploitation of residual normal action—can be, and often are, confused with each other. One reason for this confusion is that these three kinds of learning can be going on simultaneously in a therapy session, whether this is intentional or not.

6.2.9 The Problem with Comparing Motor Learning in Groups with Different Baselines

A conundrum for the field is that it is not possible to directly compare learning ability in patients either with controls or with other patients with paresis of different severity. It might not seem clear at first why this should be the case, as a difference between groups, either additive (a delta) or multiplicative (a percent change), can be computed and a statistical comparison made. Unfortunately, this approach is not valid. Consider a training-related change, using an arbitrary performance metric, from 2 to 4 in a patient compared to 6 to 10 in a control. If an additive comparison were made, then one would conclude that controls are better learners than patients (i.e., a delta of 4 versus 2, respectively). In contrast, a multiplicative comparison, which is a form of normalization, would suggest that the patient is a better learner (100 percent of the initial) than the control (66 percent of the initial). Thus, opposite conclusions would be reached based entirely on how one wants to measure the change. This, however, is not even the main problem. The fundamental problem is that the magnitude of the change in performance is *not* a proxy for the magnitude of the learning-related change in the central nervous system (CNS), which we cannot

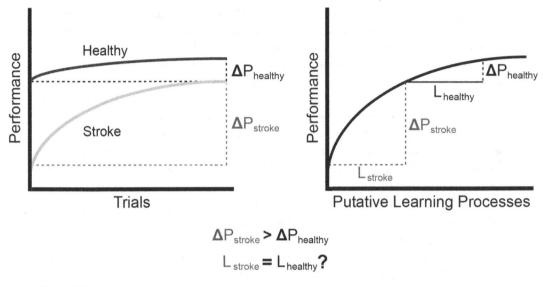

$$\Delta P_{stroke} > \Delta P_{healthy}$$
$$L_{stroke} = L_{healthy}?$$

Figure 6.1
The difficulty of comparing learning capacity when baseline performance is not matched. (A) Hypothetical plot of improvements in performance as a function of trials for a stroke patient in gray and a healthy control in black. ΔP indicates the change in performance. (B) Hypothetical plot of improvements in performance as a function of the putative processes underlying learning. It becomes apparent that (1) the healthy control has less improvement in *performance* compared to the stroke patient, and (2) this smaller change in *performance*, however, requires the same amount of *learning*. The differences in learning cannot be inferred by comparing improvements in performance when the baselines are not matched.

measure (i.e., we do not know the function relating learning and performance). It is very likely, given that learning curves follow a power law, that for a higher starting performance, compared to a lower one, more learning will be required for the same performance change (figure 6.1). This problem is insuperable, and the only way around it is to have initial performances matched, which is often not possible because by its very nature, stroke causes a performance deficit.

6.2.10   Learning a Skill versus Recovering from Paresis
Another conceptual problem arises when a task can either be used as an assay of initial impairment or be used for training, either to probe learning ability or because it may be a useful task for ADLs. The problem arises because, as we discuss in chapter 1, correlation between two variables at baseline does not mean that there will be continued covariation between them when one of them is changed (Moreau and Conway 2014), for example, through learning. This fallacy is at the heart of the fine levied by the Federal Trade Commission in early 2016 on the brain game company, Lumosity, because of their claims that improvement on a particular game would have general cognitive benefits (Span 2016). This claim was made because working memory and fluid intelligence (IQ) correlate at

baseline, but this does not mean that improving working memory in a particular game will raise IQ in general. This problem has its equivalent in motor recovery when skill learning and recovery get conflated.

The distinction between learning and recovery was examined directly by Robert Hardwick, Pablo Celnik, and colleagues. They had two groups of patients with different levels of arm paresis and a control group perform a baseline assessment on a serial visual isometric pinch task (SVIPT) (Hardwick et al. 2016), which requires skilled control of precision grip. Not surprisingly, baseline ability on the task correlated with initial severity of paresis as measured with the FMA. All three groups then trained on the task over four days, and all showed an increase in skill, measured as a shift in their speed-accuracy trade-off function. Interestingly, training on the SVIPT led the moderate-to-severe group to match the untrained (baseline) performance on the task by the mild-to-moderate group, while the trained mild-to-moderate group matched the untrained (baseline) performance of the controls. If a person who did not know about the training history of the groups were to use the SVIPT as a tool to assess severity of paresis the patients, he or she might conclude erroneously that the trained moderate-to-severe patients had the same level of paresis as the untrained mild-to-moderate patients. But as the authors show, improvement on the SVIPT did not improve patients' paresis. This is a very important result as it emphasizes the point that patients are not going to learn away their arm paresis task by task. We discussed in chapter 3 that this appears to be the case with task-based training even in the acute poststroke period, but the curse of task specificity is likely to be even more pertinent in the chronic period when plasticity levels have returned to normal.

6.3 Compensation

This section is predicated on the starting assumption that in chronic stroke—that is, when patients are six months or more poststroke—brain plasticity and the response to training are no different from what is seen in healthy subjects, with the consequence that compensatory responses are expected to dominate meaningful improvements in ADLs. Here, the term *compensation* will be restricted to changes in effector, joints, and muscles and not to use of external aids such as walkers, canes, or orthoses. In this framework, motor learning in patients with chronic hemiparesis is not different from a healthy person learning to write with his or her nondominant arm or learning to lean forward and shuffle when walking on a slippery surface. The failure to distinguish between the unique learning conditions that pertain to the sensitive period early after stroke and the ordinary motor learning that occurs during the rehabilitation of patients with chronic stroke has led, in our view, to significant conceptual confusion and the design of ill-conceived trials.

The most obvious form of compensation for arm paresis is to use the other arm, which is why constraint of the good arm has been used as a means to promote recovery of the paretic arm dating back to the primate studies early in the twentieth century discussed in chapter 3 and to the present day with CIMT, which we shall discuss in more detail below. Compensation can also occur on the affected side with perhaps the best-known strategy being trunk movements during reaching, which have been characterized in a series of studies by Mindy Levin and colleagues. Specifically, patients can improve reach speed and precision by anteriorly displacing and rotating the trunk to guide hand movement instead of using elbow extension and shoulder flexion (Roby-Brami et al. 2003). Preeti Raghavan and colleagues examined how patients with chronic stroke preshape their hand to grasp different objects. They found that the patients used a compensatory strategy by preshaping at the metacarpophalangeal joints rather than the interphalangeal joints (Raghavan et al. 2010).

The boundaries between restitution (true recovery) and compensation can blur in the case when there is muscular redundancy in the control of particular limb joints. For example, the primate wrist can be moved in a kinematically invariant way in the dimensions of ulnar/radial deviation and flexion/extension using different weightings of five muscles. In the absence of brain injury, this muscular redundancy is resolved by distributing motor commands across the redundant effectors (muscles in this case) so as to minimize effort and maximize precision (Fagg et al. 2002; Haruno and Wolpert 2005). The presence of monoarticular and biarticular muscles introduces a similar redundancy at the shoulder and elbow joints of the upper limb, which allows, for example, identical planar reach trajectories to be generated by different patterns of muscle activation (Kurtzer et al. 2006). As in the case of the wrist, the distribution of preferred torque directions of the proximal arm muscles is biased away from their preferred anatomical direction toward minimization of overall muscle activity. Thus, two types of compensatory scenarios can be envisaged after a stroke causes loss of neural control of specific muscles. In one, the effector is used but with different kinematics to reoptimize effort and accuracy with the remaining degrees of freedom. In the other scenario, a nonoptimal combination of muscles maintains the same effector kinematics. In the first scenario, there would be compensation both at the level of kinematics and at the level of muscles. In the second scenario, there would be restitution at the level of kinematics and compensation at the level of muscles. It is unclear whether differing degrees and/or types of neural reorganization are required for these two forms of compensation. This kind of fine-grained analysis has not yet been attempted in patients after stroke.

6.4   Repetitive Task-Oriented Training

The neurorehabilitation research field is proceeding in two principal directions in the quest for finding a more effective way to treat to patients with chronic hemiparesis. The first is to devise new behavioral treatments such as CIMT, robotics, functional electrical stimulation, and virtual reality (although it is questionable how "new" they really are conceptually or how truly different some of them are either from each other or task-oriented training).

The second is to increase the dose and intensity of the task-oriented training component of standard therapy and, in some cases, modulate it with either pharmacological interventions or noninvasive brain stimulation.

Repetitive task-oriented training is considered by many to be the new gold standard for neurorehabilitation and superior to both standard therapy and to the neurophysiological approach advocated by Brunnstrom and others (Winstein et al. 2014). Before we come to a definition of what task-oriented training is, it should first be admitted that it is really hard to define the content of "standard therapy," beyond it being the term we use to refer to inadequate doses of an idiosyncratic mix of neurophysiological approaches (e.g., Brunnstrom's approach described in chapter 3) and task-oriented training. Indeed, the fuzziness and vagueness of what patients actually receive is likely to cause considerable frustration to anyone surveying the field; it certainly did the authors of this book. We have already discussed the neurophysiological approach in chapters 1 and 3. The claim is often made that there are fundamental differences between this approach and task-oriented training. In our view, the difference has been consistently exaggerated, especially with regard to one being more motor learning based than the other. It is certainly true that practicing a task might be more interesting and motivating than practicing a particular joint movement, but both are practice-dependent learning. Athletes do not like drills that focus on a particular movement or stroke as much as playing full games, but they need to do both, and both are learning. The real difference, in our view, hinges on a change in emphasis for the target of training from impairment to activity.

Carolee Winstein and colleagues in a recent book chapter on task-oriented training emphasize repetition, task specificity, and skill acquisition per task (Winstein et al. 2014). They also delineate the approaches to three active ingredients: (1) the task must be challenging enough to require new learning, (2) the training must progress in difficulty and the task must be iteratively adapted to real-world relevance, and (3) the task must not lead to rote repetition but repeated attempts to solve a problem. Task-oriented training is based on various well-established empirical motor learning rules, including a focus on functional goals and the use of real-life objects, titration of difficulty, varying feedback, and distributed practice (Schmidt and Lee 1988). As we have already discussed, there is currently no adequate computational or algorithmic theory of how motor performance improves with repetitive practice in healthy humans or other animals. Thus, the conceptual framework for task-oriented training does not go much beyond the dual assumptions that practice makes perfect (with subsequent dependence on empirical work on practice schedules and task organization in healthy subjects) and that it results in beneficial plasticity in the brain.

An awkward conceptual problem when it comes to task-oriented training is that we do not have a satisfactory definition of task. We can divide motor acts into tasks from the outside, but this does not mean that these map onto distinct categories or modules inside the brain. Does practicing a particular kind of movement (e.g., reaching forward with a focus on minimizing trunk rotation and maximizing elbow extension) count as a task? Or is it only a task when the reach is incorporated into an ADL like picking up a cup? If we allow that they are both tasks but that one focuses on movement quality and the other on movement outcome, then distinctions between types of upper limb rehabilitation begin to break down. Therefore, it seems that in order for task-oriented training to remain distinctive, as some think it is (Winstein et al. 2014), then the focus must be kept on tasks relevant to ADLs. Task-oriented training implies specificity, which then brings up its opposite, generalization. The problem with testing for generalization from task A to task B is that the term *generalization* presupposes that task A and task B are different. But perhaps generalization happens because task A and task B overlap to some degree; for example, a tennis player is better at table tennis than someone who has never played a racket sport before presumably because there is overlap in perceptual and motor components. How do we establish if there is such an overlap? Test for generalization. The circularity is obvious. Thus, we need some conceptual framework that can quantify similarity between tasks *before* testing for generalization. This is not just an academic question, as it would lead to better choice of tasks in a training regimen.

As we pointed out in chapter 1, studies indicate that the amount of task-specific upper extremity therapy that is actually provided to patients during an acute rehabilitation stay is at homeopathic levels (Bernhardt et al. 2004; Lang et al. 2009). As we discuss in chapter 3, those few trials that attempted to up the dose of rehabilitation in the first three months after stroke are tricky to interpret, and there have been too few to allow any definitive conclusion about timing of rehabilitation after stroke via meta-analysis (Pollock et al. 2014; Lang et al. 2015; French et al. 2016). In contrast, however, meta-analyses of dose of task-oriented training indicate that more is indeed better regardless of time since stroke or outcome measure (Lang, Lohse, et al. 2015; Lang et al. 2016; Yagi et al. 2017). These results, in combination with a recent study by Catherine Lang and colleagues showing that it is feasible to provide a similar number of task-based repetitions (approximately 300) to patients in a one-hour session as is routinely given in animal models studies, have begun to pave the way for studies that investigate considerably higher doses of upper limb task-oriented training in chronic stroke. We shall discuss three recent studies here because they were ambitious and well conducted, and they also complement each other.

The first study by Jessica McCabe, Janis Daly, and colleagues was a randomized controlled trial that compared three kinds of very high-dose therapy (five days/week for five hours/day; sixty sessions) in patients with severe chronic arm paresis (mean of about 24/66 on the FMA). The first group ($n = 11$) received what they called motor learning-based treatment (ML), which the authors describe as containing: "movement practice as close to normal as possible" (McCabe et al. 2015). This statement is accompanied by citations to the Nudo et al. studies of prehension training that we covered in detail in chapter 3, which suggests that there was an emphasis on movement quality and not just task outcome. The authors also say that there were a high number of repetitions and training specificity. In addition to this task-based training, because of the severity of the arm paresis, initial therapy focused on "training isolated joint movement coordination of the scapula, shoulder, elbow, forearm, wrist, fingers, and thumb" and on "task component movements." The second group ($n = 12$) received ML therapy plus planar robotic assistive therapy for the elbow and shoulder. The final group ($n = 12$) received ML therapy plus functional electrical stimulation (FES) for wrist and finger flexors/extensors and forearm supinators/pronators. The main result was that all three groups showed mean gains in the FMA between 8 and 11, which is considerably higher than any previous study has reported for the FMA in chronic patients, even those with mild to moderate paresis. We should also point out that patients also showed sizable gains on a functional task called the Arm Motor Ability Test (AMAT), which is similar to the Action Research Arm Task (ARAT).

This study gives us a lot to think about. The impairment improvement was comparable in all three groups, which lends credence to the magnitude of the change in the FMA of about 9 points and suggests that it was not just a chance result. The addition of FES and robotics turned out to be a distraction from the main result—the robotics group did not show a greater proximal arm improvement, and the FES group did not show a greater distal arm improvement, even though this is what they each respectively targeted. It appears that it was the ML that mattered. Importantly, the patients in the study had severe paresis so training was employed hierarchically, beginning around a single joint followed by out-of-synergy multijoint training and then progressing to functional task-based training. Thus, we can see that the sequence is reminiscent of the neurophysiological approach taken by Brunnstrom and others, combined with motor learning based task-oriented training but with larger doses of each. It will be necessary to replicate the results and employ kinematic measures and dynamometry to ascertain how much the changes in the FMA can be attributed to true general improvements in motor control versus increases in strength (Cortes et al. 2017). The authors also addressed the feasibility of offering such a high dose of therapy by having a 1:3 ratio of therapist to patients. Overall, this study is to be commended and is the closest in spirit to what has been done to great effect in nonhuman primate models. One can only hope that this kind of study is replicated early after stroke so that it matches the promise of animal models even more closely.

It is very useful to contrast the McCabe et al. trial with another more recent phase II randomized trial conducted by Catherine Lang and colleagues, which compared four different repetition doses of task-oriented training in patients with chronic stroke. The main outcome was the ARAT activity scale, with a mean value at baseline of 32/57. An impairment measure was not obtained. Therapy was given in one-hour sessions, four days/week for eight weeks, in which patients received 3,200, 6,400, or 9,600 repetitions. An additional group had the option to go beyond eight weeks to see how much more a patient could tolerate. This group ended up receiving a median of 10,800 repetitions. The main result was that the response of the ARAT was very small, and this small change was comparable across groups. While the authors emphasize that the treatment response was dose invariant, in our view, the main result is how negligible the response was in any group; indeed, one group did not respond at all. In the discussion, the authors acknowledge the possibility that all the doses provided were too low and that animal models suggest that much higher amounts of training may be effective. They then state, however, "We are skeptical that patients could achieve another order of magnitude of dose in the context of a traditional therapy session." Now this is an odd argument: dose may indeed matter, but we cannot give larger doses in the current care environment, so dose does not matter! This is confusing the scientific with the pragmatic. As we saw above, the McCabe et al. study shows that much larger doses of arm therapy can feasibly be given and lead to impressive gains. Nick Ward and colleagues have preliminary data showing a mean FMA gain of about 10 points in 106 patients from the Queen Square intensive upper limb neurorehabilitation program who received six hours a day of upper limb–focused therapy for three weeks (personal communication 2017). Encouragingly, these gains were sustained at six months.

The surprisingly large effect sizes seen in the McCabe et al. study make it important to identify what the active ingredients in the therapy are. From the description in the paper, some aspects of the therapy are very similar to what is called the neurophysiological approach (see chapters 1 and 3), while other aspects overlap with task-oriented training approaches. Although the neurophysiological approach has a passive sensory guidance component, active movements are just as important. Brunnstrom, for example, also stressed the need for practical use of basic limb synergies (Brunnstrom 1966), which is essentially indistinguishable from task-oriented compensatory training. What the McCabe et al. study suggests is that it is *all* practice-dependent training. Training to make particular kinds of single- or multijoint movement, even if outside the context of a particular task, is still motor learning; the difference is on rewarding the quality of the movement rather than task accomplishment. Coaches in sport make this distinction all the time. Thus, what evidence is beginning to suggest is that what patients really needed all along was just a lot more of both types of training-based rehabilitation. Seen in this light, the switch from training of movement patterns to task-oriented training, along with an emphasis on compensation, was more of a pragmatic and economically motivated shortcut rather than a conceptual advance.

Given that the common denominator for better patient response to any form of therapy appears to be total number of repetitions or training trials, and therapy time is at a premium, then any way to get more learning per trial could potentially be beneficial. Noninvasive brain stimulation in the form of transcranial direct current stimulation (tDCS) appears to do exactly this for motor learning in healthy subjects when the anode is placed over the contralateral motor cortex (Reis et al. 2009; Waters-Metenier et al. 2014). In one study in which healthy subjects were trained over five days, they learned as if they experienced about 40 percent more trials of training than the controls (Reis et al. 2009). In a recent study, patients with chronic arm paresis received either anodal tDCS ($n = 11$) or sham treatment ($n = 13$) over ipsilesional primary motor cortex (M1) paired with daily, mainly task-oriented training for nine days. Outcomes were assessed right after the treatment, one week later, at one month, and at three months using the ARAT, the Wolf Motor Function Test (WMFT), and the FMA. The two activity measures showed additional improvement with anodal tDCS, but the FMA did not. One unexpected result in the study was that both groups showed unexpectedly large gains in the FMA. It would be interesting to know about the protocol that was used, which is called Graded Repetitive Arm Supplementary Program (GRASP). In the original study that used it, which also showed impressive gains, the FMA was unfortunately not an outcome measure (Harris et al. 2009). Overall, these results are entirely consistent with the idea that the ARAT and WMFT capture gains made through compensatory movements and that these movements can become more skilled with normal motor learning, which is known to be augmented by tDCS. In contrast, motor learning cannot reverse impairment in the chronic state, which is evident in the lack of improvement in the FMA.

To close, we report that the most recent Cochrane Review of the functional effectiveness of repetitive task-oriented training concluded that there was only low-quality evidence supporting its use for the arm and hand (French et al. 2016). It is clear, therefore, that task-oriented training alone is hardly a panacea and is not going to be the definitive answer to how to treat chronic upper limb paresis. It remains possible, however, that a substantial increase in dose and intensity will lead to a large nonlinear jump in gains at the levels of both impairment and activity, at least in some patients.

6.5   Constraint-Induced Movement Therapy

CIMT is often presented as a new and distinct approach to rehabilitation, but at its core, it is really just a variant on task-oriented training with the addition of the constraint component. Constraining the arm, as we have seen in chapter 3, was and has been important in primate studies to encourage use of the affected limb both when the animals are being trained and when they are alone. In humans, the constraint was conceived of a way to consolidate training gains and promote further practice when they were at home. The interest in CIMT comes both from its experimental and theoretical origins and from the fact that it was the subject of the first multicenter randomized trial in neurorehabilitation, the Extremity Constraint Induced Therapy Evaluation (EXCITE) trial (Wolf et al. 2010). The technique has two components: (1) restraint of the less affected arm and/or hand with a sling or mitten for 90 percent of waking hours and (2) task-oriented practice with the affected side using a form of training called *shaping*. The weightings for the two components and the length of the overall treatment have varied considerably in studies since the original trial. It still has not been determined what the relative contributions of the two components are to observed improvements.

It is perhaps underappreciated that the ideas that led to EXCITE originated in empirical observations in deafferented monkeys by Edward Taub and colleagues. Taub and colleagues wrote an influential paper in 1994 titled "An Operant Approach to Rehabilitation

Medicine Overcoming Learned Non-Use by Shaping" (Taub et al. 1994). In this paper, the authors presented their new rehabilitation framework based on experiments in monkeys that had been deafferented in one forelimb via dorsal rhizotomy. The key observation was that the monkeys did not resume use of the deafferented limb even after spinal shock had resolved and some use of the limb was again possible. The explanation was that early on when the limb was severely impaired, the monkeys learned that it was useless through negative reinforcement. This learning became a habit despite return of a latent capacity that was not explored. The authors discovered that the habit of nonuse could be overcome if the good limb was restrained over days. In addition to use of the restraint, the authors also retrained the limb in two different ways. In conditioned response training, the monkeys were made to make isolated repetitive movements across single joints and resist against loads. It was noted that these exercises did not generalize to functional tasks. A second, more effective training method, which they called shaping, was to incrementally reward successive approximations to a functional behavior. In essence, shaping attempted through reward to reverse the nonuse that had developed through failure. In the same paper, some promising preliminary data were presented in three patients with stroke. We can now fast forward to EXCITE, a clinical trial predicated on the ideas of restraint and shaping developed in these early studies by Taub and colleagues.

EXCITE showed that patients who received CIMT for two weeks had greater responses in a test of motor function (WMFT) and in self-report of performance quality in common daily activities. There was no assessment of motor impairment. Evidence suggests that CIMT does not lead to either significant reductions in impairment or reversion toward more normal motor control (figure 6.2) (Kitago et al. 2013). Instead, patients seem to improve via compensatory strategies. The subtle but critical point is that, unlike in the case of a monkey's recovery from spinal shock, patients are not discovering a capacity that they lost and then latently regained. Instead, compensatory strategies in the chronic state are performed with capacities that were present from the time of the stroke or recovered spontaneously in the sensitive period; they just had not been incorporated into functional tasks through practice. Thus, while it seems that an operant approach, as in deafferented monkeys, does teach useful compensatory strategies in patients after stroke, the mechanistic parallels between CIMT after stroke and after deafferentation are limited. Finally, learned nonuse has never been documented in humans (i.e., an actual plotted time course of decreasing use of the paretic limb), but this does not stop the term from being used somewhat incessantly despite it being an interpretation rather than a demonstrated fact.

Mention of plasticity and reorganization in the setting of CIMT or any other kind of task-oriented training is misleading unless these terms are thought to apply equally to healthy subjects. For example, it would presumably also occur when a healthy person's elbow is splinted into flexion so that within a few attempts, he or she flexes the trunk to make a reaching movement. To summarize, CIMT was born from studies in deafferented monkeys. Despite the original emphasis on the operant concept of shaping, which likely was more

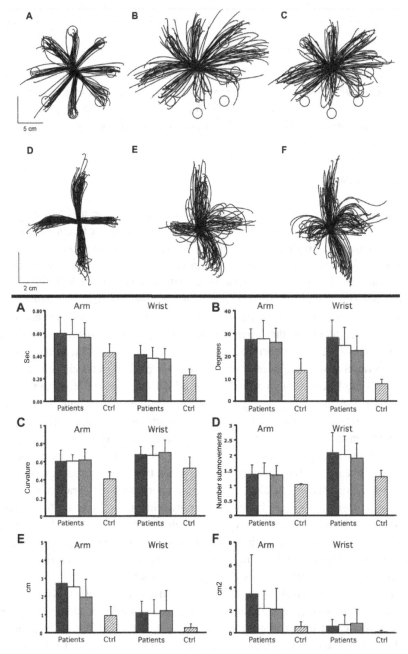

Figure 6.2
Motor control in chronic stroke does not improve with constraint-induced movement therapy (CIMT). Top Panel: Weight-supported planar reaching trajectories of the arm and vertical-plane wrist movements for randomly selected participants: (A) control, dominant arm; (B) patient, affected arm, pre-CIMT; (C) patient, affected arm, post-CIMT; (D) control subject, dominant wrist; (E) patient, affected wrist, pre-CIMT; and (F) patient, affected wrist, post-CIMT. Trajectories of healthy controls are presented for illustrative purposes. Bottom Panel: Kinematic outcomes for (A) movement time, (B) initial directional error, (C) path curvature, (D) number of submovements, (E) systematic error, and (F) variable error in the arm and wrist tasks. Scores for each session are shown: 1 (black), 2 (white), and 3 (gray). Performance of elderly controls in a separate study is shown to the right. Values are means with error bars depicting standard deviation. Adapted from Kitago et al. (2013).

important in monkeys, CIMT is a form of task-oriented training that has the expected efficacy at the activity level but not at the levels of impairment and motor control. Those studies of CIMT that have shown an effect on the FMA were conducted within the first three months after stroke, and so the possibility of an interaction with spontaneous biological recovery cannot be excluded (Kwakkel et al. 2015).

6.6   Robotic Therapy

Robots for stroke rehabilitation are considered motor learning devices that have the potential to provide a larger number of movement repetitions at higher frequency than a therapist can provide (i.e., they are fundamentally dosage and intensity machines). The first robot to be used in rehabilitation had its origin in a planar robot used in initial groundbreaking studies on force-field adaptation (Shadmehr and Mussa-Ivaldi 1994). In keeping with this adaptation origin, one training approach has robot-applied force fields make patients' trajectory errors even larger than their baseline errors (error augmentation; Patton et al. 2006). Here, the idea is that when the force field is switched off, immediate aftereffects will be more similar to normal movements. A recent trial that tested two weeks of either haptic or visual augmented error signals in patients with chronic stroke only showed a very modest beneficial effect compared with regular robotic therapy (Abdollahi et al. 2014). Although error augmentation is not likely to become the favored approach for robotic rehabilitation, it at least has the benefit of being clear with respect to the kind of motor learning it is exploiting. In contrast, it is very difficult to discern what the precise motor learning principles are that underlie the majority of robotic studies. For example, if one reads the introduction sections of the two publications reporting the results of the two largest robotic trials to date, there is no mention of the rationale behind the use of robotics or what the underlying motor learning principle is (Lo et al. 2010; Klamroth-Marganska et al. 2014). Similarly, a recent meta-analysis of robot-assisted therapy of the upper limb begins with the assumption that robots might help a patient practice (Veerbeek et al. 2016). The rather underwhelming conclusion is that the starting premise in rehabilitation robotics is that the device helps you repeat movements.

Robots allow patients to make a large number of movement repetitions with varying degrees of assistance or resistance, either with respect to gravity or along the movement trajectory. Robotic settings can be titrated in an assist-as-needed manner as patients improve, although this has to be preset before the training block—no rehabilitation robots thus far have force sensors that allow online cooperative control. In 2D, a manipulandum is sufficient to constrain movements to a kinematic path, but in 3D, an exoskeleton is required to provide the equivalent trajectory constraint. Described this way, robots are providing haptic guidance in a manner similar to a therapist applying the neurophysiological approach aimed at the impairment level, with the goal of having the patient experience a more normal movement both visually and proprioceptively.

We now return to the question of why a robotic trajectory constraint would be expected to help motor learning at all. As we discussed in section 6.2, we do not have a working theory as to why practice leads to increases in motor skill. Instruction, imitation, adaptation, and reinforcement are alternative ways to help the patient converge on the desired trajectory, but then motor commands have to iteratively improve so as to better execute the trajectory with speed and precision. In other words, knowing the kinematics and knowing the dynamics that underlie the kinematics are distinct problems. Indeed, we showed many years ago that these appear to be learned independently (Krakauer et al. 1999). A robot can be considered a way to provide a proprioceptive demonstration of the trajectory, but this does not tell a patient how to find the commands that will produce the time-varying forces that will generate the trajectory—this requires practice. Thus, the question is whether subsequent practice benefits from the continued presence of a trajectory constraint. That is to say, there is a fundamental difference between having the robot initially show you what the therapist wants you to do and having it continue to apply a constraint as you practice over hundreds if not thousands of trials. At the current time, there is no evidence that haptic guidance augments the effects of prolonged practice on motor skill acquisition in healthy subjects. Haptic guidance may speed acquisition of certain initial task requirements like timing or velocity (Bluteau et al. 2008) but not the shape of the trajectory (Liu et al. 2006; Yang et al. 2008; Sigrist et al. 2015). Robotic assistance impedes learning of novel dynamics (Lüttgen and Heuer 2012), which intuitively makes sense as it would slow down learning of one's own feedback control policy. Thus, overall, it is not clear why robotic trajectory assistance should help patients after stroke, and seen in this light, it should perhaps not be surprising that the results of robotic trials have been almost uniformly disappointing.

The largest robotics trial to date, the Veterans Affairs (VA) Robotic-Assisted Upper-Limb Neurorehabilitation in Stroke Patients study, treated forty-nine patients with chronic stroke with thirty-six one-hour sessions over twelve weeks by using the MIT-Manus device (Lo et al. 2010). The results were essentially negative: patients who received robotic therapy gained only 2 FMA points over the usual care group. A minimum meaningful effect size for the FMA is a change of 7 points (Gladstone et al. 2002). The second largest trial treated thirty-eight patients with chronic stroke with twenty-four forty-five-minute sessions over eight weeks using a 3D exoskeleton with seven degrees of freedom reported. As in the ROBOTICS study, the benefit of the robotic therapy over usual care was a trivial 0.78 points on the FMA. A recent meta-analysis of forty-four randomized controlled trials (RCTs) of robotic therapy for the upper limb after stroke reports a mean FMA change of 2 points (Veerbeek et al. 2016). Thus, the benefit of robotics for the treatment of impairment in chronic stroke appears to be minimal. This should not come as a surprise as motor learning is not repair.

Despite these unimpressive results, there are important lessons to be learned, especially from the ROBOTICS trial. First, the study showed that standard of care has no effect at all on impairment, disability, or quality of life. This observation alone cries out for the need for new treatments. Second, therapists outside of a research setting would not be able to consistently provide doses of assisted arm movements of around 1,000 per session (the average in real-world settings is twenty to forty-five). Third, there were no serious adverse events in forty-nine patients who performed 1,024 movements per session with the robot, three times a week for twelve weeks.

The recent meta-analysis did not find any evidence of added value of robotic therapy for activity level measures such as the ARAT or WMFT or for global activity measures such as the FIM or Barthel Index. This also should not be overly surprising as robotic therapy focuses more on movement training than task-oriented training. Given both the questionable value of assistance to motor learning from a theoretical standpoint and the very disappointing empirical results, should we give up on robotic therapy for stroke? Answering this question is tricky. It is not likely that assistance along prespecified trajectories in either 2D or 3D is ever going to have a meaningful impact on stroke recovery no matter how much better the technology gets or when the therapy is given. One caveat to this is that perhaps experiencing passive proprioceptive feedback very early after stroke may potentiate spontaneous recovery—this has not been tested. Weight support is, in our view, a more promising direction. There are two main reasons for this view. First, titrating weight support is a way to train for strength; indeed, the recent meta-analysis showed a strength benefit for robotic therapy (Veerbeek et al. 2016). Second, as we discuss in detail in chapters 2 and 3, strength and motor control are dissociable in terms of their neural bases and their recovery profiles. Weight support may allow for practice of motor control in 2D and 3D before strength improves in the early stroke period.

In chapter 2, we described the work done by Jules Dewald and colleagues, showing increases in the planar reaching workspace area when a robot fully supports the weight of the arm (Ellis et al. 2016). This observation formed the basis of a small study in patients with chronic arm paresis in which progressive shoulder abduction loading was used to train reaching range. The core idea is somewhat subtle: that patients could be trained to make planar reaches with reduced intrusion by a flexor synergy by having them progressively support more of the weight of their arm as robotic antigravity support was titrated away. It is important to appreciate that the goal was not strength training per se. The patients who experienced progressive shoulder abduction unloading showed a larger increase in work area than the control group despite there being no change in strength within or between the two groups. Thus, what happened here is that the patients from the intervention group learned to control their limb during a planar reach better for a given level of weight support than did the control group.

It would be of interest to extend this kind of training to 3D, as most arm movements are vertical rather than horizontal. This was demonstrated in a very interesting study by Konrad Kording, Daniel Wolpert, and colleagues of six healthy subjects who were given a wearable motion-tracking system to record their arm movements as they went about their daily life (Ingram et al. 2008). Despite the large range of possible movements, the investigators found that during most normal everyday tasks, the arms are confined to a small volume of space around the body and movements are predominantly in the vertical, not the horizontal, plane across a variety of tasks. Thus, weight support in 3D early after stroke could allow exploration and practice of movements that are useful to many everyday tasks. The ongoing Study to Enhance Motor Acute Recovery With Intensive Training After Stroke (SMARTS 2) trial is attempting to do just this (NCT02292251).

6.7 Conclusions

Overall, it appears that current training regimens, regardless of the putative motor learning mechanism recruited or therapeutic modality used, are not going to reverse impairment to any significant degree in the chronic state after stroke. It is a biological, not a technological, limit. The study by McCabe and colleagues, in which sixty sessions of five hours/day of task-oriented training of the upper limb were given, suggests that larger gains are possible if the dose and intensity of therapy are substantially increased, but these results will need to be replicated (McCabe et al. 2015). More importantly, it will be necessary to determine what the impressive gains in the FMA actually represent. There are two possibilities: 1) the higher dosage of training is truly reducing impairment, or 2) training is leading to strength gains only, which unmask latent control capacity. In essence strength increases are performing an analogous role to the antigravity support given by a robot as it was done in the studies by Jules Dewald and colleagues (see chapter 2). Kinematic measurements and dynamometry will be required to determine which of these two possibilities is occurring. Even if more therapy is beneficial, it is almost certain that biological modulators and physiological interventions will be needed to augment training effects in chronic stroke, especially if several hours of upper limb therapy a day proves to be an unrealistic option in current care environments. For chronic stroke, CIMT does not offer anything qualitatively different from comparably dosed task-oriented therapy. Robotics in chronic stroke has a real but negligible effect on the FMA, but this might relate to strength increases rather than true improvements in motor control. Robotics does not have any apparent benefit at the activity level of the International Classification of Functioning (ICF) (i.e., ADLs; Pollock et al. 2014; Veerbeek et al. 2016). Two things should be said to qualify this rather pessimistic conclusion with regard to robotics. First, it is likely that the negative results pertain more to the case when the robot provides assistance along the line of movement than when it provides antigravity support. In the latter case, supporting the weight of the arm seems to be the way to allow practice of motor control (Sukal et al. 2007), although whether expensive robots are required to just provide titrable weight support is another question. Second, the jury is still out on the use of robotics in any configuration early after stroke (first three months), but even in this timeframe antigravity support and non-specific assistance (negative damping) will likely be better than directed assistance.

A point that is often made with regard to the disappointing results of rehabilitation trials in the chronic stroke period is that there may be a subpopulation of patients that responds much better to new interventions, such as robotics, but whose positive response is diluted by nonresponders when all patients are viewed together. The implicit argument is that heterogeneity in the capacity to recover is discoverable with better biomarkers (i.e., good chronic responders can be identified and then presumably triaged to different types, doses, and intensities of therapy). The critical question, of course, is whether such a subpopulation of good responders exists.

In 2014, Steven Cramer and colleagues suggested that measures of both M1–M1 connectivity and percentage of CST injury are the best predictors of response to therapy (Burke Quinlan et al. 2015). Notably, however, the two patients in their study with the largest changes in FMA (approx. 12 points) were not distinguished by either one of these predictors. The rest of the patients made only modest gains in FMA. In the study by McCabe and colleagues, mentioned above, the mean gain in FMA was about 10 points, but there were three patients who made gains of 15, 18, and 25 points (McCabe et al. 2015). Are such patients who make large gains simply outliers in the distribution of usual recovery, or do they arise from a different distribution altogether?

Despite arguments to the contrary, the presence of large outliers in a study does not automatically imply the existence of a responsive subpopulation. Patients who make large gains may not be identifiable by baseline characteristics; the recovery process may be inherently random. Nonetheless, these studies can provide pilot data for generating hypotheses regarding the distinguishing features of chronic recoverers and non-recoverers, a process that is analogous to the efforts made to identify those patients who do and do not show proportional recovery early after stroke (chapter 3). Additional studies using the identified predictors to tailor treatment would then be necessary to confirm the features' prognostic ability. Until such studies are done, and absent a clear mechanistic explanation for large gains, the presence of small numbers of seemingly dramatic responders in a trial that otherwise shows a small effect size should be viewed with caution and not be seen as a way to make larger claims for the treatment in question.

7

Pharmacological and Cell Therapies for Recovery from Stroke

7.1 Introduction

With the discovery of basic biological principles of stroke recovery and the parallel development of the stem cell field in regenerative medicine, candidate therapies that enhance recovery from stroke are now possible. There is also clear preclinical evidence to support specific drug classes in stimulating recovery after stroke, such as tonic GABA signaling antagonists, selective serotonin reuptake inhibitors (SSRIs), phosphodiesterase inhibitors, and Rho kinase (ROCK) inhibitors. Biologics, nondrug molecules such as antibody conjugates that block myelin-associated proteins, offer another promising therapeutic approach. Finally, beyond these single-molecule therapies, the stem cell field has generated much hype in its promise of treating many different central nervous system (CNS) and systemic diseases. In aggregate, there is substantial evidence from preclinical studies that cell delivery after stroke enhances functional recovery and, recently, from promising phase I clinical trials (Kaladka et al. 2016; Steinberg et al. 2016). However, the hype surrounding stem cell studies has blurred important conceptual distinctions between different delivery methods, as well as between stem cells, neural progenitor cells, and other adult or progenitor cell types. This chapter will review candidate drugs, biologics, and cellular therapies for neural repair after stroke, and make some predictions as to where this area of neurorehabilitation is heading.

7.2 Brief Overview of Stem Cell Therapy in Stroke

The field of regenerative medicine is dominated by the concept of stem cell therapy. Stem cell therapies are in various stages of preclinical or clinical development for ischemic cardiac disease, peripheral limb ischemia, joint cartilage repair, and pancreatic islet cell replacement in diabetes, to name just a few (Cogger and Nostro 2014; Dimmeler et al. 2014; Fox et al. 2014; Pavo et al. 2014). For motor recovery after stroke, stem cell therapy has meant transplantation of cells after stroke to enhance neural repair. It is important to recognize that the brain after stroke mounts an endogenous stem or progenitor cell response to the ischemic stimulus, as reviewed in chapters 4 and 5. However, stem cell therapy in stroke has come to refer only to exogenous stem cells or their relatives, delivered to the body through brain or systemic injection. Preclinical studies have indicated a substantial effect with some of these stem cell treatments in promoting functional recovery. On the other hand, while there have been many clinical trials for stem, progenitor, and other cell types in stroke, most of these have been poorly regulated or reported, and thus details about outcome measures, cell types delivered, and analysis methods have been inadequate. Indeed, a search of clinical trial listings, such as clinicaltrials.gov, for cell or stem cell therapies in stroke generates hundreds of listed trials. The cell source and characterization of the cells is obscure in many of these trials. In many cases, these are early, phase I trials and have statistical power only to show safety, which they have demonstrated for diverse cell types and delivery approaches: both intravenous and intra-arterial systemic delivery of bone marrow progenitor cells early after stroke (Bang et al. 2005; Savitz et al. 2011; Friedrich et al. 2012; Moniche et al. 2012; Hess et al. 2014) and stereotactic injection of bone marrow progenitor or fetal neuronal progenitor cells into the brain in the chronic state after stroke (Kalladka et al. 2016; Steinberg et al. 2016). Here we provide a classification of cell therapy ranging from the most basic and straightforward cell therapy approach (Stem Cells 0.25) to the very near future when advanced stem cell therapies can be expected to be applied to stroke (Stem Cells 2.0).

7.3   What Is a Stem Cell?

A true stem cell can differentiate into any other cell type in the body, a property referred to as "pluripotency." Another property of a true stem cell is "asynchronous division": with each division, it gives rise to a daughter cell that is a copy of itself and a second cell that is on the path toward differentiation. With age and disease, stem cells may stop dividing asynchronously and instead produce two daughter cells that are committed toward a differentiated fate. In this case, the stem cell pool will become depleted over time as stem cells do not replace themselves. Examples of stem cells in regenerative medicine are embryonic stem cells (ES cells) and induced pluripotent stem cells (iPS cells). ES cells are in largest number in the preimplantation blastocyst of the embryo in humans. Obtaining human ES cells requires manipulating and in most cases destroying an embryo. This has led to an ethical debate over the use of ES cells, as those that believe that life starts at the creation of the fertilized egg recognize destruction of an embryo as the taking of this life. Other points of view recognize human life as initiating during pregnancy at some point but not at this early stage of pregnancy. The ethics of ES cells are complex, and beyond the scope of this book.

iPS cells were originally described by Shinya Yamanaka and colleagues in 2006 (Takahashi and Yamanaka 2006) and for which the Nobel Prize was awarded in 2012. iPS cells are adult cells, most often skin or lung fibroblasts, that have been reprogrammed to a pluripotent state by treatment with specific proteins or, more recently, small molecules. These cells can then be sustained as stem cells in that they will divide asynchronously and give rise to all tissues of the body. As they come from adult cells, iPS cells have not been subject to the same ethical controversies as ES cells. However, the reprogramming of the iPS cell genome from the adult cell state is not perfect; both genetic and epigenetic signatures remain in the iPS cell that harken back to the adult state and contribute to heterogeneity within iPS-derived cell lines (Cahan and Daley 2013). This issue has not hampered translational research with iPS cells; studies of differentiation protocols for iPS cells into neurons and glial cells or transplantation of iPS-derived cells into models of stroke proceed at a rapid pace. However, the fact that there may be genetic instability, mutations, or epigenetic alterations in an iPS cell as a result of reprogramming, may affect the clinical utility of these cells because such genetic alterations may lead to cell instability. For example, reprogramming an adult somatic cell (such as skin cell) into an iPS cell may alter the epigenetic code for that cell, such as DNA methylation state, which is transmissible and carries a risk in coding for a disease-related change in gene expression. Genetic instability in an iPS cell trial for retinal disease resulted in a transient hold because of genetic mutations in the iPS cells (http://www.ipscell.com/2015/07/firstipscstop/), and this led to a switch in other trials from patient-derived iPS cells to autologous iPS cells (http://www.ipscell.com/2015/11/parkinsons-ips-cell-trial-in-japan-switching-to-allogeneic/). This is further discussed below. There are other methods for the production of stem cells, such as somatic cell nuclear transfer, in which the nucleus of an oocyte (initially from a female ovary) is removed and an adult cell nucleus is inserted into the oocyte. This results in reprogramming of the adult nucleus and the generation of an egg that can produce an adult organism. This process has not seen widespread application in neural repair studies and will not be further discussed.

7.4 Types of Stem and Progenitor Cells in Brain Therapy

Cells used in stroke neural repair studies are categorized as adult- or pluripotent-derived (figures 7.1 and 7.2). Pluripotent cell types are derived from iPS or ES cells. Adult cell types are derived from adult cells, such as from bone marrow or adipose tissue. Neither pluripotent-derived nor adult-derived cells are true stem cells. In all cases, the cells used in neural repair studies are further differentiated than stem cells and cannot give rise to all tissue types in the body. This differentiation of stem cells prior to use as a brain repair therapy is a good thing, as transplantation of true stem cells into the brain or body produces tumors (Lee et al. 2013). For neural repair therapies, iPS and ES cells are differentiated from the pluripotent state into a more committed neural precursor cell (NPC) that can differentiate into all of the cell types of the CNS (neurons, astrocytes, oligodendrocytes). NPCs can be further differentiated into neuronal precursors, which only give rise

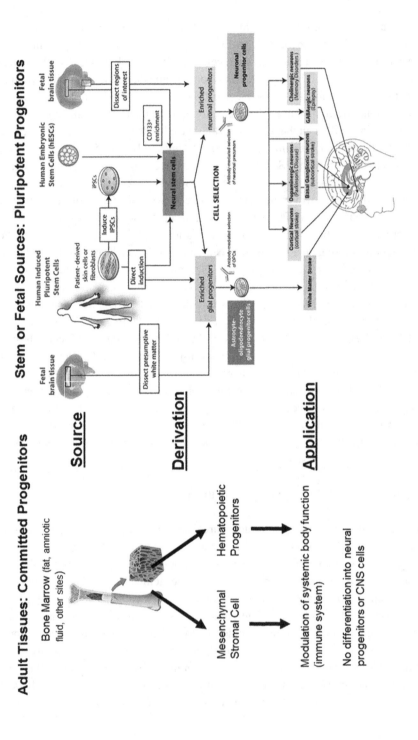

to neurons, and glial precursor cells, which only give rise to astrocytes and/or oligodendrocytes. Each of these can be even further differentiated into neuronal or glial subtypes, so that in theory, a very specific cell product can be transplanted into the brain after stroke, such as a cortical neuron into a cortical stroke (Tornero et al. 2013; Michelsen et al. 2015) (figure 7.1).

Adult progenitor cells are more restricted in their differentiation potential than pluripotent-derived progenitor cells. In general, in the body or in the lab, adult progenitor cells are termed *tissue-restricted progenitors*, as they reside in a given organ (brain, bone marrow, muscle) and can give rise to the cells of that organ when they are stimulated. In practice for stroke neural repair, adult progenitors have been taken from bone marrow, fat, placenta, or umbilical cord. These progenitor cells do not form neurons *in vivo* after transplant, nor have they been found to consistently form neurons or CNS cells *in vitro* under cell culture conditions. Thus, adult progenitor cells are very different from ES- or iPS-derived precursor cells. ES- or iPS-derived cells do differentiate into neurons or glia—indeed, that is the goal of the ES or iPS transplant field in the brain. Adult progenitor cells do not differentiate into neurons or glia and remain in their progenitor state when transplanted.

Adult progenitor cells are a very diverse group. The most commonly used adult progenitor cell is the mesenchymal progenitor cell. This is sometimes referred to as the mesenchymal "stem cell" or mesenchymal stromal cell (MSC) but it is important to stress, despite the misleading name, that these cells are not stem cells. As noted, none of the cells used for neural repair transplantation are stem cells since they are not pluripotent. MSCs are instead a progenitor cell that is related to a support cell in the bone marrow or other body tissues. MSCs are specialized to produce growth factors and extracellular matrix proteins that modify their local microenvironment. MSCs differentiate into bone, skeletal muscle, adipose tissue, cartilage, and tendon but do not consistently differentiate into any CNS cell types. In the bone marrow, MSCs support hematopoiesis (Via et al. 2012; Kfoury

Figure 7.1
Stem cell derivation and application. Adult progenitor cells are derived from bone marrow and other sources, such as amniotic fluid, umbilical cord blood, adipose tissue, and dental pulp. These cells include blood cell progenitors (hematopoietic progenitors) and mesenchymal stromal cells (MSCs). Hematopoietic progenitors and MSCs do not differentiate into neurons. When administered after stroke, these cells stimulate or alter the immune system. Stem or fetal cell sources are used to derive glial or neuronal precursors. This means that a main distinction between adult progenitors and stem or fetal progenitors is that adult progenitors will not become central nervous system (CNS) cells, but stem or fetal-derived cells are used because they can be differentiated into CNS cells. Stem or fetal cells can be sorted during the production process by antibodies that are specific to cell surface proteins that mark or identify a cell type. CD133 is a cell surface marker that recognizes CNS progenitors in a mix of fetal cells. Further antibody selection, using cell sorting such as fluorescence-activated cell sorting, may select glial cells (such as oligodendrocyte progenitor cells, OPCs) or neurons. In the most common strategies for developing cells for brain transplantation in stroke, embryonic stem cells (ES cells) or induced pluripotent stem cells (iPS cells) are taken through the neural stem cell stage and then further differentiated, as in the flow down through the figure from the top. A recent set of technologies allows direct neuronal or glial induction from the iPS or ES (pluripotent) state. This process of direct induction is not currently amenable to the scale-up necessary for production of large numbers (billions to trillions) of cells necessary for a stroke therapy. Image on left is adapted from Goldman (2016).

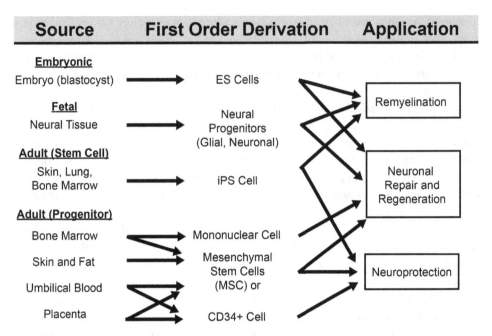

Figure 7.2
Flow diagram of cell sources and applications for stroke.

and Scadden 2015). MSCs are relatively easy to purify and grow from a bone marrow isolate and thus have been used extensively in preclinical studies of stroke neural repair (Vu
et al. 2014). MSCs have also been derived from placenta, adipose tissue, and umbilical cord
blood. These cells though from diverse sources—bone marrow, placenta, adipose tissue—
are very similar in their mesenchymal properties of differentiation and can all be classified
as "MSCs" (Via et al. 2012). Despite this common classification based on their mesenchymal differentiation pattern, there is considerable cellular and molecular heterogeneity
within MSCs across different laboratories and even within the same source (Rennerfeldt
and Van Vliet 2016; Han et al. 2017), which makes the accurate characterization and reporting of MSCs that are used in stroke recovery trials critical—so that the field knows
what is actually being transplanted. Although MSCs do not differentiate into CNS cells
(neurons or glia), they do produce growth factors, cytokines, and extracellular matrix
molecules (Tana and Sachan 2016) that likely are the mechanism behind their ability to
repair the brain after stroke.

A second cell approach in this category of adult progenitor cells is the bone marrow
mononuclear progenitor cell. Compared to MSCs, which are a support cell in the bone marrow, these cells are part of the bone marrow production of blood cells. In the studies

reported in stroke recovery trials, bone marrow mononuclear cells are immature cells in the bone marrow lineage and contain a mix of hematopoietic progenitor cells, endothelial progenitor cells, and more mature cells in the hematopoietic lineages (Giraldi-Guimarães et al. 2009). These cells also produce growth factors, cytokines, and chemokines that may provide the basis for a neural repair effect. Bone marrow mononuclear cells are relatively easy to purify from a bone marrow isolate.

A final cell type that has been applied to CNS repair studies is the fetally derived neural progenitor cell. These are cells taken from the brains of the human fetus, and as such, they are CNS progenitor cells or neural progenitor cells (NPCs) (figure 7.1). These cells can be expanded in number to produce the billions or trillions of cells that may be needed for a clinical application by genetic means. This issue of "scale-up" is a problem in the stem cell field: producing huge batches of a stem cell–derived or progenitor-derived cells for the millions of stroke patients who would need a transplant. In the example of fetally derived cells, it is not feasible, and perhaps not ethical, to isolate the large number of cells that would be needed for a stroke therapy from fetuses. Instead, an NPC that is derived from a fetal source must grow and proliferate in cell culture through several passages so that the few originally isolated cells will generate billions of cells needed for a therapy in many patients. One approach to "scale up" that has been used in this category of fetally derived cells is to insert an oncogene into the cell and use it to drive cell division to get many more cells and then to switch this gene off prior to transplantation. This is what is meant by using "genetic means" to induce scale-up. In one example in stroke neural repair, human fetal brain cells were genetically modified to contain a cancer gene that provides for cell proliferation, the *myc* oncogene. This gene is under control of a specific chemical (tamoxifen), and when this tamoxifen is removed, the *myc* gene turns off. This approach allows large-scale production of clinical-grade neural progenitors that will not be tumor forming after transplantation (Stroemer et al. 2007). Fetally derived neural precursors have also been applied to the treatment of pediatric white matter disease, such as Pelizaeus-Merzbacher disease (Gupta et al. 2012), and have been under clinical trial for other CNS diseases, such as through the work of specific biotechnology companies (http://www.stemcellsinc.com/Clinical-Programs/Overview). This genetic approach may be risky, as it requires that control of the oncogene but close to absolute: the gene should not reactivate in the cells once they are transplanted into the brain.

7.5 Mechanisms of Stem Cell Repair in Stroke

Cell therapies in stroke may repair the brain through replacement of damaged tissue, stimulation of endogenous neural repair processes, or reduction of secondary injury events. In theory, a transplanted stem/progenitor cell that is a neural precursor may differentiate into neurons and glial cells and reconstitute damaged or lost brain tissue. This possibility was one of the main initial drivers of this field. Transplantation of neural progenitor cells into the neonatal brain indicated that the transplanted cells can differentiate into neurons and integrate into the normal brain circuitry (Park et al. 2002). Recent evidence suggests that neuronal differentiation of transplanted iPS-derived neuronal precursors may generate widespread circuit integration of the donor cells (Tornero et al. 2013). Direct replacement of lost or damaged circuits after stroke occurred in these preclinical studies via novel approaches that generate very specific progenitor cells, such as cortical cells or striatal cells (Tornero et al. 2013). Despite these few studies, most of the preclinical evidence to date suggests that cell therapies stimulate endogenous repair processes rather than directly replacing damaged tissue (figure 7.2).

ES, iPS NPCs, or fetally derived NPCs are the cell types most likely to differentiate into neurons and glia and reconstitute brain tissue after stroke. These neural precursor cells are most often administered through direct injections into the region of tissue in the brain adjacent to the stroke damage. The transplanted cells then migrate to areas of damage. These cells can differentiate into more mature neurons or glial cells, as shown by their expression of markers such as neurofilament proteins, astrocyte intermediate filaments (i.e., glial fibrillary acidic protein, GFAP), and neuronal transcription factors (i.e., NeuN) (Kelly et al. 2004; Tornero et al. 2013). In some cases, these NPCs also extend axons within the damaged brain after stroke, particularly in studies with iPS-derived NPCs (Tornero et al. 2013). However, in all cases, these differentiated neurons and astrocytes do not form the type of organized brain structure that was present at the stroke site prior to the damage. In most cases, most of the transplanted cells die, and those cells that survive only differentiate in small numbers into mature neurons or astrocytes. Yet, assessments of behavioral recovery in these transplant experiments show an enhanced sensorimotor recovery. This has led to the concept that the transplanted cells are stimulating endogenous neural repair events in the brain, events which occur normally after stroke but in a much more limited fashion (Lemmens and Steinberg 2013; Zhang and Chopp 2013).

The endogenous brain repair events stimulated by stem or progenitor cell transplantation fall into the three categories: angiogenesis, neurogenesis, and axonal sprouting. Many different cell types engage in these three events after direct transplantation: ES-NPCs, iPS-NPCs, and adult progenitor cells (Lemmens and Steinberg 2013; Zhang and Chopp 2013; Savitz 2015). One problem in the field is that this work rarely goes beyond showing an association of cell transplant and a behavioral recovery effect—few studies have demonstrated causality. In showing an association, a cell line is transplanted into the brain in a rodent model of stroke, the mouse or rat experiences an improvement in its recovery from stroke, and then the process that was stimulated is considered the mechanism of recovery. In showing causality, the cell type would be characterized for the factor or factors that might be stimulating tissue repair, and these factors would then be knocked out in the cell, and the recovery effect not seen. Few studies have done the extra work to actually prove a causal role for a cell transplant–stimulated process on behavioral recovery. Thus considerable uncertainty remains about the mechanism of action of cell therapies. One exception is a study in which a human fetally-derived NPC was transplanted into a rat stroke model and the enhanced behavioral recovery was blocked by an anti-human vascular endothelial growth factor (VEGF) antibody (Horie et al. 2011). This study clearly demonstrated that human VEGF, which can only be from the transplanted human cells in this rat model, played a causal role in the recovery. In summary, cell transplantation has been associated with enhancement of many biological tissue repair events in the poststroke brain, but a true recovery-related mechanism has only been proven in one instance to date and this recovery effect is via local secretion of a growth factor.

In most cases of transplantation of adult progenitor cells, which is mainly accomplished by intravenous or intra-arterial routes, the cells do not differentiate into CNS cell types and do not enter the brain at all. These cells are mostly trapped in lung or spleen once they enter the bloodstream (Hicks and Jolkkonen 2009; Boltze et al. 2015). Yet, animal models indicate that these cells still promote functional recovery despite not getting into the brain. The mechanism may be modulation of the immune system after stroke (Eckert 2013; Boltze et al. 2015). Stroke stimulates an inflammatory response not just at the site of the stroke in the brain but also in the peripheral immune system. For example, stroke stimulates the spleen to contract and release inflammatory cells, and also to experience cell death and shrinkage. (Macrez et al. 2011; An et al. 2014). Adult progenitor cell transplantation, such as with MSCs, prevents splenic shrinkage and modulates the systemic and local brain inflammatory response in a way that may promote recovery (Scheibe et al. 2012). That said, a causal role of immune modulation in the ability of adult progenitor cells to promote stroke recovery has yet to be demonstrated definitively.

7.6   Translation of Stem Cell Preclinical Studies to the Clinic

7.6.1   Stem Cells Version 0.25

The simplest form of stem cell therapy, Version 0.25, is to take a cell that is readily available and which requires minimal treatment, culture, or manipulation and to then give this cell back to the patient. This is the case when the cell comes directly from the patient him or herself (it is "autologous"), which precludes the need to manipulate the cell in the lab prior to delivering it to the patient. This approach has been used in bone marrow transplantation for hematologic and solid tumor malignancies for over fifty years (Jenq and van den Brink 2010). In CNS disease, this approach has been used with hematopoietic progenitor cells, which are harvested in most cases from the bone marrow but may also be taken from the peripheral blood (Shyu et al. 2006). These cells are heterogeneous, containing precursor cells for most white blood cell lineages, such as lymphocytes, neutrophils, basophils, and macrophages, as well as mature cells in these lineages. A more specialized approach to

Stem Cells Version 0.25 would be to isolate a more immature or progenitor-type cell from the bone marrow preparation, such as a CD34-positive cell (Shyu et al. 2006; Sidney et al. 2014). In most preclinical and clinical studies, bone marrow is harvested and the tissue isolate is simply centrifuged and washed so as to isolate mononuclear cells (Taguchi et al. 2004; Savitz et al. 2011; Moniche et al. 2012; Moniche et al. 2015; Taguchi et al. 2015). Bone marrow mononuclear cells processed in this way can be obtained from a patient and immediately processed and delivered back to that patient, so that cell culture and manufacturing issues that may limit a scale-up of cell number or production of a specific cell type are not a concern. Delivery has been accomplished intravenously, intra-arterially, and via direct injection into the brain (intracerebral delivery). In head-to-head comparisons, there is no significant difference in functional recovery among these delivery approaches (Yang et al. 2013). Intra-arterial delivery results in more cells in the brain compared to intravenous delivery (Pendharkar et al. 2010; Yang et al. 2013). This is of interest because most bone marrow mononuclear cells delivered intravenously are filtered in the liver and spleen and do not reach the CNS in human stroke or animal models of stroke (Hicks and Jolkkonen 2009; Korhonen and Jolkkonen 2013; Khabbal et al. 2015). Those cells that do reach the vasculature of the brain, do not gain access to the actual brain parenchyma, and do not differentiate into neurons or glia (Savitz et al. 2011; Lemmens and Steinberg 2013). Clinical trials with bone marrow mononuclear cells have been small and primarily aimed at safety evaluation but have not yet found a therapeutic benefit (Prasad et al. 2014). One large trial with infusion of a bone-marrow-derived progenitor has shown promise in post-hoc analysis and is continuing to a larger phase IIb trial (Hess et al. 2017). This study found efficacy in intravenous infusion of these bone marrow-derived progenitors in the earlier time period of the treatment epoch, 24–26 hours after stroke. Future work in Stem Cells 0.25 would be to increase the study size in a phase II trial that could more definitively look at functional outcome. Finally, based on what is known about its mechanism of action, Stem Cells 0.25 treatment is likely to be restricted to the acute or very early sub-acute stage in stroke, when inflammation or secondary injury is still ongoing.

7.6.2 Stem Cells Version 0.5

A more advanced approach in cell therapy for stroke is to use a cell type that is more complex, grown in larger numbers, and is not autologous but can be used off-the-shelf for a therapy. All of these differences add complexity to the isolation and processing of stem cells for therapeutic use, placing them in a different category from Stem Cells 0.25. Compared to autologous transplantation, a non-self (allogenic) source for a cell therapy is desirable because it can be handled like a drug product—purity tested, identify confirmed, efficacy established, and the dose of cells can be administered any time. In allogenic cell therapy, the cell product is in a pre-made preparation, such as a bag or syringe, which is in the freezer and can be thawed for administration to each patient. This contrasts with an autologous source, which is isolated from the patient on an individual basis and then given back—there is no off-the-shelf product. Since autologous therapies are unique and individually derived, each patient receives a different product and responses across patients may be more varied. A concern in allogenic cell therapy is whether the cell product might be immunogenic because an allogenic cell is not "self." In terms of pure transplant biology, the placement of a cell that is from one human being into another human being is certainly an immunogenic event. However, the CNS has some immune privilege compared to systemic organs due to the blood-brain barrier, reduced lymphatics, and reduced intrinsic immune surveillance (Lampron et al. 2013). This might mean that a cell therapy for the brain may not require the same degree of immunosuppression that an organ transplant would. Also, some cell types in Stem Cells 0.5 appear to not provoke a vigorous immune response, such as MSCs. For this reason, MSCs have been transplanted both with and without immunosuppression (Jablonska et al. 2013), and in both cases, there has been demonstration of functional benefit (Eckert 2013; Lemmens and Steinberg 2013).

Cellular sources for stroke neural repair in Stem Cells 0.5 include MSCs and many MSC-like cells. These are cells that were characterized as originally isolated from the bone marrow. In their native state *in vivo*, MSCs do not produce hematopoietic cells (Stem Cells 0.25) as they are not true stem cells, but they provide support to the hematopoietic stem cells and their more differentiated progeny. MSCs secrete growth factors, cytokines, and extracellular matrix proteins in their bone marrow stem cell support role. This action provides the foundation for their beneficial effect in stroke: secreting factors that reduce inflammation and support the biological events of neural repair. As with the bone marrow progenitor cells of Stem Cells 0.25, MSCs in Stem Cells 0.5 do not differentiate into neurons or glia and do not directly replace lost cells in stroke. MSCs are found in adipose tissue, umbilical cord, placenta, amniotic fluid, dental pulp, skeletal muscle, tendons, and synovial membrane and are primarily responsible for the repair effect in transplantation from cells derived from these sources (Via et al. 2012; Kfoury and Scadden 2015). For example, an umbilical cord "stem cell" or fat-derived "stem cell" is in fact an MSC (Via et al. 2012; Kfoury and Scadden 2015). These cells are relatively easy to grow in cell culture and to expand in large numbers. They serve as the basis for many preclinical studies in stroke, where they have been found to produce recovery in sensorimotor tasks when administered in the acute and subacute phase after stroke, and in the chronic phase (more than one month after stroke in the rodent) (Eckert 2013). MSCs or cells closely related to MSCs have gone into clinical trials in human stroke (Lee et al. 2010; ; Steinberg et al. 2016; Hess et al. 2017), where they have been found to be safe. MSCs have been used in other human disease, such as graft versus host disease (Cyranoski 2012; Kfoury and Scadden 2015).

Most delivery approaches for a cell therapy in Stem Cells 0.5 have focused on intravenous or intra-arterial delivery. Preclinical studies suggest that the time window for efficacy for this approach extends into the chronic phase (months) after stroke (Eckert et al. 2013).

However, as these cells are rapidly filtered by the lungs, liver, and spleen after delivery, and most likely work via immunomodulation, it is likely that the clinical efficacy for an intravenous MSC therapy will be in the early stages, the day of or within a few days, after stroke. Efficacy at this early time point has been suggested in a recent clinical trial (Hess et al. 2017).

7.6.3   Stem Cells Version 1.0

The next tier of cell therapies in stroke uses more specialized cells that have been specifically tailored to therapy in the brain and may more specifically stimulate brain repair. Stem Cells 1.0 actually uses stem cells, whereas in Stem Cells 0.25 and 0.5 progenitor cells were used with no real stem cell association or link. ES cells and iPS cells can be easily differentiated into neural precursors with simple changes in culture conditions (Murry and Keller 2008). In fact, the differentiation of ES or iPS cells into neural precursor cells is something of a default differentiation pathway (Wonders and Anderson 2006; Maroof et al. 2013; Steinbeck and Studer 2015).

iPS-NPCs and ES-NPCs have been extensively used in preclinical stroke models. Unlike bone marrow progenitors (Stem Cells 0.25) or MSCs (Stem Cells 0.5), ES-NPCs and iPS-NPCs will differentiate into astrocytes, neurons, and, to a lesser extent, oligodendrocytes after brain transplantation. The transplanted cells migrate to areas of injury after stroke and are inhibited in their survival by local inflammation (Kelly et al. 2004; Lemmens and Steinberg 2013). iPS-NPC–derived neurons can form extensive local and long-distance connections in the brain after stroke (Oki et al. 2012; Tornero et al. 2013; Thompson and Björklund 2015).

iPS-NPCs and ES-NPCs have been delivered through direct brain injection or systemic (intravenous and intra-arterial) delivery. However, as with bone marrow progenitors and MSCs, intravenous and intra-arterial delivery does not introduce transplanted cells into the brain in appreciable numbers. This means that systemic delivery is likely to alter body-wide events, such as inflammation and its attendant secondary injury, rather than inducing primary brain tissue repair. This will be an important consideration when it comes to the timing of treatment for these cells. Systemic immunomodulation as a mechanism of action is not likely to be a viable target in late subacute and chronic stroke because most inflammatory cytokines and inflammatory cells have returned to their baseline levels around the first week in stroke (Perini et al. 2001; Suzuki et al. 2009; Sobrino et al. 2012; Kaito et al. 2013).

7.6.4   Stem Cells Version 2.0

What is coming down the road in possible stem cell therapies for stroke recovery? There are two promising developments in the field: brain region-specific stem cell therapies and bioengineered applications for stem cell therapies. Brain regions possess specialized neurons, for example Betz cells in motor cortex and medium spiny neurons in the striatum. In most preclinical studies, Stem Cells 1.0 in this chapter's terminology, a sort of generic neural precursor or NPC has been transplanted into stroke models regardless of the region that is damaged. However, it is possible in cell culture to make an ES or iPS cell differentiate into a neuron with a cortical, basal ganglia-like (Tornero et al. 2013; Anderson and Vanderhaeghen 2014; Capetian et al. 2014), or interneuron-like phenotype (Wonders and Anderson 2006; Maroof et al. 2013). When cortical neuronal progenitors are transplanted into brain regions more closely matching their phenotypes, the integration and effect on recovery appear to be better in terms of axonal connections (Tornero et al. 2013).

Stroke in humans damages a substantial proportion of white matter. This is relatively unique to the human disease, as in the human brain, white matter constitutes over 50 percent of brain volume, while in rats and mice, it is less than 10 percent (Hofman 1989). For a human stroke cell therapy, replacing lost glia may be as viable an approach as replacing lost neurons (figure 7.1). Glial cells, such as oligodendrocytes, stabilize injured axons and directly secrete neuronal growth factors (Gallo and Armstrong 2008). Astrocytes also interact with oligodendrocytes and oligodendrocyte precursor cells to promote resistance to stress and differentiation (Moore et al. 2011). A stem cell therapy for human stroke thus might be directed at glial replacement, as it is in other CNS diseases (Noble et al. 2011). New protocols are available for oligodendrocyte differentiation of ES and iPS cells, and protocols for directed astrocyte differentiation have been tested in disease models such as spinal cord injury (Noble et al. 2011; Wang et al. 2013; Xie et al. 2014). It is likely that future stem cell therapies in stroke will apply glial precursors or more differentiated glial cells as therapies.

A major hurdle in brain region-specific stem cell therapies is the inefficiency of generating these cell types in culture. Current protocols for developing region-specific neurons or oligodendrocytes take weeks to months and do not generate high numbers of cells. Yet, for a human application in stroke, it is likely that trillions of cells would be needed of a successful therapeutic line, and these would need to be developed under stringent clinical regulatory controls. An increased understanding of specific differentiation factors may help to increase efficiency of differentiation, but work is still ongoing in this area. This is one of the major challenges for Stem Cells 2.0.

Bioengineering in the stem cell field allows for tissue matrices or scaffolds to provide trophic support and for physical guidance or molecular cues to facilitate stem cell engraftment. A limitation in the stroke cell transplant field is that most of the transplanted cells die; there is substantial transplant stress. More differentiated neurons might provide better repair of a brain region (see above) but are usually more sensitive to such transplant stress (Moshayedi and Carmichael 2013; Moshayedi et al. 2016). Immature neurons or neural precursors are more resistant to transplant stress but less likely to differentiate into mature neurons and integrate *in vivo*. A solution to poor survival and lack of differentiation is to transplant stem cells within a matrix that supports survival and directs differentiation *in vivo*. Many groups are working on scaffolds or hydrogels that contain protein motifs that stem cells recognize as "home," such as blood vessel protein sequences and support survival (figure 7.3) (Kim et al. 2012; Fisher et al. 2014; Moshayedi et al. 2016). Scaffolds or hydrogels can incorporate growth factors that support differentiation and integration of transplanted cells. In stroke, the fact that the stroke itself is a cavity and can accept such a stem cell/hydrogel matrix without adjacent tissue damage represents an opportunity (Zhong et al. 2010; Zhang et al. 2011; Modo et al. 2013; Moshayedi et al. 2016). The field of bioengineering for stem cell transplantation has come further along in spinal cord injury. Recent studies indicate that iPS-NPCs transplanted within a hydrogel matrix with growth factors promote profound neuronal projections of the transplanted cells and integration into the brain (Lu et al. 2014; Tuszynski et al. 2014). Future applications for stem cell transplantation in stroke will likely combine a bioengineered matrix with a specific cell type to enhance transplant survival, integration, and functional recovery.

7.7 Pharmacological Therapies for Stroke Recovery

The growing biological understanding of neural repair mechanisms has led to the development of several categories of drug therapies to stimulate recovery after stroke. These drug therapy approaches stimulate known neuroplasticity signaling systems that have a role in memory or motor learning in the normal brain, boost the effects of the classic neuromodulatory transmitters (dopamine, serotonin, norepinephrine), alter the excitatory/inhibitory balance in peri-infarct brain tissue, influence cytoskeletal structure and axonal outgrowth, or generally stimulate brain growth programs. Each has its own unique application and potential, and as with all pharmacological therapies in the CNS, each has its own potential drawbacks. An important principle in the pharmacological therapy for stroke recovery is that these drugs likely will need to be combined with a behavioral therapy to realize their full potential in stimulating CNS tissue repair (see chapter 5).

7.7.1 Amphetamines

The pharmacological therapy of motor recovery after adult brain injury got its start in studies of amphetamines. A landmark paper by Dennis Feeney in *Science* in 1982 showed that amphetamine treatment of rats after suction ablation of the sensorimotor cortex improved behavioral recovery (beam walking) only if paired with rehabilitative activity (Feeney et al. 1982). The concept of modulation of noradrenaline and other classical neurotransmitters (serotonin, acetylcholine) in the development of normal brain function and in learning and memory started in the 1950's and gathered steam in the 1960s and 1970s (Brengelmann,

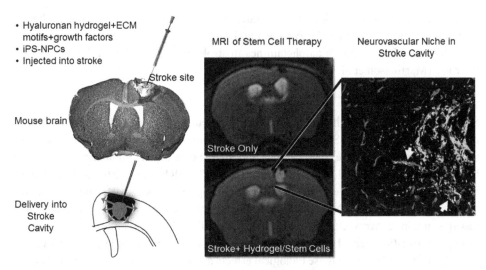

Figure 7.3

Biomaterials approach to stem cell therapy. Most stem cell transplantation approaches are not successful because the cells die after the injection into the brain. This is likely because the brain after stroke provides a harsh environment for the cells. One solution to this is to transplant stem cells within a prosurvival matrix. In this example, a hydrogel matrix was developed with hyaluronan that contains extracellular matrix (ECM) proteins and growth factors within the hydrogel. When this hydrogel is injected into the stroke cavity, it assembles once inside and supports induced pluripotent stem cell (iPS)–neural precursor cell (iPS-NPC) survival. This hydrogel is visible on magnetic resonance imaging (middle column) after stroke inside the stroke cavity, allowing it to be tracked in real time *in vivo*. Inside the stroke cavity, the hydrogel/iPS-NPC induces blood vessel in-growth into the cavity, and the transplanted cells (brighter cells) associate with the blood vessels (arrows). Adapted from Moshayedi et al. (2016).

1958; Mechner F and Latranyi, 1963; Lorens et al. 1971; Anlezark et al. 1973), leading to important findings that augmenting noradrenergic function enhanced some measures of learning and memory and that depletion of noradrenaline impaired cortical plasticity in the critical period of brain development (Kasamatsu and Pettigrew 1976). Feeney's work identified not only a pharmacological mechanism to enhance recovery in the adult brain but also the dependency of recovery in the adult on pairing drugs with behavioral therapies. This report initiated an explosion in this area. The finding of an interaction of brain injury, amphetamines, and rehabilitative therapy was replicated in many animal models (Adkins et al. 2009). Human motor learning can be enhanced by amphetamines (Bütefisch et al. 2002), and small initial trials of amphetamine or similar noradrenergic agents (methylphenidate; Tardy et al. 2006) indicated that these agents promoted recovery in stroke (Gladstone and Black 2000; Goldstein 2000). However, amphetamines have not translated to a clinical therapy for stroke recovery, as four subsequent trials indicated that this treatment provided no benefit after stroke (summarized in Goldstein 2009). It should be noted that unlike the animal studies, human trials were not paired with rehabilitation therapy of the same intensity. These drugs have powerful cardiac and vascular stimulation effects, producing hypertension, an elevated heart rate, and an enhanced tendency for arrhythmias (Martinsson and Wahlgren 2003; Westover and Halm 2012) that further limit their usefulness. A review of all data for amphetamines in stroke found insufficient data to support their use (Martinsson et al. 2003; Goldstein 2009).

7.7.2 SSRIs

SSRIs have been used in the treatment of several neurological and psychiatric diseases, most notably depression. SSRIs alter fundamental neurobiological processes in the brain, such as neurogenesis in the hippocampus and the balance of excitation and inhibition in brain circuits (Mostert et al. 2008; Méndez et al. 2012). SSRIs increase brain-derived neurotrophic factor (BDNF) levels and activate the transcription factor cAMP response element-binding protein (CREB) (Mostert et al. 2008), two molecular events involved in motor learning and in recovery after stroke (Ménard et al. 2015). The SSRI fluoxetine (Prozac®) can open the critical period of plasticity in the visual system of the rat in the adult (Vetencourt et al. 2008). Based on these biological effects and the relevance of their mechanisms to stroke recovery, many animal studies have tested the effect of SSRIs in motor recovery after stroke. The results have been mixed. Several well-conducted and rigorous preclinical studies have found no effect of SSRIs in enhancing motor recovery after stroke (Jolkkonen et al. 2000; Windle and Corbett 2005; Sun et al. 2016), while others have found a positive effect on recovery (Li et al. 2009; Ng et al. 2015). Important differences in these studies relate to the experimental model of stroke, the pairing of SSRI administration with concomitant poststroke training, and the use of different types of motor outcome measures. Thus, the preclinical data are not uniformly supportive of a positive effect on stroke recovery with SSRIs, and the reasons behind these are not easily explained by simple experimental differences across these studies. One recent study found that fluoxetine prolonged or reopened the period after stroke in which neurorehabilitative training produces its most substantial effect on stroke (Ng et al. 2015). This fluoxetine effect was associated with decreases in markers of intracortical inhibitory neurons and appears similar to the effect of fluoxetine on brain inhibition and opening of the critical period in visual cortex (Vetencourt et al. 2008; Méndez et al. 2012).

In the fluoxetine for motor recovery after acute ischemic stroke (FLAME) trial (Chollet et al. 2011), the SSRI fluoxetine enhanced motor recovery after stroke. This was a well-designed but still small phase IIb trial, with fifty-seven patients in the fluoxetine group. Patients with moderate to severe motor impairments after stroke were administered fluoxetine (20 mg once a day) beginning at five to ten days after stroke for ninety days while receiving traditional neurorehabilitation. Patients with a pre-specified cutoff on a formal depression metric were excluded. Patients on fluoxetine had a 10 point improvement in the upper extremity Fugl-Meyer Motor Assessment (FMA), as well as an impressive improvement in the National Institutes of Health Stroke Scale (NIHSS), and an increase in the percentage of patients with modified Rankin scores of 0 to 2 (none to minor disability).

Patients also improved in depression scores with fluoxetine treatment after stroke. This aspect of depression score is important because the trial treated depression as a binary element: a score above a certain threshold in the Montgomery-Åsberg Depression Rating Scale was defined as depressed and a score below this threshold was defined as not depressed. It is more likely that depression after stroke is a continuum and not a binary function, and that an effect of fluoxetine on mild depression after stroke may enhance motivation, participation in neurorehabilitation and greater incorporation of affected limbs into activities of daily living, and lead to greater recovery, independent of a direct motor effect. However, of the possible mechanisms of action, fluoxetine most likely had a direct effect on motor recovery rather than an indirect one through reduction in depression. There are two reasons why we say this. First, a back of the envelope calculation reveals that the average recovery in the group in the FLAME trial that received fluoxetine was exactly what would be predicted by the proportional recovery rule. As we have already pointed out, standard rehabilitation has no impact on proportional recovery (chapter 3). Second, in the mouse model of stroke (Ng et al. 2015), fluoxetine's effect on motor recovery was dependent on the timing of when it was given—it is not clear why this should be true for motivating effects (i.e., once they kick in, an animal should recover at any time point after stroke).

7.7.3 Cerebrolysin

Cerebrolysin is a mixture of peptides and amino acids derived from pig brain. The production process and constituents of this mixture have not been published. The first reports on this mixture/brain extract date back to 1972 (Harrer et al. 1972), where it was referred to as "brain hydrolysate." Most early studies are in the German- or Russian-language literature. The first English-language report on the effects of cerebrolysin was in 1990 (Kofler et al. 1990). Cerebolysin has been applied to many different conditions over the past fifty-five years, including Raynaud's phenomenon, as an immune stimulant, for postsurgical conditions, vascular dementia, perinatal encephalopathy, hyperacusis, diabetic neuropathy, Alzheimer disease, and "organic brain syndrome." There have been many small studies *in vitro* and *in vivo* investigating its potential to promote neuroprotection in stroke. The application of cerebrolysin to CNS disease reflects evolving intellectual trends in this scientific field over the past half-century. Cerebrolysin was first studied for its growth factor effects in the 1970s to 1990s, during the heyday of brain growth factor discovery and application. It was studied for neuroprotective effects in the mid-1990s to early 2000s during the peak of interest of neuroprotection for stroke. More recently, interest has shifted toward a cerebrolysin effect on repair and recovery in stroke, with a parallel shift in focus to more nuanced molecular signaling systems, such as sonic hedgehog effects in neurogenesis (Zhang et al. 2013).

The largest trial of cerebrolysin in stroke recovery indicated a beneficial effect of this peptide/amino acid mixture when administered beginning one to three days after acute stroke (Muresanu et al. 2016). Cerebolysin was given as a daily intravenous infusion for twenty-one days in conjunction with traditional neurorehabilitation, although the amount and intensity of neurorehabilitation were not recorded. The Action Research Arm Test (ARAT) was the primary outcome measure and showed a statistically significant improvement in recovery with cerebrolysin ($n = 104$) compared to a placebo ($n = 104$) infusion at ninety days after stroke. It was also stated that there was a significant effect on the NIHSS and other measures, but the actual scores were not reported. A Cochrane review of all cerebrolysin trials in stroke found faults in trial structure and reporting, including previous trials in which trial analysis personnel were employed by the company that makes this peptide/amino acid mixture and an excess of side effects in cerebrolysin recipients compared to controls (Ziganshina et al. 2016). Thus, at this stage, although there is some promising preliminary work to support a larger trial, current data do not justify use of this mixture in clinical practice.

Outside of questions of efficacy and related concerns with the clinical trials to date, cerebrolysin is problematic as a drug product. First, the active ingredient in the mixture has not been identified. Second, whether elements of this mixture permeate the blood-brain barrier has not yet been established. Studies demonstrate clear biological effects of cerebrolysin on brain tissue, such as by inducing specific effects in neural progenitors in the brain after stroke (Zhang et al. 2013), suggesting that some element of this mixture gets into the brain to mediate a biological effect. Mechanistic and pharmacokinetic studies to determine bioavailability and mechanism of action are possible, such as with biological labeling of peptides or amino acids in the cerebrolysin mixture, but these have not been done. Third, the batch-to-batch variability of a brain lysate has not been reported. In a clinical delivery scenario, what happens when a group of patients does not experience an expected benefit from the mixture? Is this because that particular batch differed from a previous one? If so, did it differ in a specific bioactive molecule? What if side effects occur in a particular batch—how will these be attributed to particular elements in the Cerebrolysin mixture?

7.7.4   Dopamine

Dopamine inputs to motor cortex contribute to motor learning (Molina-Luna et al. 2009; Hosp et al. 2011). Dopaminergic drugs are used extensively in Parkinson disease and have well-established dosing and side effect profiles. It is a natural evolution of these two facts: a common clinical usage and known direct brain action, that stroke recovery would be considered a target for clinical trials of using drugs such as L-dopa or direct dopamine agonists.

Despite some initial promise, dopaminergic drugs have not been shown to cause improved recovery after stroke. Initial optimism for efficacy of L-dopa in stroke recovery came from a well-designed but small trial published in *The Lancet* in 2001 (Scheidtmann et al. 2001). Beginning three to six weeks after stroke, fifty-three patients were given

L-dopa plus a peripheral dopa-decarboxylase inhibitor (which reduces the peripheral adverse effects of L-dopa) or placebo. These patients were then followed for a three-week washout phase of no L-dopa (or placebo) while still receiving neurorehabilitation to test if treatment effects persisted. The neurorehabilitation was not described in detail but appeared to consist of a standard poststroke regimen. The patients had a significant improvement in the Rivermead Motor Assessment at three and six weeks compared to placebo, suggesting recovery persisted after treatment. Subsequent clinical trials of L-dopa, however, have not been positive (Sonde and Lökk 2007; Lokk et al. 2011). The largest study of dopamine in stroke recovery is the DARS (Dopamine Augmented Rehabilitation in Stroke) trial. This trial aimed to recruit 572 patients, five to forty-two days poststroke, and to administer L-dopa/dopa decarboxylase inhibitor for six weeks. L-dopa was given in conjunction with standard neurorehabilitation but not closely paired with it, as was done in *The Lancet* trial (Bhakta et al. 2014). The results of this trial have not been published, but oral presentations at stroke meetings have indicated a negative outcome.

Outside of stimulating dopamine production and release with L-dopa, studies with direct dopamine agonists have also been conducted. Unfortunately, these followed the same general pattern as the L-dopa studies; larger studies have been negative (Cramer et al. 2009), even though small trials, such as of neglect, have been positive (Gorgoraptis et al. 2012). With the results from medium to large trials showing no efficacy of L-dopa or direct dopamine agonists, it appears that the dopamine hypothesis will be abandoned.

7.7.5 Manipulating the Excitatory/Inhibitory Balance in Peri-Infarct Tissue

Stroke produces a hypoexcitable state in brain tissue that borders the infarct. In a sense, stroke "stuns" adjacent cortex and reduces signaling in pyramidal neurons in this area. This occurs because reactive astrocytes downregulate their GABA uptake systems, increasing the extracellular levels of this inhibitory neurotransmitter. Increased GABA activates tonic (extrasynaptic) GABA receptors, which increase the "shunt current" of these cells. The shunt current establishes the baseline level that a neuron needs to depolarize to fire an action potential. An increase in tonic GABA signaling means an increased shunt current, and this in turn means that pyramidal neurons require greater activation to reach the threshold for action potential discharge. This hypoexcitable state can be reversed by a tonic GABA receptor blocker. Tonic GABA receptors have a unique structure, with incorporation of the $GABA_A$ $\alpha 5$ or δ subunits (Glykys and Mody, 2007). Preclinical studies demonstrate that genetic knockdown of either subunit improves recovery (Clarkson et al. 2010). Similarly, $GABA_A$ $\alpha 5$ inverse agonists bind to tonic GABA receptors and decrease their function. These promote recovery after stroke, with an interesting efficacy profile: their positive effect on motor function is almost immediate after administration in rats and mice with experimental stroke. The positive effects of $GABA_A$ $\alpha 5$ inverse agonists in stroke have been replicated in multiple species and multiple models (Clarkson et al. 2010; Lake et al. 2015). There is one ongoing clinical trial for GABA blockade to enhance stroke recovery, using an α5 antagonist (NCT02877615).

An alternative to manipulating inhibitory tone to increase cellular excitability after stroke is to increase excitatory glutamatergic signaling. Most excitatory signaling in the brain occurs via the glutamate AMPA receptor. The number of AMPA receptors in the postsynaptic membrane is tightly controlled and modulated up or down depending on synaptic activity to modulate the strength of the excitatory synapse (Henley and Wilkinson 2013). Potentiation of glutamate signaling and an increase in cell surface AMPA receptors is one of the means of producing long-term potentiation, or LTP, which has long been thought to underlie learning and memory formation (Henley and Wilkinson 2013; Arai and Kessler 2007; Bassani et al. 2013; Park et al. 2015). Based on these findings and reasoning, treatments or experimental manipulations that enhance AMPA receptor signaling have frequently been tried to enhance learning memory in the normal condition and in CNS disease (Pirotte et al. 2013; Park et al. 2015).

In stroke, this approach walks a fine line, because glutamate signaling plays a major role in inducing neuronal cell death via excitotoxicity in the early phases after stroke (see chapter 4), which of course would make stroke recovery worse by enlarging the initial zone of stroke cell death. However, enhancing AMPA receptor function through a positive allosteric modulator—a drug that only increases ionic conductance when glutamate is bound—increases recovery of motor function in stroke. This effect of increasing recovery of function after stroke with an AMPA receptor-positive allosteric modulator was effective if initiated three days after the infarct. However, if administered before this three-day delay, the approach made the infarct larger, consistent with the excitotoxic effects of glutamate signaling in acute stroke. The mechanism of AMPA receptor-mediated improvement in functional recovery is dependent on the downstream induction of BDNF, as AMPA-positive allosteric modulators that do not induce BDNF do not promote recovery of function, and recovery of function with these drugs can be blocked by antagonizing BDNF action (Clarkson et al. 2011). Like tonic GABA receptor blockers, inhibitors, or inverse agonists, this approach of enhancing neuronal network excitability after stroke works with a built-in delay—a time period in the first few days after stroke in which initial cell death must subside. AMPA receptor-positive allosteric modulators remain a promising target in stroke recovery, with only preclinical data at present to support the approach.

7.7.6 Phosphodiesterase Inhibitors

Cyclic nucleotides (cAMP and cGMP) are critical intracellular signaling molecules that are required for brain development and normal function (Xu et al. 2011; Sanderson and Sher 2013). cAMP and cGMP signaling are related through distinct pathways to long-term potentiation; they play a role in many models of learning and memory and also promote axonal sprouting (Xu et al. 2011; Heine et al. 2012; Sanderson and Sher 2013). cAMP and cGMP are degraded by 3′,5′-cyclic nucleotide phosphodiesterases (PDEs) (Francis et al. 2011; Xu et al. 2011). cAMP, through activation of cAMP-dependent protein kinase, stimulates the transcription factor CREB, which regulates the transcription of many genes associated with learning and memory and axonal outgrowth, such as BDNF (Xu et al. 2011). Both CREB and BDNF have been implicated in recovery of function after stroke (Clarkson 2015). cGMP signaling can be enhanced to promote learning and memory, through nitric oxide synthase and other pathways (Xu et al. 2011; Sanderson and Sher 2013). The signaling cascades of cyclic nucleotide action in neurons thus places them within several molecular systems that might have a role in functional recovery.

Eleven families of PDEs (PDE1–11) are encoded by twenty-one genes with over 100 different proteins expressed. Each of the twenty-one PDE isoforms shows a unique tissue expression profile (Kelly et al. 2010; Xu et al. 2011; Stephenson et al. 2012; Kelly et al. 2014), including unique profiles within regions of the brain, making them attractive drug targets. Targeting PDEs with small molecules or biologics may be one way to modulate cyclic nucleotides in a brain region–specific manner while minimizing undesirable side effects in undamaged brain regions. In stroke, studies investigating PDE4 and PDE5 inhibitors have shown improvement in motor performance. The PDE4 inhibitors rolipram and HT-0712 enhance motor performance when paired with rehabilitative therapy in the rat (MacDonald et al. 2007). However, PDE4 inhibitors have been closely associated with vomiting, which may be triggered at doses equivalent to levels that produce motor recovery after stroke, hindering their clinical translation. The vomiting action is due to the presence of PDE4 in brainstem centers that trigger vomiting. Other PDE isoforms are not present in this location.

Sildenafil, a PDE5 inhibitor, and other experimental PDE5 inhibitors also improve functional recovery in preclinical stroke models (Zhang et al. 2012). Sildenafil (Viagra®) is approved by the Food and Drug Administration (FDA) and has well-established safety. The mechanism of action of PDE5 inhibitors in brain repair is not clear. There has been a reported effect of sildenafil on neural progenitor cells (Zhang 2012), but a lack of significant expression of PDE5 in neurons in the brain has been reported (Menniti et al. 2009). Clinical trials for PDEs have been completed in Alzheimer disease and have found no efficacy (PF-04447943; Schwam et al. 2014) or have not yet been reported (MK0952). Cilostazol is a PDE3 inhibitor that also blocks platelet function and has been tested for its effect in secondary prevention of stroke but not for a role in stroke recovery. In general, the application of PDE inhibitors appears promising but requires both greater preclinical modeling and access to isoform-specific PDE inhibitors.

7.7.7 Axonal Growth Inhibitors

In the adult brain, axons interact with their environment through extracellular receptors that sense myelin (oligodendrocyte) membrane proteins, extracellular matrix proteins, and astrocyte membrane proteins. These can cause growth cone collapse and block the formation of new connections after CNS injury (Benowitz and Carmichael 2010; Sharma et al. 2012; Fawcett et al. 2012). Several members of these classes of axonal growth inhibitors have been implicated in functional recovery after stroke. These include the myelin proteins Nogo, OMgp, and MAG; the chondroitin sulfate proteoglycans (CSPGs); and astrocyte protein ephrin A5. Blocking these proteins promotes the formation of new connections within the ischemic hemisphere and between cortex and spinal cord (Li et al. 2010; Overman et al. 2012; Wahl et al. 2014). This is reviewed in chapter 4. Although mechanistically promising, therapeutic strategies to target these signaling systems are in their infancy. These have initially used antibody or small-molecule blockade of the molecule itself or of its receptor. The Nogo receptor, NgR1, binds Nogo, MAG, and OMgp. CSPGs bind leukocyte common antigen receptors, such as PTPsigma. These can be targeted with protein motifs from the Nogo protein or of CSPG molecules, either administered as the protein motif (peptide) or as a conjugate of these protein motifs to an immunoglobulin chain, which aids in production of the molecule (Fiedler et al. 2002; Li et al. 2004; Robak et al. 2009; Lang et al. 2015). This approach improves motor performance in stroke (Fang et al. 2010; Zai et al. 2011; Barbay et al. 2015).

Making an antibody molecule with protein motifs for the growth inhibitory protein or its receptor is relatively easy to do technically and generates a potential biological therapy for stroke. It has been applied to ephrin A5 (Overman et al. 2012) and MAG (Barbay et al. 2015) and has moved into clinical trials with an anti-MAG therapy (Cramer et al. 2013). A problem is that this approach generates a large molecule, with limited or no penetration of the blood-brain barrier (BBB). When the BBB is disrupted in stroke, this might allow brain penetration, but this has not been routinely tested in preclinical models or in patients.

An alternative approach to inhibiting the binding or signaling of CSPGs after stroke is to actually degrade them. This is most commonly done by degrading or digesting their proteoglycan side chains. These side chains, and not the core protein, are responsible for most of the growth inhibitory activity of CSPGs. The CSPGs side chains can be degraded after stroke using a bacterial enzyme that digests the side chains (chondroitinase ABC). This has limits because the molecule is large and must be directly administered into the brain, and it is antigenic, likely provoking some degree of inflammatory response in the brain.

An alternative to generating a biologic therapy against an axonal growth inhibitory system is to target common downstream signaling pathways for axonal growth inhibitors. Most of the axonal growth inhibitory receptors, such as Nogo receptor 1 and the CSPG receptors, signal through rho kinase and thereby directly alter cytoskeletal dynamics (Monnier et al. 2003; Fujita and Yamashita 2014). Rho kinase, or ROCK, can be blocked by small molecules (Takekazu et al. 2007). Small-molecule inhibitors of ROCK are neuroprotective in stroke and promote enhanced recovery (Ding et al. 2009; Hyun Lee et al. 2014). Fasudil is a small-molecule inhibitor of ROCK, along with other actions. It has been shown to promote recovery of function in preclinical stroke models (Huang et al. 2008; Liu et al. 2014). A problem with the approach of selectively targeting ROCK is that it is involved in other cellular processes than axonal growth, such as angiogenesis. The iso-form ROCK II is more neuron specific and may represent a target for blockade to promote axonal growth and recovery of function in stroke, but it has not been fully studied in the stroke context to date.

7.7.8 Growth Factors

The simplest approach to promoting recovery of function after stroke is to give a mole-cule that activates many cellular growth programs. With the idea of "the more growth ac-tivated the better," delivering a growth factor or growth-promoting cytokine after stroke could be considered the nuclear bomb of neural repair. Growth factors or growth-promoting cytokines, such as EGF, FGF, GDNF, EPO, G-CSF, and GDF10, will activate most cells of the CNS, including neurons, astrocytes, oligodendrocytes, and endothe-lial cells. These may stimulate a host of processes that are active in tissue turnover after stroke, such as angiogenesis, neurogenesis, axonal sprouting, myelination, and glial re-modeling (Maurer et al. 2008; Gutiérrez-Fernández et al. 2012; Greenberg and Jin 2013). However, as discussed in the chapter on secondary principles of stroke neural repair, sys-temic delivery of a growth factor will activate many biological processes that are not exactly ideal targets for neural repair, such as hematopoiesis, renal fibrosis, altered liver synthetic and excretory function, glucose intolerance, and, in many sites, cell prolifera-tion. These actions will lead to effects that are not "off-target effects" because they are legitimate and specific actions of the growth factor, but they will complicate a potential stroke therapy. Examples include white blood cell stimulation in the G-CSF trial and thrombogenic complications in the EPO trial in stroke (Ehrenreich et al. 2009; Bath et al. 2013; Ringelstein et al. 2013).

A solution to these systemic complications is to focus on local delivery of growth factors. This approach requires a delivery platform and usually some degree of invasiveness for its placement in the brain. Such approaches have been applied with bioengineered scaf-folds or hydrogels that would release growth factors locally into the peri-infarct tissue from the stroke cavity (Moshayedi and Carmichael 2013; Wang et al. 2013; Cook et al. 2016; Moshayedi et al. 2016). These are an exciting area of development but will require a great deal more preclinical work before they are translated into therapies.

7.8 Conclusions: Molecular Neurorehabilitation

Neurorehabilitation in stroke and other CNS diseases uses primarily a behavioral ap-proach. Physical, occupational, and cognitive/speech therapies promote recovery. How-ever, neurorehabilitation as a field is unique in that it has no medical therapies to treat its constituent diseases. The revolution of molecular medicine has changed all other med-ical specialties, such as oncology, cardiology, gastrointestinal, and rheumatological dis-ease areas. In neurology, new drugs reach specific molecular targets in the subspecialties of movement disorders, neuro-oncology, and epilepsy. Despite the past focus on purely be-havioral approaches to stroke recovery, research in the field of neural repair in stroke is at a pivot point at which promising preclinical results can be taken into clinical trials. This process applies to distinct classes of pharmacological therapies, including drugs that en-hance synaptic plasticity. Older and readily available drugs that manipulate neuromodula-tors, such as SSRIs, have been tested in initial clinical trials (Chollet et al. 2011; Cramer 2015) and shown to be effective in small sample sizes. Widespread translation into clinical practice will need larger trials to prove efficacy. The stem cell field and its popular consid-eration has boosted research across many different cell types, and the preclinical research for many of these looks positive. Clinical trials for cell therapy are more complex than for a small-molecule drug, but initial progenitor cell therapy trials in stroke have just been completed with promising results (Kalladka et al. 2016; Steinberg et al. 2016). The next five to ten years should see novel medical therapies move into the clinic as the field of neurorehabilitation takes its place in the modern pantheon of medical specialties and in fact moves beyond them as it uniquely demonstrates the fundamental interaction between the nervous system and the body as a whole, and the action of drugs and cells on its con-stituent parts during behavior.

8

A Future Approach to Neurorehabilitation after Stroke
If Humans Had Wings

An inference from the above traced course of evolution of the vertebrate brain is that the freeing of a limb-pair for more manifold use as "tool," while the other limb-pair still assured efficient land locomotion gave an impulsion for cerebral development which was of decisive importance in the evolution of pallial (cortical) growth and function. This inference raises the surmise that, had wings arisen in the vertebrates, as actually in the insect, without cost of a limb-pair to co-exist with "land-locomotor" leg and "tool" arm, the consequent additional experience and exploitation of a great three-dimensional medium (containing, unlike water, ample oxygen) would have evolved a brain of wider components and fuller lines than is the human.

—C. S. Sherrington (1910)

8.1 Introduction

We started this book on arm and hand recovery after stroke describing the work of early pioneers in neurology and neurophysiology, for example, Sir Charles Scott Sherrington and his elegant experiments on spinal reflexes and decerebrate rigidity. The principles detailed in this book suggest that future therapies that stimulate brain plasticity while immersing the body and brain in behaviorally stimulating and more enriching environments will promote greater recovery after stroke. In this final chapter, we will briefly touch upon four areas that have the potential for improving motor recovery after stroke. These are video gaming and virtual reality, invasive brain stimulation, reopening the sensitive period, and precision medicine.

8.2 Video Games and Virtual Reality

As we have discussed in chapter 3, enriched environments promote motor recovery in animal models (Biernaskie et al. 2004). The question that then arises is, how can one apply the principle of enrichment to patients? Video games and virtual reality experiences have the potential to create more immersive, stimulating, and mood-enhancing experiences than conventional therapy. They also offer the possibility for general aesthetic enhancement of the clinical environment and may provide the opportunity for multiplayer interpersonal engagement—patients spend a lot of time in their rooms alone or with bored relatives and friends. There are two aspects to this issue—one pertains to making clinical and home environments more stimulating and thereby encouraging motor activity outside of therapy sessions, and the other is to making physical therapy itself more immersive and motivating. The mechanisms of enrichment are not precisely known but likely work through promoting more time on task, adding task variety, and increasing the gain on skill learning and retention through the provision of reward, perhaps mediated through the modulatory effects of neurotransmitters such as dopamine (Johansson and Ohlsson 1996; Biernaskie and Corbett 2001; Hosp et al. 2011). In chapter 6, we defined one aspect of skill as a shift in the speed-accuracy trade-off. It has been shown in both nonhuman primates and humans that reward can lead to an instantaneous shift in the speed-accuracy trade-off for a task (Xu-Wilson et al. 2009; Manohar et al. 2015; Wong et al. 2015), a phenomenon that could potentially interact with larger practice-induced gains in skill (Hikosaka et al. 2002). Thus, the hope is that video games and virtual reality will have potentiating effects on motor recovery in humans in the same way that enriched environments do in animal models (Lewis and Rosie 2012). Almost everyone knows what a video game is, whether played on a smartphone, tablet console, or at an arcade. Virtual and augmented reality, however, are relatively new to the health care space, and the differences between them are not always appreciated. Virtual reality (VR) is defined in the *Oxford English Dictionary* as "the computer-generated simulation of a three-dimensional image or environment that can be interacted within a seemingly real or physical way by a person using special electronic equipment, such as a helmet with a screen inside or gloves fitted with sensors" (OED online 2017). Augmented reality (AR), in contrast, does not take the user out of the real world but instead superimposes a computer-generated image on a user's view of the real world. The recent Pokémon Go craze was a simple version of this (Lovelace 2016). In addition to the "fun" aspect of VR and AR, they also provide a means to control and weight multimodal feedback and allow patients to perform activities that might be unsafe in reality. At the current time, most of the high-quality VR experiences co-opt existing commercial platforms, although there is increasing interest in systems specifically designed for stroke patients. A recent survey of patients and therapists revealed that at the current time, patients prefer fun-oriented commercial games to existing rehabilitation-specific games, which are considered for the most part too boring (Hung et al.2016). This stance might change as game production values improve in the medical space. Patients also expressed a preference for large displays and physical sensors, which again speaks to the appeal of immersion. A recent Cochrane review reports that at the current time, the quality of the evidence for the effectiveness of VR and video gaming is low but nevertheless concludes that they may be beneficial in improving upper limb activity-level performance measures and global activities of daily living (ADLs) when used as an adjunct to usual care or when compared to equivalently dosed conventional therapy (Laver et al. 2015). As with other new technologies such as robotics, it will be necessary to have a better conceptual framework for what games and VR uniquely offer so that they go beyond just being a way to provide more of the same kind of therapy. Since the Cochrane review, a multicenter trial called Effectiveness of Virtual Reality Exercises in STroke rehabilitation (EVREST) was conducted in patients within three months of stroke, which compared two kinds of two-week intervention: gaming with the Nintendo Wii versus recreational therapy—playing cards, bingo, music, and so on (Saposnik et al. 2016). Both interventions were added on to conventional therapy. The result was that there was no difference in upper extremity outcome assessed with the Wolf Motor Function Test (WMFT) at four weeks. The authors of the study concluded that the type of task used in rehabilitation might not matter as long it is given at high intensity and is task specific. We have been here before—a new technology is introduced but there is no new conceptual framework to come along with it, and so its use is subsumed under an existing one. After all, giving regular task-oriented therapy in virtual reality is not likely to be that different from giving it in reality. A new paradigm— or, in the parlance of this chapter—a future direction in neurorehabilitation is to use VR for a qualitatively different rehabilitation experience, rather than a game to simply entice more of the same. Enriched environments for rodents provide full-field stimulation of social, physical, and perceptual needs, a far cry from repetitive reaching for sugar pellets. The promise of these new technologies has yet to be realized, but we suspect that we are on the cusp of a revolution in rehabilitation where the richness of the cognitive and motor experiences that the technology will allow, with limitless mappings between movements and their played-out consequences, will be unprecedented. This will require imaginative work on the part of clinicians; we must partake in designing these experiences, rather than waiting for the technology to tell us what to do. As an example of how training can be dramatically accelerated in controlled virtual environments, the Nissan academy takes people who are good at video games and, in a matter of months, turns them into professional drivers through use of a racing simulation game (Nissan GT Academy 2017). The critical point is that the training experience can be varied and intensified without risk to the participant and thereby allows freer and more playful exploration and experimentation (Lewis and Rosie 2012).

8.3 Direct Physiological Interventions

The book has laid heavy emphasis on behavioral and pharmacological interventions. As with the advent of deep brain stimulation (DBS) for movement disorders, there is growing interest in invasive brain stimulation methods, especially for the treatment of severe hemiparesis in chronic stroke. A very strong case, however, will need to be made that invasive approaches will have large effect sizes, given both their attendant greater risks and the less-than-impressive success so far for noninvasive brain stimulation methods. That is to say, the justification for invasive brain stimulation needs to be based on strong conceptual frameworks and high-quality preclinical evidence. Here we will offer a brief survey of current trends, but a few general points will be made beforehand. First, as we have discussed in chapters 3 and 6, improving on a specific task with training does not indicate general motor recovery—it just means that motor learning can occur in patients within the envelope of their performance deficit. By analogy, anyone can improve their handwriting in their nondominant hand with practice—this does not mean they have recovered from "nondominance." This confusion between learning and recovery is ubiquitous in studies in animal models and in patients after stroke. Second, some papers advocating for invasive brain stimulation often justify the approach based solely on the claim that reorganization has occurred perilesionally, as revealed by intracortical microstimulation. Again, as we point out in chapters 3 and 4, perilesional changes in cortical maps are neither necessary nor sufficient for recovery (i.e. these changes are not evidence for *functional* reorganization but may just be a short-term marker of learning). Arguably, these two kinds of misunderstanding; equating either learning or map plasticity with recovery, along with exaggerated enthusiasm for the results of noninvasive brain stimulation studies, led to a premature human cortical stimulation stroke trial, which unsurprisingly failed (Levy et al. 2016). Invasive brain stimulation is in its infancy, and it is too early to predict how useful these approaches will be clinically. There are other ways to perhaps benefit from the potentiating effects of stimulation paired with training, without the need for epidural or subdural electrodes. For example, vagus nerve stimulation (VNS) is approved by the Food and Drug Administration (FDA) for some types of refractory epilepsy and requires that a device be placed under the skin of the chest wall, and a wire runs from it to the vagus nerve in the neck. Studies in healthy rats show that pairing VNS with motor behavior, such as a lever press task, increased motor cortical map plasticity (Khodaparast et al. 2014). As we have cautioned at other points in this book, however, whether this kind of reorganization is of behavioral relevance remains an open question.

8.3.1   Cortical Stimulation

The general premise of this approach is to improve "recovery" from stroke through enhancement of plasticity in response to training by pairing subthreshold electrical stimulation of ipsilesional residual motor cortex, epidurally or subdurally, with rehabilitative interventions. Initial enthusiasm for this approach was generated by studies in rats (Adkins-Muir and Jones 2003; Teskey et al. 2003; Kleim et al. 2004) and in the squirrel monkey (Plautz et al. 2003) in the early 2000s. These animal studies, along with phase II results in patients (Brown et al. 2006; Levy et al. 2008), led to a phase III trial called Everest, in which epidural electrical stimulation centered on a motor cortical target identified with functional magnetic resonance imaging (fMRI) was combined with task-oriented upper limb motor rehabilitation (NCT00170716; Levy et al. 2016). The investigational and control groups received six weeks of therapy, which consisted of 2.5 hours of therapy a day divided into two sessions. The two primary outcome measures were the upper extremity Fugl-Meyer Motor Assessment (FMA) and the Arm Motor Ability Test (AMAT), the latter tests activity-level performance. There was no significant difference between groups on either primary measure at four weeks. More important, the changes in both groups, for both the impairment and activity measures, were very small. Once again, there are some hard lessons to be learned here. First, why would task-oriented training be expected to change impairment? Second, the effects of training alone on both primary outcome measures were very small, so modulating a very small effect is unlikely to lead to an effect large enough to justify an invasive procedure. This is a recurring theme in stroke recovery research—if the behavioral intervention alone barely has an impact, then modulating it is not likely to lead to a magical amplification effect.

8.3.2 Deep Cerebellar Stimulation

The core idea underlying this approach for stroke recovery is that deep brain stimulation of the dentate nucleus will lead to increased excitability and reorganization of the contralateral perilesional motor cortex via the dentatothalamocortical pathway, which in turn will lead to gains in motor function (Machado et al. 2013). A number of questionable assumptions form the basis for this rationale. First, that increased cortical excitability is behaviorally relevant—see Bestmann and Krakauer (2015) for a critique of this overly simplistic position. Second, that there is convincing evidence that crossed-cerebellar diaschisis is associated with worse stroke outcomes—there is not; see chapter 2. Third, that changes in perilesional cortical maps indicate that functionally significant reorganization has occurred (Cooperrider et al. 2014)—evidence does not support this; see chapter 2. The original work suggesting a potential stroke recovery benefit for deep cerebellar stimulation was done in rats, and reported only modest behavioral gains when chronic 20- or 30-Hz stimulation of the lateral cerebellar nucleus was given for several weeks with and without concomitant training. Indeed, the gains seen are comparable to what was seen with cortical stimulation protocols in animal models, which does not bode well for a human study that has just begun, called Electrical Stimulation of the Dentate Nucleus Area (EDEN, NCT02835443). The objective of this study ($n = 12$) is to document the safety and patient outcomes of electrical stimulation of the dentate nucleus area for the management of chronic, moderate, to severe upper extremity hemiparesis due to ischemic stroke. Results are expected in 2019.

8.3.3 Bidirectional Brain Computer Interfaces (BBCI)

The core recovery-related idea underlying interest in bidirectional brain computer interfaces (BCCI) is that these interfaces (so called "neurochips") can strengthen weak synaptic connections by inducing spike-timing dependent plasticity (STDP) (Fetz 2015). A particularly interesting example of behaviorally relevant synaptic strengthening through BCCI was provided by Eberhard Fetz and colleagues in an experiment in monkeys (Nishimura et al. 2013). Specifically, they demonstrated that they could strengthen the connections between the motor cortex and spinal cord by interposing a recurrent neural interface that delivered electrical stimuli at fixed delays to the spinal cord when triggered by the output of monosynaptic corticomotorneuronal cells during free behavior. This conditioning was provided to the monkeys overnight in their home cages. As we have seen in chapter 3, spontaneous recovery via the proportional recovery rule is predicated on the state of the corticospinal tract (CST). It would indeed be very exciting if spared corticospinal connections could be strengthened through the potentiating effect of a BCCI on spontaneous cortical firing. There is of course a great deal of work to be done before such an invasive approach could be considered in humans but the authors of the monkey BCCI study suggest that the findings of their invasive study in monkeys could potentially be clinically translatable if adapted to a noninvasive form (Nishimura et al. 2013). There is some irony in the fact that the potentially most promising approach for reducing impairment after stroke is the one that currently seems the one least likely to lead to an invasive human trial.

Overall, the promise of invasive approaches will ultimately depend on first developing behavioral interventions with a good effect size on their own. Physiological approaches would then modulate these new training paradigms. To hope that invasive approaches will substitute for or enhance inadequate behavioral interventions is not in our view a promising direction.

8.4 Reopening the Sensitive Period

The sensitive period is a compelling concept in stroke recovery. As discussed extensively in previous chapters, this is the period within which most spontaneous recovery occurs, and extends over the first three months after stroke in humans. In this sensitive period, motor, sensory, language, and other brain functions improve spontaneously in almost all stroke patients, even those with the most severe impairments, and do so in a way that is largely independent of the type, duration, and intensity of neurorehabilitative therapies (Duncan et al. 1994; Jørgensen et al. 1995; Zeiler and Krakauer 2013; van Kordelaar et al. 2014). A sensitive period for spontaneous recovery implies that stroke triggers innate short-lived biological repair processes in the adult brain. By definition the notion of a sensitive period after stroke means that the window of enhanced plasticity driving spontaneous biological recovery must close, after which the chronic post-stroke phase is entered. There are over seven million patients with chronic stroke in the United States and only very modest gains in motor function can be expected with current rehabilitation practice (Lo et al. 2010; Lang et al. 2016; Winstein et al. 2016b).

The sensitive period after stroke shares similarities with critical periods in brain development; the time during early brain development when brain function is uniquely plastic and most susceptible to shaping by environmental inputs. The earliest scientific description of the critical period is the famous finding of imprinting in young goslings to the mother goose in Konrad Lorenz's work (figure 8.1). David Hubel and Torsten Wiesel (1970) characterized the critical period in the visual cortex by detailing physiological and structural

Mechanisms for Opening the Critical Period in the Adult Brain

Mechanism	Signaling System	Evidence	Possible Therapy in Stroke
Perineuronal Net	chondroitin sulfate proteoglycans	Visual system	Chondroitinase ABC
Nogo	NogoA, NgR1	Visual and Auditory systems	Function-blocking antibodies
PirB	MHC1, Nogo, MAG, OMgp	Visual system	Function-blocking antibodies
SSRIs	cortical inhibition	Visual system	Fluoxetine
Cortical inhibition	GABA agonists, antagonists	Visual and Auditory systems	Tonic GABA antagonists, phasic GABA agonists
HDAC inhibition	epigenetic modification of gene expression	Visual and Auditory systems	Valproate
Cell Transplantation	excitatory/inhibitory balance	Visual system	Inhibitory neuron transplantation

Figure 8.1
Reopening the critical period and recovery in chronic stroke. Konrad Lorenz showed that geese have imprinted on him during the critical period of development. The critical period for maternal imprinting in goslings is 17 hours and in humans critical periods for sensory, motor and language functions extend to 2 years. Activating the sensitive period for stroke recovery in the chronic phase of stroke is metaphorically like the initiation of herding in adult geese. The table shows cellular and molecular systems that influence closure of the developmental critical period or have been shown to open the critical period in the adult. Note that most of the studies on this system are in the visual system, and have not been shown in other brain systems, such as other primary sensory areas (auditory, somatosensory) or the motor system (Hensch, 2015; Erzurumlu and Gaspar, 2012). Manipulation of the epigenetic state, such as histone acetylation (HDAC inhibition), promotes learning and memory and can alter the critical period. Current drugs that affect histone acetylation can affect many other signaling systems, alter gene expression in a widespread way, and have substantial cognitive and systemic side effects. These drugs, such as valproate, are not candidates for a selective or targeted effect in stroke and are not further discussed. HDAC=histone deacetylation.

changes occurring in response to altered retinal input. During this critical period, diminished visual input from one eye allows the input from the intact eye to drive connections in the visual cortex so that these connections control a greater number of neurons and produce a greater response to that eye's input. This plasticity to altered visual experience becomes greatly reduced beyond the neonatal period. This is the first feature of the critical period: substantial changes in brain circuits in response to altered environmental inputs.

The second feature of the critical period is an enhanced ability to revert to the normal pattern of organization if the reorganized brain circuits are reexposed to further environmental alteration. In the Hubel and Wiesel studies in the visual system, reducing visual input from one eye caused the inputs from the "good" eye to take over real estate in the cortex that was previously devoted to the now-impaired eye (first element of a critical period). When the impaired eye was then made the "good" eye and the "good" eye was made impaired, the initial shift in cortical responses could then be mostly reversed—as long as the young animal was still in the critical period. The initial plasticity phase of the critical period and the second recovery phase of the critical period are lost or very much reduced after the first weeks of life in most experimental animals. Critical periods have been described for other sensory modalities in animals and humans, and for cognitive processes such as language (Kandel et al. 2012). In humans, the critical period can extend for several years (Takesian et al. 2013).

The similarity between the critical period in brain development and the sensitive period in stroke appears obvious: after stroke there is an initial loss of brain structure or function followed by partial recovery that is greatest early after stroke and then diminishes over time. If stroke triggers a period of initially rapid neuroplasticity and recovery, it can be assumed that there must be a signal or signals that activate this sensitive period. The logical follow-on to this hypothesis is that if stroke activates an enhanced time-limited phase of plasticity and recovery then a second stroke, administered after this sensitive period has closed, should reopen it. This is indeed what occurred in mice that were subjected to paradigm that induced strokes sequentially (Zeiler et al. 2016) (see chapter 3). The clinical utility of such an approach is limited, but the concept of a stroke-initiated sensitive period is validated.

The similarity between the critical period in brain development and the sensitive period for recovery in stroke is only partial. There are multiple critical periods across different sensory and cognitive modalities in development (Kandel et al. 2012) and even between different circuit elements within the same sensory processing unit in the brain (Tagawa et al. 2005). There is a critical period in the motor cortex (Huntley 1997), but the cellular and molecular events that underlie it have not yet been determined. Even in the better characterized visual system, timing differs between the initial phase of the critical period, in which a sensory perturbation leads to widespread circuit change, and the recovery phase, in which additional sensory perturbations lead to partial or full restoration (Daw 2003).

These complexities in developmental critical periods should be borne in mind because it suggests that there will be similar heterogeneity in sensitive periods for recovery after stroke, for example with respect to cognitive modality (sensory, motor, language), by circuit (cortical, thalamic, striatal), and their time course.

8.4.1 Can We Reopen the Sensitive Period in Stroke?

A comparison of the recovery that is achieved in neurorehabilitation trials in chronic stroke with the level of recovery that occurs spontaneously in the sensitive period after stroke, suggests that motor recovery gains in chronic stroke are approximately one-tenth of those seen during the sensitive period (Lo et al. 2010; Duncan 2013; Lang et al. 2016; Winstein et al. 2016b). Furthermore, the small gains seen in these trials are produced in the unique patient cohorts within these trials: highly motivated, undergoing intense rehabilitative training regimens, and are not likely representative of most chronic stroke patients. Patients with chronic stroke are usually elderly, have physical impairments from their stroke, have accompanying comorbidities that affect physical activity (such as obesity and osteoarthritis), and so likely did not exercise much before their stroke. Thus, outside of the first three months after stroke—the sensitive period for recovery—chronic stroke patients not only do not show substantial further recovery but also usually show further declines in their level of impairment and disability (Kernan et al. 2005; Thorsén et al. 2006; Dhamoon et al. 2012). This decline is likely due to non-use, for example of the paretic arm, and overall inactivity.

If the molecular and cellular elements that trigger the sensitive period after stroke were identified, then new therapies that open brain plasticity in chronic stroke, or stimulate it even further in the sensitive period. A future goal of neurorehabilitative therapy is to reproduce the molecular events of the sensitive period in chronic stroke. Several molecular and cellular events reopen the critical period of brain plasticity in the adult and what we know about these events might provide clues as to how to initiate a sensitive period in patients with chronic stroke. These include blocking the signaling between the cell surface molecule Nogo and its receptors (Nogo receptor 1 and PirB), digesting perineuronal nets or altering chondroitin sulfate proteoglycans on these nets, changing the excitatory/inhibitory balance in cortex, and administering the drug fluoxetine. All of these manipulations have been shown to increase visual cortex plasticity as seen with monocular deprivation with the resultant changes in cortical maps in the non-deprived eye (Levelt and Hubner, 2012). Fluoxetine administration reopens a critical period for the visual cortex in mice (Vetencourt et al. 2008) and maintains it in early stages after stroke (Ng et al. 2015). It has also shown efficacy in enhancing motor recovery or possibly in treating post-stroke depression in the subacute phase of stroke in a small trial—the FLAME trial (see chapter 7) (Chollet et al. 2011). However, a recent animal study by Steven Zeiler and colleagues suggests that fluoxetine or other selective serotonin reuptake inhibitors (SSRIs) may not be effective in reopening the sensitive period in chronic stroke (Ng et al. 2015). Specifically they showed that fluoxetine had to be started early after stroke to have an effect on the sensitive period (see chapter 3). This result in stroke is consistent with a recent clinical trial that reported that fluoxetine could not reopen a critical period for recovery from amblyopia, thus failing to reproduce the promising effect first seen in rodents (Evaluate Group 2013).

Other manipulations such as altering myelin or growth inhibitory signaling systems are plausible as mechanistic approaches to opening up the sensitive period in chronic stroke. The myelin protein Nogo signals through a receptor system that includes the receptors paired immunoglobulin-like receptor B (PirB) and Nogo receptor 1 (NgR1). In addition to Nogo, NgR1 and PirB bind to chondroitin sulfate proteoglycans, which themselves increase around neurons during the critical period and function to terminate it (see below). These molecular relationships suggest that there is an "interactome" of molecules that signal through NgR1 and PirB, which develops at the end of the period of brain maturation to close the critical period and establish the adult state of limited brain plasticity (Akbik et al. 2012; Willi and Schwab 2013; Stephany et al. 2016). Blocking NgR1 signaling does indeed enhance cortical plasticity in local and corticospinal projections and increases recovery after stroke during the sensitive period in rodent stroke models (Li et al. 2010; Wahl et al. 2014). In white matter stroke, blocking NgR1 signaling enhances oligodendrocyte progenitor responses and myelin repair and motor recovery, even in the chronic period, months after stroke in rodent models (Sozmen et al. 2016). These studies suggest that targeting the NgR1/PirB interactome may re-establish a sensitive period for recovery in chronic stroke. A major hurdle to trying this approach is that currently there are no published small-molecule drugs that will selectively target this system.

A third approach to opening the sensitive period in chronic stroke involves manipulation of perineuronal nets. Perineuronal nets consist of chondroitin sulfate proteoglycans that form most densely around inhibitory interneurons in cortex. As these nets form, the critical period closes (Pizzorusso et al. 2002). Digestion of these nets with an enzyme that removes the side chains of the chondroitin sulfate proteoglycans (chondroitinase ABC) opens the critical period in the adult (Pizzorusso et al. 2002). Perineuronal nets are normally digested in the first week after stroke, most likely by local activation of matrix metalloproteinases. These nets re-form after one month (Carmichael et al. 2005), indicating that in the early weeks after stroke, pathological digestion of perineuronal nets produces a condition that is similar to the experimental digestion of perineuronal nets that reopens the critical period. In addition to the rather gross manipulation of enzymatic digestion of perineuronal nets, altering the sulfation pattern of the chondroitin sulfate side chains can also prolong the critical period (Miyata et al. 2012). There are two main limitations to attempting to manipulate the perineuronal net for reopening the sensitive period after stroke. First, the data in support of a role for perineuronal nets in the closure of the critical period are established only for one aspect of visual function in visual cortex and may involve a molecular mechanism that is unique to the visual system's critical period (Beurdeley et al. 2012). Second, as noted, the most common approach to altering perineuronal nets is enzymatic digestion of the chondroitin sulfate side chains with chondroitinase ABC. This is a large and highly immunogenic bacterial enzyme that is not suited for therapeutic development in stroke, and so other approaches would need to be developed.

A fourth approach to re-opening the sensitive period in chronic stroke is through manipulations of excitatory or inhibitory neuronal signaling. During the critical period, there is a two-phase evolution of cortical inhibition that first establishes the critical period with increased GABAergic intracortical inhibition, then closes it as this system further develops (Hensch 2005). Indeed, delivering GABAergic neurons via cell transplantation can open a critical period in the visual cortex in the adult (Southwell et al. 2010), suggesting that transplantation of cortical inhibitory neurons may be a way to improve recovery after stroke by altering local inhibition. After stroke, complex changes in GABAergic signaling may be exploited to prolong the sensitive period or enhance recovery in the chronic phases of stroke. Stroke enhances extrasynaptic GABAergic/tonic GABA signaling (Clarkson et al. 2010; Lake et al. 2015) (see chapter 4), and this can be blocked with tonic GABA signaling antagonists, such as $GABA_A$ $\alpha 5$ inverse agonists to enhance recovery (described in chapter 7). This approach is currently the focus of a clinical trial in subacute stroke (NCT02928393). More generally, blocking GABA signaling in stroke in the subacute phase (Alia et al. 2016) and manipulations that reduce intracortical GABAergic cell function (Ng et al. 2015) enhance motor performance after stroke. Conversely, stroke also increases synaptic levels of $GABA_A$ $\alpha 1$ receptors, which can be activated by a more selected agonist such as zolpidem to enhance recovery in experimental stroke (Hiu et al. 2015). Anecdotal clinical experience with patients on zolpidem in subacute and chronic stroke, which is commonly prescribed for insomnia, does not suggest a large effect on stroke recovery, but there have been no systematic studies of optimal dosing or of pairing the drug with behavioral intervention. Overall, preclinical evidence suggests that treatments designed to alter local inhibition in the brain after stroke hold some promise and could be translated to clinical trials with relative speed.

Reopening the sensitive period for stroke recovery in chronic stroke is a Promethean task of bringing light to a darkened world. This did not work out so well for Prometheus and may also come at a cost for the adult brain after stroke. Closing the critical period produces the adult brain state in which long-term memories and sensory perceptions are mostly invariant, allowing synaptic networks to be subject to only minor modifications that build learning or experience-dependent modifications onto a stabilized construct. Inducing greater brain plasticity may destabilize memories, learned responses, and motor and sensory representations or lead to abnormalities of malformed circuits, such as epilepsy. As an example, digesting perineuronal nets in the adult mouse can erase stored fear memories that are usually resistant to elimination (Gogolla et al. 2009). Congenital mutation or alterations of the molecular systems that end the critical period are associated with neurodevelopmental disorders, including schizophrenia and autism (Bitanihirwe et al. 2016). It is also clear that any induced state of heightened plasticity after stroke will need to be paired with the proper training regimens (see chapter 5).

8.5   Precision Health and Recovery after Stroke

Health care is increasingly focused on individual variations between patients and how these relate to diagnosis, treatment, and overall progression of disease. This focus on the unique features of patients in their health or disease state is termed *precision medicine*, *precision health*, or *personalized medicine*. Most commonly, precision health focuses on the genotype of the patient, but the focus could also be on epigenetics or in theory other large "omics" data sets, such as proteomics or metabolomics. The core idea is that a set of genes or molecules will determine if and how a patient develops a disease or responds to a drug treatment. Precision health as applied to motor recovery after stroke, or in the larger context neurorehabilitation, would comprise a focus on the genes and molecules that influence spontaneous motor recovery and the brain's response to rehabilitative therapies.

8.5.1   Pharmacogenetics in Personalized Health

The most immediate application of precision health in stroke is in the area of patient responses to drugs; genetic variation in the metabolism of drugs will affect the amount of drug that is acting on its target. A good example of this relates to clopidogrel, which is prescribed for primary or secondary prevention of stroke. It is an inhibitor of the platelet P2Y12 adenosine diphosphate (ADP) receptor that is metabolized by the liver in the cytochrome P450 system into its active form. Genetic variation in this P450 system in the *CYP2C19* gene reduces the metabolism to active drug. Carriers of the loss-of-function CYP2C19 alleles (CYP2C19*2 and CYP2C19*3) have been associated with poor responsiveness to clopidogrel (Shuldiner et al. 2009; Campo et al. 2011). This means that a patient with this allele in the P450 system who is prescribed clopidogrel will have an increased risk of stroke because the drug is less active compared to a patient without this allele. There are genetic variations associated with altered metabolism or effects for many of the drugs used to manage secondary occurrence of stroke, such as warfarin, statins, aspirin, and many blood pressure drugs (Zaiou and El Amri 2016). As the effects of drugs on stroke recovery become better characterized, precision medicine in this area will likely begin with genotyping of the patient for the purpose of establishing how they metabolize a candidate recovery drug. For example, individual variations in metabolism of selective serotonin reuptake inhibitors are likely to produce large variations in their drug levels (Chang et al. 2014). The SSRI fluoxetine is the only drug at the moment with a proven association with stroke recovery (Chollet et al. 2011), and information on the individual metabolism of this drug would be useful in dosing it or in making the decision to try a different medication.

8.5.2 Discovery Science: Precision Health and the Genetics of Stroke Recovery

Individual genetic variations in brain signaling systems may lead to differences in stroke recovery. Genetic variations in several genes have been associated with distinct outcomes after stroke, including brain-derived neurotrophic factor (BDNF), ApoE4, a serotonin transporter, and the dopamine metabolizing enzyme, catechol-O-methyltransferase (COMT). BDNF is strongly linked to stroke recovery in preclinical models of stroke (Berretta et al. 2014). Increases in BDNF, via exercise, task-specific activity, direct delivery, secondary elevations from drugs, or from bioengineered materials, promote recovery (see chapters 4 and 5). A naturally occurring human mutation in the *BDNF* gene reduces BDNF function; a valine is replaced with methionine at position 66, which results in 18 to 30 percent less activity-dependent secretion of the BDNF protein (Chen et al. 2004). This val/met mutation is associated with reduced levels of cortical excitability and motor learning, but conflicting evidence exists with regard to its relevance to recovery from brain injury (Pearson-Fuhrhop et al. 2012; Di Pino et al. 2016). For example, a study of patients with penetrating traumatic brain injury (TBI) found that the BDNF Met allele was correlated with *improved* executive function, compared with the Val allele (Krueger et al. 2011). The lack of a clear role for a BDNF loss-of-function mutation highlights the challenges in precision health for stroke recovery.

Several other gene systems have links to stroke recovery. The *apolipoprotein E* gene has four alleles, and the *ApoE4* allele is associated with poorer outcome in stroke, as seen in improvement in the National Institutes of Health Stroke Scale over the first month after stroke (Cramer and Procaccio 2012). This is not surprising as ApoE4 is also associated with worse outcomes in TBI, and it is associated with worse outcome in other CNS pathologies, such as in the development of Alzheimer disease (Pearson-Fuhrhop et al. 2012). A mutation with a single amino acid substitution in the dopamine metabolizing enzyme COMT disrupts dopamine clearance, an effect that produces an increase in synaptic levels of dopamine (Käenmäki et al. 2010). Although dopamine agonists and L-dopa have not panned out as validated drugs that enhance recovery after stroke (chapter 7), initial small-scale studies suggested that these drugs might promote recovery. One study screened patients admitted to a single inpatient neurorehabilitation facility and found that this valine-to-methionine amino acid change at position 108/158 of the COMT was associated with the severity of the initial functional deficit in stroke but did not have an effect on recovery (Liepert et al. 2013). Polymorphisms exist in other signaling systems that play a role in excitatory neuronal transmission and set cortical excitability levels, such as in the NMDA subunits NR1 and NR2B and TRPV1 (Mori et al. 2011; Mori et al. 2012). There are strong suggestions that variations in specific gene systems may alter how humans recover after stroke. Progress in this area will require large sample sizes and accurate outcome measures over time.

8.5.3   Epigenetics in Stroke Recovery

Epigenetic changes in the expression of genes refers to DNA methylation, histone modification, and RNA-based mechanisms (microRNA and other noncoding RNAs). Most epigenetic changes are acquired in life, although these can be passed on to subsequent generations (Sales et al. 2017). This adds a second level to the study of the individual variation in health and disease that is at heart of precision health: individuals may vary not just in their genetic code but also in how this code is modified over the lifetime of an individual. For example, smoking (Gao et al. 2015), diabetes (Raghuraman et al. 2016), and environmental exposures alter DNA methylation. These alterations affect gene expression and protein levels. Many studies have reported that epigenetic alterations can lead to the development of atherosclerosis (Zaiou and El Amri 2016). It is likely that epigenetic changes may affect signaling systems that play a role in recovery. The field of precision medicine will likely expand to the characterization of DNA methylation changes and histone methylation state. A major limitation in this expansion, however, will be that, unlike a mutation or allelic variation in the genome, which is present in all cells, an epigenetic change is cell specific, and the effect of this epigenetic change is related to the cell in which this change has occurred. For stroke recovery, most of the important cellular changes are in the brain or spinal cord. Epigenetic changes in cortical neurons or astrocytes will likely be most important for stroke recovery, but these cells are not amenable to large-scale epigenetic screening. Most human samples will come from saliva (good for whole-exome sequencing) or blood (for genotyping and for epigenetic studies). However, for stroke recovery, this is not going to be good enough. Since brain biopsies will almost certainly not become the new norm for precision health in neurology, specific epigenetic modifications might need to be established in peripheral blood cells and then validated as surrogate markers for alterations in the central nervous system, either through cell sampling in CSF or through detailed correlation in animal models.

8.6   Conclusions

Enhancing recovery after stroke literally means changing the brain. Spontaneous recovery in the first three months only achieves 70 percent of the maximum possible on average. In chronic stroke, further true recovery is minimal. Promising research studies in humans and in experimental models of stroke indicate that environmental enrichment, certain kinds of drug and cell therapy, and intense training may enhance recovery after stroke in both the acute and chronic phases. As can be seen from the quote that opens the introduction to this chapter, in Sherrington's estimation, the development of the human motor system has the cost of a limb-pair: a "land-locomotor" leg and "tool" arm. After stroke, there is an additional cost of damage to the motor system, and the land-locomotor and tool appendages do not fully recover. The future of stroke recovery will require that we give humans wings.

References

Abdollahi, F., E. D. Case Lazarro, M. Listenberger, R. V. Kenyon, M. Kovic, R. A. Bogey, D. Hedeker, B. D. Jovanovic, and J. L. Patton (2014). "Error augmentation enhancing arm recovery in individuals with chronic stroke: a randomized crossover design." *Neurorehabilitation and Neural Repair* 28(2): 120–128.

Ada, L., N. O'Dwyer, J. Green, W. Yeo, and P. Neilson (1996). "The nature of the loss of strength and dexterity in the upper limb following stroke." *Human Movement Science* 15(5): 671–687.

Adkins, D. L., T. Schallert, and L. B. Goldstein (2009). "Poststroke treatment." *Stroke* 40(1): 8–9.

Adkins-Muir, D. L., and T. A. Jones (2003). "Cortical electrical stimulation combined with rehabilitative training: enhanced functional recovery and dendritic plasticity following focal cortical ischemia in rats." *Neurological Research* 25(8): 780–788.

Akbik, F., W. B. J. Cafferty, and S. M. Strittmatter (2012). "Myelin associated inhibitors: A link between injury-induced and experience-dependent plasticity." *Experimental Neurology* 235(1): 43–52.

Alia, C., C. Spalletti, S. Lai, A. Panarese, S. Micera, and M. Caleo (2016). "Reducing GABAA-mediated inhibition improves forelimb motor function after focal cortical stroke in mice." *Scientific Reports* Nov. 29; 6:37823:1–15.

Alstermark, B., and T. Isa (2012). "Circuits for skilled reaching and grasping." *Annual Review of Neuroscience* 35: 559–578.

Alstermark, B., T. Isa, Y. Ohki, and Y. Saito (1999). "Disynaptic pyramidal excitation in forelimb motoneurons mediated via C3–C4 propriospinal neurons in the Macaca fuscata." *Journal of Neurophysiology* 82(6): 3580–3585.

Alstermark, B., and J. Ogawa (2004). "In vivo recordings of bulbospinal excitation in adult mouse forelimb motoneurons." *Journal of Neurophysiology* 92(3): 1958–1962.

Alstermark, B., J. Ogawa, and T. Isa (2004). "Lack of monosynaptic corticomotoneuronal EPSPs in rats: disynaptic EPSPs mediated via reticulospinal neurons and polysynaptic EPSPs via segmental interneurons." *Journal of Neurophysiology* 91(4): 1832–1839.

Alt Murphy, M., C. Willén, and K. S. Sunnerhagen (2012). "Movement kinematics during a drinking task are associated with the activity capacity level after stroke." *Neurorehabilitation and Neural Repair* 26(9): 1106–1115.

An, C., Y. Shi, P. Li, X. Hu, Y. Gan, R. A. Stetler, R. K. Leak, Y. Gao, B.-L. Sun, P. Zheng, and J. Chen (2014). "Molecular dialogs between the ischemic brain and the peripheral immune system: Dualistic roles in injury and repair." *Progress in Neurobiology* 115: 6–24.

Anderson, M. A., J. E. Burda, Y. Ren, Y. Ao, T. M. O'Shea, R. Kawaguchi, G. Coppola, B. S. Khakh, T. J. Deming, and M. V. Sofroniew (2016). "Astrocyte scar formation aids central nervous system axon regeneration. " *Nature* 532(7598):195–200.

Anderson, S., and P. Vanderhaeghen (2014). "Cortical neurogenesis from pluripotent stem cells: complexity emerging from simplicity." *Current Opinion in Neurobiology* 27: 151–157.

Anlezark, G. M., T. J. Crow, and A. P. Greenway (1973). "Impaired learning and decreased cortical norepinephrine after bilateral locus coeruleus lesions." *Science* 181(4100): 682–684.

Arai, A. C., and M. Kessler (2007). "Pharmacology of ampakine modulators: From AMPA receptors to synapses and behavior." *Current Drug Targets* 8(5): 583–602.

Armatas, C. A., J. J. Summers, and J. L. Bradshaw (1994). "Mirror movements in normal adult subjects." *Journal of Clinical and Experimental Neuropsychology* 16(3): 405–413.

Ashby, P., and D. Burke (1971). "Stretch reflexes in the upper limb of spastic man." *Journal of Neurology, Neurosurgery & Psychiatry* 34(6): 765–771.

Astrup, J., B. K. Siesjö, and L. Symon (1981). "Thresholds in cerebral ischemia—the ischemic penumbra." *Stroke* 12(6): 723–725.

The Avert Trial Collaboration Group (2015). "Efficacy and safety of very early mobilisation within 24 h of stroke onset (AVERT): A randomised controlled trial." *The Lancet* 386: 46–55.

Ay, H., E. M. Arsava, L. Gungor, D. Greer, A. B. Singhal, K. L. Furie, W. J. Koroshetz, and A. G. Sorensen (2008). "Admission international normalized ratio and acute infarct volume in ischemic stroke." *Annals of Neurology* 64(5): 499–506.

Baker, S. N. (2011). "The primate reticulospinal tract, hand function and functional recovery." *The Journal of Physiology* 589(23): 5603–5612.

Bakheit, A. M. O., A. F. Thilmann, A. B. Ward, W. Poewe, J. Wissel, J. Muller, R. Benecke, C. Collin, F. Muller, and C. D. Ward (2000). "A randomized, double-blind, placebo-controlled, dose-ranging study to compare the efficacy and safety of three doses of botulinum toxin type A (Dysport) with placebo in upper limb spasticity after stroke." *Stroke* 31(10): 2402–2406.

Bang, O. Y., J. S. Lee, P. H. Lee, and G. Lee (2005). "Autologous mesenchymal stem cell transplantation in stroke patients." *Annals of Neurology* 57(6): 874–882.

Barbay, S., E. J. Plautz, E. Zoubina, S. B. Frost, S. C. Cramer, and R. J. Nudo (2015). "Effects of postinfarct myelin-associated glycoprotein antibody treatment on motor recovery and motor map plasticity in squirrel monkeys." *Stroke* 46(6): 1620–1625.

Barde, Y.-A. (1989). "Trophic factors and neuronal survival." *Neuron* 2(6): 1525–1534.

Baron, J. C., M. G. Bousser, D. Comar, N. Duquesnoy, J. Sastre, and P. Castaigne (1981). "Crossed cerebellar diaschisis: A remote functional depression secondary to supratentorial infarction of man." *Journal of Cerebral Blood Flow & Metabolism* 1(suppl 1): S500–501.

Barreto, G. E., X. Sun, L. Xu, and R. G. Giffard (2011). "Astrocyte proliferation following stroke in the mouse depends on distance from the infarct." *PLoS ONE* 6(11): e27881.

Barth, A. L., M. McKenna, S. Glazewski, P. Hill, S. Impey, D. Storm, and K. Fox (2000). "Upregulation of cAMP response element-mediated gene expression during experience-dependent plasticity in adult neocortex." *The Journal of Neuroscience* 20(11): 4206–4216.

Bassani, S., A. Folci, J. Zapata, and M. Passafaro (2013). "AMPAR trafficking in synapse maturation and plasticity." *Cellular and Molecular Life Sciences* 70(23): 4411–4430.

Bath, P. M. W., N. Sprigg, and T. England (2013). "Colony stimulating factors (including erythropoietin, granulocyte colony stimulating factor and analogues) for stroke." *Cochrane Database of Systematic Reviews.* Jun 24 (6):CD005207: 1–44.

Bavelier, D., C. S. Green, A. Pouget, and P. Schrater (2012). "Brain plasticity through the life span: learning to learn and action video games." *Annual Review of Neuroscience* 35: 391–416.

Beck, C. H., and W. W. Chambers (1970). "Speed, accuracy, and strength of forelimb movement after unilateral pyramidotomy in rhesus monkeys." *Journal of Comparative and Physiological Psychology* 70(2, Pt. 2): 1–22.

Beebe, J. A., and C. E. Lang (2008). "Absence of a proximal to distal gradient of motor deficits in the upper extremity early after stroke." *Clinical Neurophysiology* 119(9): 2074–2085.

Beebe, J. A., and C. E. Lang (2009). "Relationships and responsiveness of six upper extremity function tests during the first 6 months of recovery after stroke." *Journal of Neurologic Physical Therapy: JNPT* 33(2): 96.

Beer, R. F., J. P. A. Dewald, and W. Z. Rymer (2000). "Deficits in the coordination of multijoint arm movements in patients with hemiparesis: evidence for disturbed control of limb dynamics." *Experimental Brain Research* 131(3): 305–319.

Beer, R. F., M. D. Ellis, B. G. Holubar, and J. Dewald (2007). "Impact of gravity loading on post-stroke reaching and its relationship to weakness." *Muscle & Nerve* 36(2): 242–250.

Benakis, C., L. Garcia-Bonilla, C. Iadecola, and J. Anrather (2015). "The role of microglia and myeloid immune cells in acute cerebral ischemia." *Frontiers in Cellular Neuroscience* 8:461:1–16.

Benecke, R., B. U. Meyer, and H. J. Freund (1991). "Reorganisation of descending motor pathways in patients after hemispherectomy and severe hemispheric lesions demonstrated by magnetic brain stimulation." *Experimental Brain Research* 83(2): 419–426.

Benjamin, E. J., M. J. Blaha, S. E. Chiuve, M. Cushman, S. R. Das, R. Deo, S. D. de Ferranti, J. Floyd, M. Fornage, C. Gillespie, C. R. Isasi, M. C. Jiménez, L. C. Jordan, S. E. Judd, D. Lackland, J. H. Lichtman, L. Lisabeth, S. Liu, C. T. Longenecker, R. H. Mackey, K. Matsushita, D. Mozaffarian, M. E. Mussolino, K. Nasir, R. W. Neumar, L. Palaniappan, D. K. Pandey, R. R. Thiagarajan, M. J. Reeves, M. Ritchey, C. J. Rodriguez, G. A. Roth, W. D. Rosamond, C. Sasson, A. Towfighi, C. W. Tsao, M. B. Turner, S. S. Virani, J. H. Voeks, J. Z. Willey, J. T. Wilkins, J. H. Y. Wu, H. M. Alger, S. S. Wong, and P. Muntner (2017). "Heart disease and stroke statistics—2017 update: A report from the American Heart Association." *Circulation.* Mar 7;135(10):e146–e603.

Bennett, D. J. (1994). "Stretch reflex responses in the human elbow joint during a voluntary movement." *The Journal of Physiology* 474(2): 339.

Benowitz, L. I., and S. T. Carmichael (2010). "Promoting axonal rewiring to improve outcome after stroke." *Neurobiology of Disease* 37(2): 259–266.

Benson, M. D., M. I. Romero, M. E. Lush, Q. R. Lu, M. Henkemeyer, and L. F. Parada (2005). "Ephrin-B3 is a myelin-based inhibitor of neurite outgrowth." *Proceedings of the National Academy of Sciences of the United States of America* 102(30): 10694–10699.

Bernhardt, J., H. Dewey, A. Thrift, and G. Donnan (2004). "Inactive and alone physical activity within the first 14 days of acute stroke unit care." *Stroke* 35(4): 1005–1009.

Berretta, A., Y. C. Tzeng, and A.N. Clarkson (2014). "Post-stroke recovery: the role of activity-dependent release of brain-derived neurotrophic factor." Expert Review of Neurotherapeutics 14(11):1335–1344.

Bestmann, S., and J. W. Krakauer (2015). "The uses and interpretations of the motor-evoked potential for understanding behaviour." *Experimental Brain Research* 233(3): 679–689.

Beurdeley, M., J. Spatazza, H. H. C. Lee, S. Sugiyama, C. Bernard, A. A. D. Nardo, T. K. Hensch, and A. Prochiantz (2012). "Otx2 binding to perineuronal nets persistently regulates plasticity in the mature visual cortex." *Journal of Neuroscience* 32(27): 9429–9437.

Bhakta, B. B., J. A. Cozens, J. M. Bamford, and M. A. Chamberlain (1996). "Use of botulinum toxin in stroke patients with severe upper limb spasticity." *Journal of Neurology, Neurosurgery & Psychiatry* 61(1): 30–35.

Bhakta, B. B., S. Hartley, I. Holloway, J. A. Couzens, G. A. Ford, D. Meads, C. M. Sackley, M. F. Walker, S. P. Ruddock, and A. J. Farrin (2014). "The DARS (Dopamine Augmented Rehabilitation in Stroke) trial: protocol for a randomised controlled trial of Co-careldopa treatment in addition to routine NHS occupational and physical therapy after stroke." *Trials* 15: 316.

Biernaskie, J., and D. Corbett (2001). "Enriched rehabilitative training promotes improved forelimb motor function and enhanced dendritic growth after focal ischemic injury." *Journal of Neuroscience* 21:5272–5280.

Biernaskie, J., G. Chernenko, and D. Corbett (2004). "Efficacy of rehabilitative experience declines with time after focal ischemic brain injury." *The Journal of Neuroscience* 24(5): 1245–1254.

Birkenmeier, R. L., E. M. Prager, and C. E. Lang (2010). "Translating animal doses of task-specific training to people with chronic stroke in 1-hour therapy sessions: a proof-of-concept study." *Neurorehabil Neural Repair* 24(7): 620–635.

Bitanihirwe, B. K. Y., S. A. Mauney, and T.-U. W. Woo (2016). "Weaving a net of neurobiological mechanisms in schizophrenia and unraveling the underlying pathophysiology." *Biological Psychiatry* 80(8): 589–598.

Bland, S. T., R. N. Pillai, J. Aronowski, J. C. Grotta, and T. Schallert (2001). "Early overuse and disuse of the affected forelimb after moderately severe intraluminal suture occlusion of the middle cerebral artery in rats." *Behavioural Brain Research* 126(1–2): 33–41.

Bland, S. T., T. Schallert, R. Strong, J. Aronowski, and J. C. Grotta (2000). "Early exclusive use of the affected forelimb after moderate transient focal ischemia in rats." *Stroke* 31(5): 1144–1152.

Blasi, F., M. J. Whalen, and C. Ayata (2015). "Lasting pure–motor deficits after focal posterior internal capsule white-matter infarcts in rats." *Journal of Cerebral Blood Flow & Metabolism* 35(6): 977–984.

Blicher, J. U., J. Near, E. Næss-Schmidt, C. J. Stagg, H. Johansen-Berg, J. F. Nielsen, L. Østergaard, and Y.-C. L. Ho (2015). "GABA levels are decreased after stroke and GABA changes during rehabilitation correlate with motor improvement." *Neurorehabilitation and Neural Repair* 29(3): 278–286.

Block, F., M. Dihné, and M. Loos (2005). "Inflammation in areas of remote changes following focal brain lesion." *Progress in Neurobiology* 75(5): 342–365.

Bluteau, J., S. Coquillart, Y. Payan, and E. Gentaz (2008). "Haptic guidance improves the visuo-manual tracking of trajectories." *PLoS One* 3(3): e1775.

Boake, C., E. A. Noser, T. Ro, S. Baraniuk, M. Gaber, R. Johnson, E. T. Salmeron, T. M. Tran, J. M. Lai, and E. Taub (2007). "Constraint-induced movement therapy during early stroke rehabilitation." *Neurorehabilitation and Neural Repair* 21(1): 14–24.

Bobath, B. (1990). *Adult hemiplegia: Evaluation and treatment.* 3rd Edition. Oxford: Butterworth-Heinemann.

Bobath, K., and B. Bobath (1984). "The neuro-developmental treatment." In: Scrutton D, editor. *Management of the Motor Disorders of Children with Cerebral Palsy.* Clinics in Developmental Medicine No. 90. London: Spastics International Medical Publications.

Boekhoorn, K., M. Joels, and P. J. Lucassen (2006). "Increased proliferation reflects glial and vascular-associated changes, but not neurogenesis in the presenile Alzheimer hippocampus." *Neurobiology of Disease* 24(1): 1–14.

Bogey, R., and T. G. Hornby (2007). "Gait Training strategies utilized in poststroke rehabilitation: Are we really making a difference?" *Topics in Stroke Rehabilitation* 14(6): 1–8.

Boltze, J., A. Arnold, P. Walczak, J. Jolkkonen, L. Cui, and D.-C. Wagner (2015). "The dark side of the force—constraints and complications of cell therapies for stroke." *Frontiers in Neurology* 6:155:1–21.

Bonita, R., and R. Beaglehole (1988). "Recovery of motor function after stroke." *Stroke* 19(12): 1497–1500.

Borod, J. C., P. M. Fitzpatrick, N. Helm-Estabrooks, and H. Goodglass (1989). "The relationship between limb apraxia and the spontaneous use of communicative gesture in aphasia." *Brain and Cognition* 10(1): 121–131.

Bosch, M., and Y. Hayashi (2012). "Structural plasticity of dendritic spines." *Current Opinion in Neurobiology* 22(3):383–388.

Bosse, F., K. Hasenpusch-Theil, P. Küry, and H. W. Müller (2006). "Gene expression profiling reveals that peripheral nerve regeneration is a consequence of both novel injury-dependent and reactivated developmental processes." *Journal of Neurochemistry* 96(5): 1441–1457.

Bourbonnais, D., S. V. Noven, K. M. Carey, and W. Z. Rymer (1989). "Abnormal spatial patterns of elbow muscle activation in hemiparetic human subjects." *Brain* 112(1): 85–102.

Bradnam, L. V., C. M. Stinear, P. A. Barber, and W. D. Byblow (2012). "Contralesional hemisphere control of the proximal paretic upper limb following stroke." *Cerebral Cortex* 22(11): 2662–2671.

Brashear, A., M. F. Gordon, E. Elovic, V. D. Kassicieh, C. Marciniak, M. Do, C.-H. Lee, S. Jenkins, and C. Turkel (2002). "Intramuscular injection of botulinum toxin for the treatment of wrist and finger spasticity after a stroke." *New England Journal of Medicine* 347(6): 395–400.

Brengelmann, J. C. (1958). "D-amphetamine and amytal. I. Effects on memory and expressive movement." *Journal of Mental Science* 104(434):153–159.

Brodal, A. (1973). "Self-observations and neuro-anatomical considerations after a stroke." *Brain* 96(4): 675–694.

Broderick, J. P. (2004). "William M. Feinberg lecture: Stroke therapy in the year 2025." *Stroke* 35(1): 205–211.

Brown, C. E., P. Li, J. D. Boyd, K. R. Delaney, and T. H Murphy (2007). "Extensive turnover of dendritic spines and vascular remodeling in cortical tissues recovering from stroke." *Journal of Neuroscience* 27(15):4101–4109.

Brown, C. E., K. Aminoltejari, H. Erb, I. R. Winship, and T. H. Murphy (2009). "In vivo voltage-sensitive dye imaging in adult mice reveals that somatosensory maps lost to stroke are replaced over weeks by new structural and functional circuits with prolonged modes of activation within both the peri-infarct zone and distant sites." *The Journal of Neuroscience* 29(6): 1719–1734.

Brown, C. E., J. D. Boyd, and T. H. Murphy (2010). "Longitudinal in vivo imaging reveals balanced and branch-specific remodeling of mature cortical pyramidal dendritic arbors after stroke." *Journal of Cerebral Blood Flow & Metabolism* 30(4): 783–791.

Brown, C. E., C. Wong, and T. H. Murphy (2008). "Rapid morphologic plasticity of peri-infarct dendritic spines after focal ischemic stroke." *Stroke* 39(4): 1286–1291.

Brown, J. A., H. L. Lutsep, M. Weinand, and S. C. Cramer (2006). "Motor cortex stimulation for the enhancement of recovery from stroke: A prospective, multicenter safety study." *Neurosurgery* 58(3): 464–473.

Brown-Séquard, C. E. (1875). "On localization of functions in the brain." *Boston Medical and Surgical Journal* 93:119–124.

Brumm, A. J., and S. T. Carmichael (2012). "Not just a rush of blood to the head." *Nature Medicine* 18(11): 1609–1610.

Brunnstrom, S. (1966). "Motor testing procedures in hemiplegia: based on sequential recovery stages." *Physical Therapy* 46(4): 357–375.

Buchkremer-Ratzmann, I., M. August, G. Hagemann, and O. W. Witte (1996). "Electrophysiological transcortical diaschisis after cortical photothrombosis in rat brain." *Stroke* 27(6): 1105–1111.

Buchkremer-Ratzmann, I., and O. W. Witte (1997). "Extended brain disinhibition following small photo-thrombotic lesions in rat frontal cortex." *Neuroreport* 8(2): 519–522.

Buetefisch, C. M. (2015). "Role of the contralesional hemisphere in post-stroke recovery of upper extremity motor function." *Frontiers in Neurology* 6:214:1–10

Burke, R. E. (2003). "Postnatal developmental programmed cell death in dopamine neurons." *Annals of the New York Academy of Sciences* 991:69–79.

Burke, D., C. J. Andrews, and J. W. Lance (1972). "Tonic vibration reflex in spasticity, Parkinson's disease, and normal subjects." *Journal of Neurology, Neurosurgery & Psychiatry* 35(4): 477–486.

Burke, D., L. Knowles, C. Andrews, and P. Ashby (1972). "Spasticity, decerebrate rigidity and the clasp-knife phenomenon: an experimental study in the cat." *Brain: A Journal of Neurology* 95(1):31–48.

Burke Quinlan, E., L. Dodakian, J. See, A. McKenzie, V. Le, M. Wojnowicz, B. Shahbaba and S. C. Cramer (2015). "Neural function, injury, and stroke subtype predict treatment gains after stroke." *Annals of Neurology* 77(1): 132–145.

Burkhardt, R. W. (2005). *Patterns of behavior: Konrad Lorenz, Niko Tinbergen, and the founding of ethology.* Chicago: University of Chicago Press.

Burne, J. A., V. L. Carleton, and N. J. O'Dwyer (2005). "The spasticity paradox: movement disorder or disorder of resting limbs?" *Journal of Neurology, Neurosurgery, and Psychiatry* 76(1): 47–54.

Bütefisch, C. M., B. C. Davis, D. Sawaki, D. Waldvogel, J. Classen, L. Kopylev, and L. G. Cohen (2002). "Modulation of use-dependent plasticity by d-amphetamine." *Annals of Neurology* 51(1): 59–68.

Button, K. S., J. P. A. Ioannidis, C. Mokrysz, B. A. Nosek, J. Flint, E. S. J. Robinson, and M. R. Munafò (2013). "Power failure: Why small sample size undermines the reliability of neuroscience." *Nature Reviews Neuroscience* 14(5): 365–376.

Buxbaum, L. J., S. H. Johnson-Frey, and M. Bartlett-Williams (2005). "Deficient internal models for planning hand–object interactions in apraxia." *Neuropsychologia* 43(6): 917–929.

Byblow, W. D., C. M. Stinear, P. A. Barber, M. A. Petoe, and S. J. Ackerley (2015). "Proportional recovery after stroke depends on corticomotor integrity." *Annals of Neurology* 78(6): 848–859.

Cahan, P., and G. Q. Daley (2013). "Origins and implications of pluripotent stem cell variability and heterogeneity." *Nature Reviews Molecular Cell Biology* 14(6): 357–368.

Campo, G., M. Miccoli, M. Tebaldi, J. Marchesini, L. Fileti, M. Monti, M. Valgimigli, and R. Ferrari (2011). "Genetic determinants of on-clopidogrel high platelet reactivity." *Platelets* 22(6): 399–407.

Capetian, P., M. G. Pauly, L. M. Azmitia, and C. Klein (2014). "Striatal cholinergic interneurons in isolated generalized dystonia—rationale and perspectives for stem cell-derived cellular models." *Frontiers in Cellular Neuroscience* 8:205:1–9.

Caracciolo, L., Y. Sano, S. J. Alcino, and S. T. Carmichael (2012) "Role of CREB in neural repair and recovery after focal stroke." Program No. 350.06/O3 2012 *Neuroscience Meeting Planner*. New Orleans, LA: Society for Neuroscience.

Caracciolo, L., M. Marosi, Y. Sano, A. J. Silva, C. Portera-Cailliau, and S. T. Carmichael (2014). "CREB facilitates recovery and reorganization of forelimb sensory map after a photothrombotic stroke in mice." Program No. 716.23/DD4 2014 *Neuroscience Meeting Planner*. Washington D.C.: Society for Neuroscience.

Caracciolo, L., A. Hamade, A. Bulfone, A. Guzner, Y. Sano, A. J. Silva, and S. T. Carmichael (2015). "CREB/ DREADD system: switching on/off recovery of motor function after stroke." Program No. 324.25/N18 2015 *Neuroscience Meeting Planner*. Chicago, IL: Society for Neuroscience.

Carmichael, S. T. (2012). "Brain excitability in stroke: The yin and yang of stroke progression." *Archives of Neurology* 69(2): 161–167.

Carmichael, S. T. (2014). "Chapter 14: Cellular mechanisms of plasticity after brain lesons." in: *Textbook of Neural Repair and Rehabilitation: Volume 1, Neural Repair and Plasticity*. Eds: M. Selzer, S. Clarke, L. Cohen, G. Kwakkel, and R. Miller. Cambridge University Press, New York, USA.

Carmichael, S. T. (2016a). "The 3 Rs of stroke biology: radial, relayed, and regenerative." *Neurotherapeutics* 13(2): 348–359.

Carmichael, S. T. (2016b). "Emergent properties of neural repair: Elemental biology to therapeutic concepts." *Annals of Neurology* 79(6): 895–906.

Carmichael, S. T., I. Archibeque, L. Luke, T. Nolan, J. Momiy, and S. Li (2005). "Growth-associated gene expression after stroke: Evidence for a growth-promoting region in peri-infarct cortex." *Experimental Neurology* 193(2): 291–311.

Carmichael, S. T., and M.-F. Chesselet (2002). "Synchronous neuronal activity is a signal for axonal sprouting after cortical lesions in the adult." *Journal of Neuroscience* 22(14): 6062–6070.

Carmichael, S. T., K. Tatsukawa, D. Katsman, N. Tsuyuguchi, and H. I. Kornblum (2004). "Evolution of diaschisis in a focal stroke model." *Stroke* 35(3): 758–763.

Carmichael, S. T., L. Wei, C. M. Rovainen, and T. A. Woolsey (2001). "New patterns of intracortical projections after focal cortical stroke." *Neurobiology of Disease* 8(5): 910–922.

Carr, J. H. (1987). *Movement science: Foundations for physical therapy in rehabilitation*. Rockville, Maryland: Aspen Publishers.

Carrera, E., and G. Tononi (2014). "Diaschisis: Past, present, future." *Brain* 137(9): 2408–2422.

Castro-Alamancos, M. A., and J. Borrell (1995). "Functional recovery of forelimb response capacity after forelimb primary motor cortex damage in the rat is due to the reorganization of adjacent areas of cortex." *Neuroscience* 68(3): 793–805.

Celio, M. R., R. Spreafico, S. De Biasi, and L. Vitellaro-Zuccarello (1998). "Perineuronal nets: Past and present." *Trends in Neurosciences* 21(12): 510–515.

Cenci, M. A., I. Q. Whishaw, and T. Schallert (2002). "Animal models of neurological deficits: How relevant is the rat?" *Nature Reviews Neuroscience* 3(7): 574–579.

Centonze, D., S. Rossi, A. Tortiglione, B. Picconi, C. Prosperetti, V. De Chiara, G. Bernardi, and P. Calabresi (2007). "Synaptic plasticity during recovery from permanent occlusion of the middle cerebral artery." *Neurobiology of Disease* 27(1): 44–53.

Chaboub, L. S., and B. Deneen (2012). "Developmental origins of astrocyte heterogeneity: The final frontier of CNS development." *Developmental Neuroscience* 34(5): 379–388.

Chang, M., G. Tybring, M.-L. Dahl, and J. D. Lindh (2014). "Impact of cytochrome P450 2C19 polymorphisms on citalopram/escitalopram exposure: A systematic review and meta-analysis." *Clinical Pharmacokinetics* 53(9): 801–811.

Cheetham, C. E. J., M. S. L. Hammond, R. McFarlane, and G. T. Finnerty (2008). "Altered sensory experience induces targeted rewiring of local excitatory connections in mature neocortex." *The Journal of Neuroscience* 28(37): 9249–9260.

Chen, P., D. E. Goldberg, B. Kolb, M. Lanser, and L. I. Benowitz (2002). "Inosine induces axonal rewiring and improves behavioral outcome after stroke." *Proceedings of the National Academy of Sciences* 99(13): 9031–9036.

Chen, Z., L. C. Trotman, D. Shaffer, H.-K. Lin, Z. A. Dotan, M. Niki, J. A. Koutcher, H. I. Scher, T. Ludwig, W. Gerald, C. Cordon-Cardo, and P. Paolo Pandolfi (2005). "Crucial role of p53-dependent cellular senescence in suppression of Pten-deficient tumorigenesis." *Nature* 436(7051): 725–730.

Chen, Z.-Y., P. D. Patel, G. Sant, C.-X. Meng, K. K. Teng, B. L. Hempstead, and F. S. Lee (2004). "Variant brain-derived neurotrophic factor (BDNF) (Met66) alters the intracellular trafficking and activity-dependent secretion of wild-type BDNF in neurosecretory cells and cortical neurons." *Journal of Neuroscience* 24(18): 4401–4411.

Chollet, F., J. Tardy, J.-F. Albucher, C. Thalamas, E. Berard, C. Lamy, Y. Bejot, S. Deltour, A. Jaillard, P. Niclot, B. Guillon, T. Moulin, P. Marque, J. Pariente, C. Arnaud, and I. Loubinoux (2011). "Fluoxetine for motor recovery after acute ischaemic stroke (FLAME): A randomised placebo-controlled trial." *The Lancet Neurology* 10(2): 123–130.

Chu, X., X. Fu, L. Zou, C. Qi, Z. Li, Y. Rao, and K. Ma (2007). "Oncosis, the possible cell death pathway in astrocytes after focal cerebral ischemia." *Brain Research* 1149: 157–164.

Cicinelli, P., P. Pasqualetti, M. Zaccagnini, R. Traversa, M. Oliveri, and P. M. Rossini (2003). "Interhemispheric asymmetries of motor cortex excitability in the postacute stroke stage a paired-pulse transcranial magnetic stimulation study." *Stroke* 34(11): 2653–2658.

Cirstea, M. C., and M. F. Levin (2000). "Compensatory strategies for reaching in stroke." *Brain: A Journal of Neurology* 123(Pt. 5): 940–953.

Clarke, P. G. H. (1985). "Neuronal death in the development of the vertebrate nervous system." *Trends in Neurosciences* 8: 345–349.

Clarkson, A. N. (2015). "Combined ampakine and BDNF treatments enhance poststroke functional recovery in aged mice via AKT-CREB signaling." *Journal of Cerebral Blood Flow and Metabolism* 35(8): 1272–1279.

Clarkson, A. N., B. S. Huang, S. E. MacIsaac, I. Mody, and S. T. Carmichael (2010). "Reducing excessive GABA-mediated tonic inhibition promotes functional recovery after stroke." *Nature* 468(7321): 305–309.

Clarkson, A. N., H. E. López-Valdés, J. J. Overman, A. C. Charles, K. C. Brennan, and S. T. Carmichael (2013). "Multimodal examination of structural and functional remapping in the mouse photothrombotic stroke model." *Journal of Cerebral Blood Flow & Metabolism* 33(5): 716–723.

Clarkson, A. N., J. J. Overman, S. Zhong, R. Mueller, G. Lynch, and S. T. Carmichael (2011). "AMPA receptor-induced local brain-derived neurotrophic factor signaling mediates motor recovery after stroke." *Journal of Neuroscience* 31(10): 3766–3775.

Classen, J., J. Liepert, S. P. Wise, M. Hallett, and L. G. Cohen (1998). "Rapid plasticity of human cortical movement representation induced by practice." *Journal of Neurophysiology* 79(2): 1117–1123.

Classen, J., A. Schnitzler, F. Binkofski, K. J. Werhahn, Y. S. Kim, K. R. Kessler, and R. Benecke (1997). "The motor syndrome associated with exaggerated inhibition within the primary motor cortex of patients with hemiparetic." *Brain* 120(4): 605–619.

Cogger, K., and M. C. Nostro (2014). "Recent advances in cell replacement therapies for the treatment of type 1 diabetes." *Endocrinology* 156(1): 8–15.

Colebatch, J. G., and S. C. Gandevia (1989). "The distribution of muscular weakness in upper motor neuron lesions affecting the arm." *Brain* 112(3): 749–763.

Conover, J. C., and K. L. Todd (2016). "Development and aging of a brain neural stem cell niche." *Experimental Gerontology* epub ahead of print.

Cook, D. J., C. Nguyen, H. N. Chun, I. L Llorente, A. S. Chiu, M. Machnicki, T. I. Zarembinski, and S. T. Carmichael (2016). "Hydrogel-delivered brain-derived neurotrophic factor promotes tissue repair and recovery after stroke." *Journal of Cerebral Blood Flow & Metabolism* 37(3): 1030–1045.

Cooperrider, J., H. Furmaga, E. Plow, H.-J. Park, Z. Chen, G. Kidd, K. B. Baker, J. T. Gale, and A. G. Machado (2014). "Chronic deep cerebellar stimulation promotes long-term potentiation, microstructural plasticity, and reorganization of perilesional cortical representation in a rodent model." *Journal of Neuroscience* 34(27): 9040–9050.

Cortes, J. C., J. Goldsmith, M. D. Harran, J. Xu, N. Kim, H. Schambra, A. R. Luft, P. A. Celnik, J. W. Krakauer, and T. Kitago (2017). "A short and distinct time window for recovery of arm motor control early after stroke revealed with a global measure of trajectory kinematics." *Neurorehabilitation and Neural Repair* epub ahead of print.

Coulthard, M. G., M. Morgan, T. M. Woodruff, T. V. Arumugam, S. M. Taylor, T. C. Carpenter, M. Lackmann, and A. W. Boyd (2012). "Eph/Ephrin signaling in injury and inflammation." *The American Journal of Pathology* 181(5): 1493–1503.

Cramer, S. C. (2015). "Drugs to enhance motor recovery after stroke." *Stroke* 46(10): 2998–3005.

Cramer, S. C., B. Abila, N. E. Scott, M. Simeoni, and L. A. Enney (2013). "Safety, pharmacokinetics, and pharmacodynamics of escalating repeat doses of GSK249320 in patients with stroke." *Stroke* 44(5): 1337–1342.

Cramer, S. C., and M. Chopp (2000). "Recovery recapitulates ontogeny." *Trends in Neurosciences* 23(6): 265–271.

Cramer, S. C., B. H. Dobkin, E. A. Noser, R. W. Rodriguez, and L. A. Enney (2009). "Randomized, placebo-controlled, double-blind study of ropinirole in chronic stroke." *Stroke* 40(9): 3034–3038.

Cramer, S. C., and V. Procaccio (2012). "Correlation between genetic polymorphisms and stroke recovery: Analysis of the GAIN Americas and GAIN International Studies." *European Journal of Neurology* 19(5): 718–724.

Cyranoski, D. (2012). "Canada approves stem cell product." *Nature Biotechnology* 30(7): 571–571.

Dancause, N., S. Barbay, S. B. Frost, E. J. Plautz, D. Chen, E. V. Zoubina, A. M. Stowe, and R. J. Nudo (2005). "Extensive cortical rewiring after brain injury." *Journal of Neuroscience* 25(44): 10167–10179.

Dancause, N., B. Touvykine, and B. K. Mansoori (2015). Inhibition of the contralesional hemisphere after stroke: reviewing a few of the building blocks with a focus on animal models. *Progress in Brain Research* 218: 361–387.

Danilov, C. A., and O. Steward (2015). "Conditional genetic deletion of PTEN after a spinal cord injury enhances regenerative growth of CST axons and motor function recovery in mice." *Experimental Neurology* 266: 147–160.

Darling, W. G., M. A. Pizzimenti, and R. J. Morecraft (2011). "Functional recovery following motor cortex lesions in non-human primates: Experimental implications for human stroke patients." *Journal of Integrative Neuroscience* 10(3): 353–384.

Das, T. K., and D. M. Park (1989). "Effect of treatment with botulinum toxin on spasticity." *Postgraduate Medical Journal* 65(762): 208–210.

Davidson, A. G., and J. A. Buford (2004). "Motor outputs from the primate reticular formation to shoulder muscles as revealed by stimulus-triggered averaging." *Journal of Neurophysiology* 92(1): 83–95.

Davidson, A. G., M. H. Schieber, and J. A. Buford (2007). "Bilateral spike-triggered average effects in arm and shoulder muscles from the monkey pontomedullary reticular formation." *The Journal of Neuroscience* 27(30): 8053–8058.

Davis, S., and G. A. Donnan (2014). "Time is penumbra: Imaging, selection and outcome." *Cerebrovascular Diseases* 38(1): 59–72.

Daw, N. W. (2003). "Critical periods in the visual system." in: *Neurobiology of Infant Vision,* ed. Hopkins, B. and Johnson, S. P.W, Westport, CT, USA.

De Renzi, E., and F. Lucchelli (1988). "Ideational apraxia." *Brain* 111(5): 1173–1185.

de Rugy, A., G. Loeb, and T. Carroll (2013). "Are muscle synergies useful for neural control?" *Frontiers in Computational Neuroscience* 7: 19.

Deguchi, K., M. Takaishi, T. Hayashi, A. Oohira, S. Nagotani, F. Li, G. Jin, I. Nagano, M. Shoji, M. Miyazaki, K. Abe, and N.-h. Huh (2005). "Expression of neurocan after transient middle cerebral artery occlusion in adult rat brain." *Brain Research* 1037(1–2): 194–199.

Denny-Brown, D. (1964). "The extrapyramidal system and postural mechanisms." *Clinical Pharmacology & Therapeutics* 5(6, Pt. 2): 812–827.

DeVetten, G., S. B. Coutts, M. D. Hill, M. Goyal, M. Eesa, B. O'Brien, A. M. Demchuk, and A. Kirton (2010). "Acute corticospinal tract Wallerian degeneration is associated with stroke outcome." *Stroke* 41(4): 751–756.

Dewald, J. P., P. S. Pope, J. D. Given, T. S. Buchanan, and W. Z. Rymer (1995). "Abnormal muscle coactivation patterns during isometric torque generation at the elbow and shoulder in hemiparetic subjects." *Brain: A Journal of Neurology* 118(Pt. 2): 495–510.

Dhamoon, M. S., Y. P. Moon, M. C. Paik, R. L. Sacco, and M. S. V. Elkind (2012). "Trajectory of functional decline before and after ischemic stroke." *Stroke* 43(8): 2180–2184.

Di Filippo, M., A. Tozzi, C. Costa, V. Belcastro, M. Tantucci, B. Picconi, and P. Calabresi (2008). "Plasticity and repair in the post-ischemic brain." *Neuropharmacology* 55(3): 353–362.

Di Pino, G., G. Pellegrino, G. Assenza, F. Capone, F. Ferreri, D. Formica, F. Ranieri, M. Tombini, U. Ziemann, J. C. Rothwell, and V. Di Lazzaro (2014). "Modulation of brain plasticity in stroke: A novel model for neurorehabilitation." *Nature Reviews Neurology* 10(10): 597–608.

Di Pino, G., G. Pellegrino, F. Capone, G. Assenza, L. Florio, E. Falato, F. Lotti, and V. Di Lazzaro (2016). "Val66Met BDNF polymorphism implies a different way to recover from stroke rather than a worse overall recoverability." *Neurorehabilitation and Neural Repair* 30(1): 3–8.

Dickson, T. C., R. S. Chung, G. H. McCormack, J. A. Staal, and J. C. Vickers (2007). "Acute reactive and regenerative changes in mature cortical axons following injury." *Neuroreport* 18(3): 283–288.

Diedrichsen, J., O. White, D. Newman, and N. Lally (2010). "Use-dependent and error-based learning of motor behaviors." *Journal of Neuroscience* 30(15): 5159–5166.

Diedrichsen, J., T. Wiestler, and J. W. Krakauer (2013). "Two distinct ipsilateral cortical representations for individuated finger movements." *Cerebral Cortex (New York, N.Y.: 1991)* 23(6): 1362–1377.

Dijkhuizen, R. M., A. B. Singhal, J. B. Mandeville, O. Wu, E. F. Halpern, S. P. Finklestein, B. R. Rosen, and E. H. Lo (2003). "Correlation between brain reorganization, ischemic damage, and neurologic status after transient focal cerebral ischemia in rats: A functional magnetic resonance imaging study." *Journal of Neuroscience* 23(2): 510–517.

Dimmeler, S., S. Ding, T. A. Rando, and A. Trounson (2014). "Translational strategies and challenges in regenerative medicine." *Nature Medicine* 20(8): 814–821.

Ding, J., J.-Z. Yu, Q.-Y. Li, X. Wang, C.-Z. Lu, and B.-G. Xiao (2009). "Rho kinase inhibitor Fasudil induces neuroprotection and neurogenesis partially through astrocyte-derived G-CSF." *Brain, Behavior, and Immunity* 23(8): 1083–1088.

Dobkin, B. H. (2003). *The clinical science of neurologic rehabilitation.* Oxford, UK: Oxford University Press.

Dobkin, B. H., and S. T. Carmichael (2016). "The specific requirements of neural repair trials for stroke." *Neurorehabilitation and Neural Repair* 30(5): 470–478.

Dong, Y., T. Green, D. Saal, H. Marie, R. Neve, E. J. Nestler, and R. C. Malenka (2006). "CREB modulates excitability of nucleus accumbens neurons." *Nature Neuroscience* 9(4): 475–477.

Doughty, C., J. Wang, W. Feng, D. Hackney, E. Pani, and G. Schlaug (2016). "Detection and predictive value of fractional anisotropy changes of the corticospinal tract in the acute phase of a stroke." *Stroke* 47(6): 1520–1526.

Doya, K. (1999). "What are the computations of the cerebellum, the basal ganglia and the cerebral cortex?" *Neural Networks* 12(7): 961–974.

Dromerick, A. W., M. A. Edwardson, D. F. Edwards, M. L. Giannetti, J. Barth, K. P. Brady, E. Chan, M. T. Tan, I. Tamboli, R. Chia, M. Orquiza, R. M. Padilla, A. K. Cheema, M. E. Mapstone, M. S. Fiandaca, H. J. Federoff, and E. L. Newport (2015). "Critical periods after stroke study: translating animal stroke recovery experiments into a clinical trial." *Frontiers in Human Neuroscience* 9: 231.

Dromerick, A. W., C. E. Lang, R. L. Birkenmeier, J. M. Wagner, J. P. Miller, T. O. Videen, W. J. Powers, S. L. Wolf, and D. F. Edwards (2009). "Very Early Constraint-Induced Movement during Stroke Rehabilitation (VECTORS): A single-center RCT." *Neurology* 73(3): 195–201.

Dudek, F. E., and T. P. Sutula (2007). Epileptogenesis in the dentate gyrus: A critical perspective. *Progress in Brain Research* 163: 755–773.

Duering, M., R. Righart, E. Csanadi, E. Jouvent, D. Hervé, H. Chabriat, and M. Dichgans (2012). "Incident subcortical infarcts induce focal thinning in connected cortical regions." *Neurology* 79(20): 2025–2028.

Dum, R. P., and P. L. Strick (1991). "The origin of corticospinal projections from the premotor areas in the frontal lobe." *Journal of Neuroscience* 11(3): 667–689.

Duncan, P. W. (2013). "Outcome measures in stroke rehabilitation." Handbook of Clinical Neurology 110:105–111.

Duncan, P. W., L. B. Goldstein, R. D. Horner, P. B. Landsman, G. P. Samsa, and D. B. Matchar (1994). "Similar motor recovery of upper and lower extremities after stroke." *Stroke* 25(6): 1181–1188.

Duncan, P. W., L. B. Goldstein, D. Matchar, G. W. Divine, and J. Feussner (1992). "Measurement of motor recovery after stroke: Outcome assessment and sample size requirements." *Stroke* 23(8): 1084–1089.

Duncan, P. W., S. M. Lai, and J. Keighley (2000). "Defining post-stroke recovery: Implications for design and interpretation of drug trials." *Neuropharmacology* 39(5): 835–841.

Duncan, P. W., M. Propst, and S. G. Nelson (1983). "Reliability of the Fugl-Meyer assessment of sensorimotor recovery following cerebrovascular accident." *Physical Therapy* 63(10): 1606–1610.

Dusart, I., A. Ghoumari, R. Wehrle, M. P. Morel, L. Bouslama-Oueghlani, E. Camand, and C. Sotelo (2005). "Cell death and axon regeneration of Purkinje cells after axotomy: Challenges of classical hypotheses of axon regeneration." *Brain Research Reviews* 49(2): 300–316.

Eckert, M. A. (2013). "Evidence for high translational potential of mesenchymal stromal cell therapy to improve recovery from ischemic stroke." *Journal of Cerebral Blood Flow and Metabolism* 33(9): 1322–1334.

Egorov, A. V., and A. Draguhn (2013). "Development of coherent neuronal activity patterns in mammalian cortical networks: Common principles and local hetereogeneity." *Mechanisms of Development* 130(6–8): 412–423.

Ehrenreich, H., K. Weissenborn, H. Prange, D. Schneider, C. Weimar, K. Wartenberg, P. D. Schellinger, M. Bohn, H. Becker, M. Wegrzyn, P. Jähnig, M. Herrmann, M. Knauth, M. Bähr, W. Heide, A. Wagner, S. Schwab, H. Reichmann, G. Schwendemann, R. Dengler, A. Kastrup, and C. Bartels (2009). "Recombinant human erythropoietin in the treatment of acute ischemic stroke." *Stroke* 40(12): e647–e656.

Eisner-Janowicz, I., S. Barbay, E. Hoover, A. M. Stowe, S. B. Frost, E. J. Plautz, and R. J. Nudo (2008). "Early and late changes in the distal forelimb representation of the supplementary motor area after injury to frontal motor areas in the squirrel monkey." *Journal of Neurophysiology* 100(3): 1498–1512.

Ejaz, N., J. Xu, M. Branscheidt, B. Hertler, H. Schambra, M. Widmer, A.V. Faria, M.D. Harran, J.C. Cortes, N. Kim, T. Kitago, P.A. Celnik, A. Luft, J.W. Krakauer, and J. Diedrichsen (2016). "Finger recruitment patterns during mirror movements suggest two systems for hand recovery after stroke." *bioRxiv*:129510 epub ahead of print.

Ekdahl, C. T., Z. Kokaia, and O. Lindvall (2009). "Brain inflammation and adult neurogenesis: The dual role of microglia." *Neuroscience* 158(3): 1021–1029.

Ekonomou, A., M. Johnson, R. H. Perry, E. K. Perry, R. N. Kalaria, S. L. Minger, and C. G. Ballard (2012). "Increased neural progenitors in individuals with cerebral small vessel disease." *Neuropathology and Applied Neurobiology* 38(4): 344–353.

Ellis, M. D., J. Drogos, C. Carmona, T. Keller, and J. P. A. Dewald (2012). "Neck rotation modulates flexion synergy torques, indicating an ipsilateral reticulospinal source for impairment in stroke." *Journal of Neurophysiology* 108(11): 3096–3104.

Ellis, M. D., Y. Lan, J. Yao, and J. P. A. Dewald (2016). "Robotic quantification of upper extremity loss of independent joint control or flexion synergy in individuals with hemiparetic stroke: A review of paradigms addressing the effects of shoulder abduction loading." *Journal of NeuroEngineering and Rehabilitation* 13: 95.

Ellis, M. D., T. Sukal-Moulton, and J. P. A. Dewald (2009). "Progressive shoulder abduction loading is a crucial element of arm rehabilitation in chronic stroke." *Neurorehabilitation and Neural Repair* 23(8): 862–869.

Ellis, M. D., I. Schut, and J. P. A. Dewald (2017). "Flexion synergy overshadows flexor spasticity during reaching in chronic moderate to severe hemiparetic stroke." *Clinical Neurophysiology* 128(7):1308–1314.

Elsner, B., J. Kugler, M. Pohl, and J. Mehrholz (2016). "*Transcranial direct current stimulation (tDCS) for improving activities of daily living, and physical and cognitive functioning, in people after stroke.*" The Cochrane Database of Systematic Reviews 21(3):CD009645.

Ergul, A., A. Alhusban, and S. C. Fagan (2012). "Angiogenesis: a harmonized target for recovery after stroke." *Stroke* 43(8): 2270–2274.

Ernst, A., K. Alkass, S. Bernard, M. Salehpour, S. Perl, J. Tisdale, G. Possnert, H. Druid, and J. Frisén (2014). "Neurogenesis in the striatum of the adult human brain." *Cell* 156(5): 1072–1083.

Esposito, M. S., P. Capelli, and S. Arber (2014). "Brainstem nucleus MdV mediates skilled forelimb motor tasks." *Nature* 508(7496): 351–356.

Evaluate Group (2013). "Hermo pharma reports topline data with HER-801 from clinical study in adult amblyopia." September 5, 2013. http://evaluategroup.com/Universal/View.aspx?type=Story&id =453937.

Evarts, E. V. (1968). "Relation of pyramidal tract activity to force exerted during voluntary movement." *Journal of Neurophysiology* 31(1): 14–27.

Evarts, E. V. (1981). "Role of motor cortex in voluntary movements in primates." *Comprehensive Physiology* Supplement 2: Handbook of Physiology, The Nervous System, Motor Control: 1083–1120.

Everson-Hock, E. S., M. A. Green, E. C. Goyder, R. J. Copeland, S. H. Till, B. Heller, and O. Hart (2016). "Reducing the impact of physical inactivity: evidence to support the case for targeting people with chronic mental and physical conditions." *Journal of Public Health (Oxford)* 38(2):343–351.

Fagg, A. H., A. Shah, and A. G. Barto (2002). "A computational model of muscle recruitment for wrist movements." *Journal of Neurophysiology* 88(6): 3348–3358.

Fang, P.-c., S. Barbay, E. J. Plautz, E. Hoover, S. M. Strittmatter, and R. J. Nudo (2010). "Combination of NEP 1–40 treatment and motor training enhances behavioral recovery after a focal cortical infarct in rats." *Stroke* 41(3): 544–549.

Fawcett, J. W., M. E. Schwab, L. Montani, N. Brazda, and H. W. Müller (2012). "Defeating inhibition of regeneration by scar and myelin components." *Handbook of Clinical Neurology* 109:503–22.

Feeney, D. M., A. Gonzalez, and W. A. Law (1982). "Amphetamine, haloperidol, and experience interact to affect rate of recovery after motor cortex injury." *Science* 217(4562): 855–857.

Fellows, S. J., C. Kaus, and A. F. Thilmann (1994). "Voluntary movement at the elbow in spastic hemiparesis." *Annals of Neurology* 36(3): 397–407.

Fellows, S. J., H. F. Ross, and A. F. Thilmann (1993). "The limitations of the tendon jerk as a marker of pathological stretch reflex activity in human spasticity." *Journal of Neurology, Neurosurgery, and Psychiatry* 56(5): 531–537.

Feng, W., J. Wang, P. Y. Chhatbar, C. Doughty, D. Landsittel, V. A. Lioutas, S. A. Kautz, and G. Schlaug (2015). "Corticospinal tract lesion load: An imaging biomarker for stroke motor outcomes." *Annals of Neurology* 78(6): 860–870.

Fernández-Klett, F., J. R. Potas, D. Hilpert, K. Blazej, J. Radke, J. Huck, O. Engel, W. Stenzel, G. Genové, and J. Priller (2013). "Early loss of pericytes and perivascular stromal cell-induced scar formation after stroke. " *Journal of Cerebral Blood Flow and Metabolism* 33(3):428–439.

Fetz, E. E. (2015). "Restoring motor function with bidirectional neural interfaces." *Progress in Brain Research* 218: 241–252.

Fiedler, M., C. Horn, C. Bandtlow, M. E. Schwab, and A. Skerra (2002). "An engineered IN-1 Fab fragment with improved affinity for the Nogo-A axonal growth inhibitor permits immunochemical detection and shows enhanced neutralizing activity." *Protein Engineering* 15(11): 931–941.

Fiehler, J., J.-m. Boulanger, W. Kakuda, J. S. Kim, D. S. Liebeskind, T. Neumann-haefelin, N. Nighoghossian, S. Pedraza, A. Rovira, and P. D. Schellinger (2006). "Bleeding risk analysis in stroke by T2*-weighted imaging before thrombolysis (BRASIL): a multicenter study of 600 patients of the MR Stroke Collaborative Group." *Stroke* 37(2): 636.

Filippi, M., M. A. Rocca, D. M. Mezzapesa, A. Ghezzi, A. Falini, V. Martinelli, G. Scotti, and G. Comi (2004). "Simple and complex movement-associated functional MRI changes in patients at presentation with clinically isolated syndromes suggestive of multiple sclerosis." *Human Brain Mapping* 21(2): 108–117.

Fisher, M., and G. W. Albers (2013). "Advanced imaging to extend the therapeutic time window of acute ischemic stroke." *Annals of Neurology* 73(1): 4–9.

Fisher, S. A., R. Y. Tam, and M. S. Shoichet (2014). "Tissue mimetics: Engineered hydrogel matrices provide biomimetic environments for cell growth." *Tissue Engineering Part A* 20(5–6): 895–898.

Fox, I. J., G. Q. Daley, S. A. Goldman, J. Huard, T. J. Kamp, and M. Trucco (2014). "Use of differentiated pluripotent stem cells in replacement therapy for treating disease." *Science* 345(6199): 1247391.

Fox, M. D., M. A. Halko, M. C. Eldaief, and A. Pascual-Leone (2012). "Measuring and manipulating brain connectivity with resting state functional connectivity magnetic resonance imaging (fcMRI) and transcranial magnetic stimulation (TMS)." *Neuroimage* 62(4): 2232–2243.

Franceschini, M., F. La Porta, M. Agosti, M. Massucci, and I. C. R. Group (2010). "Is health-related-quality of life of stroke patients influenced by neurological impairments at one year after stroke?" *European Journal of Physical Rehabilitation Medicine* 46(3): 389–399.

Francis, S. H., M. A. Blount, and J. D. Corbin (2011). "Mammalian cyclic nucleotide phosphodiesterases: Molecular mechanisms and physiological functions." *Physiological Reviews* 91(2): 651–690.

Franke, H., and P. Illes (2014). "Nucleotide signaling in astrogliosis." *Neuroscience Letters* 565: 14–22.

Fregni, F., P. S. Boggio, C. G. Mansur, T. Wagner, M. J. L. Ferreira, M. C. Lima, S. P. Rigonatti, M. A. Marcolin, S. D. Freedman, and M. A. Nitsche (2005). "Transcranial direct current stimulation of the unaffected hemisphere in stroke patients." *Neuroreport* 16(14): 1551–1555.

Fregni, F., P. S. Boggio, A. C. Valle, R. R. Rocha, J. Duarte, M. J. L. Ferreira, T. Wagner, S. Fecteau, S. P. Rigonatti, and M. Riberto (2006). "A sham-controlled trial of a 5-day course of repetitive transcranial magnetic stimulation of the unaffected hemisphere in stroke patients." *Stroke* 37(8): 2115–2122.

French, B., L. H. Thomas, J. Coupe, N. E. McMahon, L. Connell, J. Harrison, C. J. Sutton, S. Tishkovskaya, and C. L. Watkins (2016). "Repetitive task training for improving functional ability after stroke." Cochrane Database *of Systematic Reviews* (11):CD006073.

Friedrich, M. A. G., M. P. Martins, M. D. Araújo, C. Klamt, L. Vedolin, B. Garicochea, E. F. Raupp, J. Sartori El Ammar, D. C. Machado, J. C. da Costa, R. G. Nogueira, P. H. Rosado-de-Castro, R. Mendez-Otero, and G. R. de Freitas (2012). "Intra-arterial infusion of autologous bone marrow mononuclear cells in patients with moderate to severe middle cerebral artery acute ischemic stroke." *Cell Transplantation* 21(1): S13–S21.

Friel, K. M., S. Barbay, S. B. Frost, E. J. Plautz, A. M. Stowe, N. Dancause, E. V. Zoubina, and R. J. Nudo (2007). "Effects of a rostral motor cortex lesion on primary motor cortex hand representation topography in primates." *Neurorehabilitation and Neural Repair* 21(1): 51–61.

Fu, M., and Y. Zuo (2011). "Experience-dependent structural plasticity in the cortex." *Trends in Neurosciences* 34(4): 177–187.

Fugl-Meyer, A. R., L. Jääskö, I. Leyman, S. Olsson, and S. Steglind (1975). "The post-stroke hemiplegic patient. 1. A method for evaluation of physical performance." *Scandinavian Journal of Rehabilitation Medicine* 7(1): 13–31.

Fujita, Y., and T. Yamashita (2014). "Axon growth inhibition by RhoA/ROCK in the central nervous system." *Frontiers in Neuroscience* 8:338.

Galea, J. M., E. Mallia, J. Rothwell, and J. Diedrichsen (2015). "The dissociable effects of punishment and reward on motor learning." *Nature Neuroscience* 18(4): 597–602.

Gallo, V., and R. Armstrong (2008). "Myelin repair strategies: A cellular view." *Current Opinion in Neurology* 21(3): 278–283.

Gao, X., M. Jia, Y. Zhang, L. P. Breitling, and H. Brenner (2015). "DNA methylation changes of whole blood cells in response to active smoking exposure in adults: a systematic review of DNA methylation studies." *Clinical Epigenetics* 7: 113.

Garcia, A. D. R., R. Petrova, L. Eng, and A. L. Joyner (2010). "Sonic hedgehog regulates discrete populations of astrocytes in the adult mouse forebrain." *Journal of Neuroscience* 30(41): 13597–13608.

Gentner, R., S. Gorges, D. Weise, K. aufm Kampe, M. Buttmann, and J. Classen (2010). "Encoding of motor skill in the corticomuscular system of musicians." *Current Biology* 20(20): 1869–1874.

Gertz, K., G. Kronenberg, R. E. Kälin, C. Baldinger, C. Werner, M. Balkaya, G. D. Eom, J. Hellmann-Regen, J. Kröber, K. R. Miller, U. Lindauer, U. Laufs, U. Dirnagl, F. L. Heppner, and M. Endres (2012). "Essential role of interleukin-6 in post-stroke angiogenesis." *Brain* 135(6): 1964–1980.

Ghajar, C. M., H. Peinado, H. Mori, I. R. Matei, K. J. Evason, H. Brazier, D. Almeida, A. Koller, K. A. Hajjar, D. Y. R. Stainier, E. I. Chen, D. Lyden, and M. J. Bissell (2013). "The perivascular niche regulates breast tumour dormancy." *Nature Cell Biology* 15(7): 807–817.

Gharbawie, O. A., J. M. Karl, and I. Q. Whishaw (2007). "Recovery of skilled reaching following motor cortex stroke: Do residual corticofugal fibers mediate compensatory recovery?" *European Journal of Neuroscience* 26(11): 3309–3327.

Ghilardi, M. F., C. Moisello, G. Silvestri, C. Ghez, and J. W. Krakauer (2009). "Learning of a sequential motor skill comprises explicit and implicit components that consolidate differently." *Journal of Neurophysiology* 101(5): 2218–2229.

Giger, R. J., E. R. Hollis, and M. H. Tuszynski (2010). "Guidance molecules in axon regeneration." *Cold Spring Harbor Perspectives in Biology* 2(7): a001867.

Gillies, J. D., D. J. Burke, and J. W. Lance (1971). "Tonic vibration reflex in the cat." *Journal of Neurophysiology* 34(2):252–262.

Giraldi-Guimarães, A., M. Rezende-Lima, F. P. Bruno, and R. Mendez-Otero (2009). "Treatment with bone marrow mononuclear cells induces functional recovery and decreases neurodegeneration after sensorimotor cortical ischemia in rats." *Brain Research* 1266: 108–120.

Göritz, C., D. O. Dias, N. Tomilin, M. Barbacid, O. Shupliakov, J. Frisén (2011). "A pericyte origin of spinal cord scar tissue. " *Science* 333(6039):238–242.

Gladstone, D., and S. Black (2000). "Enhancing recovery after stroke with noradrenergic pharmacotherapy: A new frontier?" *The Canadian Journal of Neurological Sciences* 27(2): 97–105.

Gladstone, D. J., C. J. Danells, and S. E. Black (2002). "The Fugl-Meyer assessment of motor recovery after stroke: A critical review of its measurement properties." *Neurorehabilitation and Neural Repair* 16(3): 232–240.

Glazewski, S., A. L. Barth, H. Wallace, M. McKenna, A. Silva, and K. Fox (1999). "Impaired experience-dependent plasticity in barrel cortex of mice lacking the alpha and delta isoforms of CREB." *Cerebral Cortex* 9(3): 249–256.

Gleichman, A. J., and S. T. Carmichael (2014). "Astrocytic therapies for neuronal repair in stroke." *Neuroscience Letters* 565: 47–52.

Glykys, J., and I. Mody (2007). "Activation of GABAA receptors: Views from outside the synaptic cleft." *Neuron* 56(5): 763–770.

Gogolla, N., P. Caroni, A. Lüthi, and C. Herry (2009). "Perineuronal nets protect fear memories from erasure." *Science* 325(5945): 1258–1261.

Goldenberg, G. (2013). *Apraxia: The cognitive side of motor control.* Oxford, UK: Oxford University Press.

Goldman, S. A., and Z. Chen (2011). "Perivascular instruction of cell genesis and fate in the adult brain." *Nature Neuroscience* 14(11): 1382–1389.

Goldman, S. A. (2016). "Stem and Progenitor Cell-Based Therapy of the Central Nervous System: Hopes, Hype, and Wishful Thinking." *Cell Stem Cell* 18(2):174–188.

Goldsmith, J., and T. Kitago (2016). "Assessing systematic effects of stroke on motorcontrol by using hierarchical function-on-scalar regression." *Journal of the Royal Statistical Society. Series C, Applied Statistics* 65(2): 215–236.

Goldstein, L. B. (2000). "Effects of amphetamines and small related molecules on recovery after stroke in animals and man." *Neuropharmacology* 39(5): 852–859.

Goldstein, L. B. (2009). "Amphetamine trials and tribulations." *Stroke* 40(3, Suppl. 1): S133–S135.

Gonzalez-Perez, O., F. Gutierrez-Fernandez, V. Lopez-Virgen, J. Collas-Aguilar, A. Quinones-Hinojosa, and J. M. Garcia-Verdugo (2012). "Immunological regulation of neurogenic niches in the adult brain." *Neuroscience* 226: 270–281.

Gordon, J. (1987). "Assumptions underlying physical therapy intervention: Theoretical and historical perspectives." In *Movement science: Foundations for physical therapy in rehabilitation*, edited by J. Carr and R. Sheppard. Aspen Publishers, Rockville, MD.

Gorgoraptis, N., Y.-H. Mah, B. Machner, V. Singh-Curry, P. Malhotra, M. Hadji-Michael, D. Cohen, R. Simister, A. Nair, E. Kulinskaya, N. Ward, R. Greenwood, and M. Husain (2012). "The effects of the dopamine agonist rotigotine on hemispatial neglect following stroke." *Brain* 135(8): 2478–2491.

Graham Brown, T., and C. S. Sherrington (1913). "Note on the functions of the cortex cerebri." *Journal of Physiology (London)* 46: xxii.

Graziano, M. S. A., C. S. R. Taylor, and T. Moore (2002). "Complex movements evoked by microstimulation of precentral cortex." *Neuron* 34(5): 841–851.

Greenberg, D. A., and K. Jin (2013). "Vascular endothelial growth factors (VEGFs) and stroke." *Cellular and Molecular Life Sciences* 70(10): 1753–1761.

Grefkes, C., and G. R. Fink (2014). "Connectivity-based approaches in stroke and recovery of function." *The Lancet Neurology* 13(2): 206–216.

Gresham, G. E., T. E. Fitzpatrick, P. A. Wolf, P. M. McNamara, W. B. Kannel, and T. R. Dawber (1975). "Residual disability in survivors of stroke—the Framingham study." *New England Journal of Medicine* 293(19): 954–956.

Grimaldi, G., G. P. Argyropoulos, A. Bastian, M. Cortes, N. J. Davis, D. J. Edwards, R. Ferrucci, F. Fregni, J. M. Galea, M. Hamada, M. Manto, R. C. Miall, L. Morales-Quezada, P. A. Pope, A. Priori, J. Rothwell, S. P. Tomlinson, and P. Celnik (2016). "Cerebellar Transcranial Direct Current Stimulation (ctDCS): A Novel Approach to Understanding Cerebellar Function in Health and Disease." *Neuroscientist* 22(1):83–97.

Grutzendler, J., N. Kasthuri, and W.-B. Gan (2002). "Long-term dendritic spine stability in the adult cortex." *Nature* 420(6917): 812–816.

Gulati, T., S. J. Won, D. S. Ramanathan, C. C. Wong, A. Bodepudi, R. A. Swanson, and K. Ganguly (2015). "Robust neuroprosthetic control from the stroke perilesional cortex." *The Journal of Neuroscience* 35(22): 8653–8661.

Gupta, N., R. G. Henry, J. Strober, S.-M. Kang, D. A. Lim, M. Bucci, E. Caverzasi, L. Gaetano, M. L. Mandelli, T. Ryan, R. Perry, J. Farrell, R. J. Jeremy, M. Ulman, S. L. Huhn, A. J. Barkovich, and D. H. Rowitch (2012). "Neural stem cell engraftment and myelination in the human brain." *Science Translational Medicine* 4(155): 155ra137.

Gutiérrez-Fernández, M., B. Fuentes, B. Rodríguez-Frutos, J. Ramos-Cejudo, M. T. Vallejo-Cremades, and E. Díez-Tejedor (2012). "Trophic factors and cell therapy to stimulate brain repair after ischaemic stroke." *Journal of Cellular and Molecular Medicine* 16(10): 2280–2290.

Haaland, K. Y., and D. L. Harrington (1989). "Hemispheric control of the initial and corrective components of aiming movements." *Neuropsychologia* 27(7): 961–969.

Hadjiargyrou, M., and R. J. O'Keefe (2014). "The convergence of fracture repair and stem cells: Interplay of genes, aging, environmental factors and disease." *Journal of Bone and Mineral Research* 29(11): 2307–2322.

Hagemann, G., C. Redecker, T. Neumann-Haefelin, H.-J. Freund, and O. W. Witte (1998). "Increased long-term potentiation in the surround of experimentally induced focal cortical infarction." *Annals of Neurology* 44(2): 255–258.

Hall, M. (1833). "On the reflex function of the medulla oblongata and medulla spinalis." *Philosophical Transactions of the Royal Society of London* 123: 635–665.

Han, J.-H., S. A. Kushner, A. P. Yiu, C. J. Cole, A. Matynia, R. A. Brown, R. L. Neve, J. F. Guzowski, A. J. Silva, and S. A. Josselyn (2007). "Neuronal competition and selection during memory formation." *Science* 316(5823): 457–460.

Han, M.-H., C. A. Bolaños, T. A. Green, V. G. Olson, R. L. Neve, R.-J. Liu, G. K. Aghajanian, and E. J. Nestler (2006). "Role of cAMP response element-binding protein in the rat locus ceruleus: Regulation of neuronal activity and opiate withdrawal behaviors." *Journal of Neuroscience* 26(17): 4624–4629.

Han, Z.C., W. J. Du, Z. B. Han, and L. Liang (2017). "New insights into the heterogeneity and functional diversity of human mesenchymal stem cells." *Biomedical Materials and Engineering* 28(s1):S29–S4.

Hao, Z., D. Wang, Y. Zeng, and M. Liu (2013). "Repetitive transcranial magnetic stimulation for improving function after stroke." *The Cochrane Database of Systematic Reviews* 5: CD008862.

Harding, D. I., L. Greensmith, M. Mason, P. N. Anderson, and G. Vrbová (1999). "Overexpression of GAP-43 induces prolonged sprouting and causes death of adult motoneurons." *European Journal of Neuroscience* 11(7): 2237–2242.

Hardwick, R. M., V. A. Rajan, A. J. Bastian, J. W. Krakauer, and P. A. Celnik (2016). "Motor learning in stroke: Trained patients are not equal to untrained patients with less impairment." *Neurorehabilitation and Neural Repair* 31(2):178–189.

Harrer, G., U. Melnizky, and F. Wagner (1972). "Wirkweise und therapeutische Wirksamkeit eines Gehirnhydrolysates." *Arzneimittel Forschung (Drug Reearch)* 22(5): 3–15.

Harris, J. E., J. J. Eng, W. C. Miller, and A. S. Dawson (2009). "A self-administered Graded Repetitive Arm Supplementary Program (GRASP) improves arm function during inpatient stroke rehabilitation." *Stroke* 40(6): 2123–2128.

Haruno, M., and D. M. Wolpert (2005). "Optimal control of redundant muscles in step-tracking wrist movements." *Journal of Neurophysiology* 94(6): 4244–4255.

Haslinger, B., P. Erhard, F. Weilke, A. O. Ceballos-Baumann, P. Bartenstein, H. Gräfin von Einsiedel, M. Schwaiger, B. Conrad, and H. Boecker (2002). "The role of lateral premotor-cerebellar-parietal circuits in motor sequence control: A parametric fMRI study." *Brain Research: Cognitive Brain Research* 13(2): 159–168.

Hayashi, T., N. Noshita, T. Sugawara, and P. H. Chan (2003). "Temporal profile of angiogenesis and expression of related genes in the brain after ischemia." *Journal of Cerebral Blood Flow & Metabolism* 23(2): 166–180.

Hayward, K. S., J. Schmidt, K. R. Lohse, S. Peters, J. Bernhardt, N. A. Lannin, and L. A. Boyd (2017). "Are we armed with the right data? Pooled individual data review of biomarkers in people with severe upper limb impairment after stroke." *NeuroImage: Clinical* 13: 310–319.

Heine, C., K. Sygnecka, N. Scherf, A. Berndt, U. Egerland, T. Hage, and H. Franke (2012). "Phosphodiesterase 2 inhibitors promote axonal outgrowth in organotypic slice co-cultures." *Neurosignals* 21(3–4): 197–212.

Heiss, W.-D. (2012). "The ischemic penumbra: how does tissue injury evolve?" *Annals of the New York Academy of Sciences* 1268(1): 26–34.

Henley, J. M., and K. A. Wilkinson (2013). "AMPA receptor trafficking and the mechanisms underlying synaptic plasticity and cognitive aging." *Dialogues in Clinical Neuroscience* 15(1): 11–27.

Henley, J. M., and K. A. Wilkinson (2016). "Synaptic AMPA receptor composition in development, plasticity and disease." *Nature Reviews Neuroscience* 17(6):337–350.

Hensch, T. K. (2005). "Critical period plasticity in local cortical circuits." *Nature Reviews Neuroscience* 6(11): 877–888.

Herbert, W. J., K. Powell, and J. A. Buford (2015). "Evidence for a role of the reticulospinal system in recovery of skilled reaching after cortical stroke: Initial results from a model of ischemic cortical injury." *Experimental Brain Research* 233(11): 3231–3251.

Hermann, D. M., and M. Chopp (2012). "Promoting brain remodelling and plasticity for stroke recovery: Therapeutic promise and potential pitfalls of clinical translation." *The Lancet Neurology* 11(4): 369–380.

Hess, D. C., C. A. Sila, A. J. Furlan, L. R. Wechsler, J. A. Switzer, and R. W. Mays (2014). "A double-blind placebo-controlled clinical evaluation of MultiStem for the treatment of ischemic stroke." *International Journal of Stroke* 9(3): 381–386.

Hess, D. C., L. R. Wechsler, W. M. Clark, S. I. Savitz, G. A. Ford, D. Chiu, D. R. Yavagal, K. Uchino, D. S. Liebeskind, A. P. Auchus, S. Sen, C. A. Sila, J. D. Vest, and R. W. Mays (2017). "Safety and efficacy of multipotent adult progenitor cells in acute ischaemic stroke (MASTERS): a randomised, double-blind, placebo-controlled, phase 2 trial." *Lancet Neurology* 16(5):360–368.

Hicks, A., and J. Jolkkonen (2009). "Challenges and possibilities of intravascular cell therapy in stroke." *Acta Neurobiologiae Experimentalis* 69(1): 1–11

Hikosaka, O., K. Nakamura, K. Sakai, and H. Nakahara (2002). "Central mechanisms of motor skill learning." *Current Opinion in Neurobiology* 12(2): 217–222.

Hill, J. J., K. Jin, X. O. Mao, L. Xie, and D. A. Greenberg (2012). "Intracerebral chondroitinase ABC and heparan sulfate proteoglycan glypican improve outcome from chronic stroke in rats." *Proceedings of the National Academy of Sciences* 109(23): 9155–9160.

Hinman, J. D., M. D. Lee, S. Tung, H. V. Vinters, and S. T. Carmichael (2015). "Molecular disorganization of axons adjacent to human lacunar infarcts." *Brain* 138(Pt 3):736–745.

Hinman, J. D., M. N. Rasband, and S. T. Carmichael (2013). "Remodeling of the axon initial segment after focal cortical and white matter stroke." *Stroke* 44(1): 182–189.

Hiu, T., Z. Farzampour, J. T. Paz, E. H. J. Wang, C. Badgely, A. Olson, K. D. Micheva, G. Wang, R. Lemmens, K. V. Tran, Y. Nishiyama, X. Liang, S. A. Hamilton, N. O'Rourke, S. J. Smith, J. R. Huguenard, T. M. Bliss, and G. K. Steinberg (2015). "Enhanced phasic GABA inhibition during the repair phase of stroke: a novel therapeutic target." *Brain* 139(Pt 2):468–480.

Hobohm, C., A. Günther, J. Grosche, S. Roßner, D. Schneider, and G. Brückner (2005). "Decomposition and long-lasting downregulation of extracellular matrix in perineuronal nets induced by focal cerebral ischemia in rats." *Journal of Neuroscience Research* 80(4): 539–548.

Hofman, M. A. (1989). "On the evolution and geometry of the brain in mammals." *Progress in Neurobiology* 32(2): 137–158.

Horie, N., M. P. Pereira, K. Niizuma, G. Sun, H. Keren-Gill, A. Encarnacion, M. Shamloo, S. A. Hamilton, K. Jiang, S. Huhn, T. D. Palmer, T. M. Bliss, and G. K. Steinberg (2011). "Transplanted stem cell-secreted vascular endothelial growth factor effects poststroke recovery, inflammation, and vascular repair." *Stem Cells* 29(2): 274–285.

Hosp, J. A., A. Pekanovic, M. S. Rioult-Pedotti, and A. R. Luft (2011). "Dopaminergic projections from midbrain to primary motor cortex mediate motor skill learning." *Journal of Neuroscience* 31(7): 2481–2487.

Hsieh, Y.-W., C.-Y. Wu, K.-C. Lin, Y.-F. Chang, C.-L. Chen, and J.-S. Liu (2009). "Responsiveness and validity of three outcome measures of motor function after stroke rehabilitation." *Stroke: A Journal of Cerebral Circulation* 40(4): 1386–1391.

Hu, Q., X. Liang, D. Chen, Y. Chen, D. Doycheva, J. Tang, J. Tang, and J. H. Zhang (2014). "Delayed hyperbaric oxygen therapy promotes neurogenesis through reactive oxygen species/hypoxia-inducible factor-1α/β-catenin pathway in middle cerebral artery occlusion rats." *Stroke* 45(6): 1807–1814.

Huang, C., L.-P. Cen, L. Liu, S. G. Leaver, A. R. Harvey, Q. Cui, C. P. Pang, and M. Zhang (2013). "Adeno-associated virus-mediated expression of growth-associated protein-43 aggravates retinal ganglion cell death in experimental chronic glaucomatous injury." *Molecular Vision* (19):1422–32.

Huang, L., Z. He, L. Guo, and H. Wang (2008). "Improvement of cognitive deficit and neuronal damage in rats with chronic cerebral ischemia via relative long-term inhibition of rho-kinase." *Cellular and Molecular Neurobiology* 28(5):757–768.

Huang, V. S., A. Haith, P. Mazzoni, and J. W. Krakauer (2011). "Rethinking motor learning and savings in adaptation paradigms: Model-free memory for successful actions combines with internal models." *Neuron* 70(4): 787–801.

Hubel, D. H., and T. N. Wiesel (1970). "The period of susceptibility to the physiological effects of unilateral eye closure in kittens." *The Journal of Physiology* 206(2): 419–436.

Huberdeau, D. M., J. W. Krakauer, and A. M. Haith (2015). "Dual-process decomposition in human sensorimotor adaptation." *Current Opinion in Neurobiology* 33: 71–77.

Humm, J. L., D. A. Kozlowski, D. C. James, J. E. Gotts, and T. Schallert (1998). "Use-dependent exacerbation of brain damage occurs during an early post-lesion vulnerable period." *Brain Research* 783(2): 286–292.

Humm, J. L., D. A. Kozlowski, S. T. Bland, D. C. James, and T. Schallert (1999). "Use-dependent exaggeration of brain injury: is glutamate involved?" *Experimental Neurology* 157(2):349–358.

Huntley, G. W. (1997). "Differential effects of abnormal tactile experience on shaping representation patterns in developing and adult motor cortex." *Journal of Neuroscience* 17(23): 9220–9232.

Huttner, H. B., O. Bergmann, M. Salehpour, A. Rácz, J. Tatarishvili, E. Lindgren, T. Csonka, L. Csiba, T. Hortobágyi, G. Méhes, E. Englund, B. W. Solnestam, S. Zdunek, C. Scharenberg, L. Ström, P. Ståhl, B. Sigurgeirsson, A. Dahl, S. Schwab, G. Possnert, S. Bernard, Z. Kokaia, O. Lindvall, J. Lundeberg, and J. Frisén (2014). "The age and genomic integrity of neurons after cortical stroke in humans." *Nature Neuroscience* 17(6): 801–803.

Hyun Lee, J., Y. Zheng, D. von Bornstadt, Y. Wei, A. Balcioglu, A. Daneshmand, N. Yalcin, E. Yu, F. Herisson, Y. B. Atalay, M. H. Kim, Y.-J. Ahn, M. Balkaya, P. Sweetnam, O. Schueller, M. V. Poyurovsky, H.-H. Kim, E. H. Lo, K. L. Furie, and C. Ayata (2014). "Selective ROCK2 inhibition in focal cerebral ischemia." *Annals of Clinical and Translational Neurology* 1(1): 2–14.

Iadecola, C., and J. Anrather (2011). "The immunology of stroke: From mechanisms to translation." *Nature Medicine* 17(7): 796–808.

Imbrosci, B., E. Ytebrouck, L. Arckens, and T. Mittmann (2014). "Neuronal mechanisms underlying transhemispheric diaschisis following focal cortical injuries." *Brain Structure and Function* 220(3): 1649–1664.

Ingram, J. N., K. P. Körding, I. S. Howard, and D. M. Wolpert (2008). "The statistics of natural hand movements." *Experimental Brain Research* 188(2): 223–236.

Ivey, F. M., R. F. Macko, A. S. Ryan, and C. E. Hafer-Macko (2005). "Cardiovascular health and fitness after stroke." *Topics in Stroke Rehabilitation* 12(1):1–16.

Jablonska, A., M. Janowski, and B. Lukomska (2013). "Different methods of immunosuppresion do not prolong the survival of human cord blood-derived neural stem cells transplanted into focal brain-injured immunocompetent rats." *Acta Neurobiologiae Experimentalis* 73(1): 88–101.

Jackson, J. H. (1881). "Remarks on dissolution of the nervous system as exemplified by certain post-epileptic conditions." *Medical Press and Circular* 1: 329–347.

Jackson, J. H. (1884). "Evolution and dissolution of the nervous system." *Selected Writings of John Hughlings Jackson,* London, Hodder and Stoughton, New York, USA

Jakobsson, F., L. Grimby, and L. Edström (1992). "Motoneuron activity and muscle fibre type composition in hemiparesis." *Scandinavian Journal of Rehabilitation Medicine* 24(3): 115–119.

Jasinska, M., E. Siucinska, A. Cybulska-Klosowicz, E. Pyza, D. N. Furness, M. Kossut, and S. Glazewski (2010). "Rapid, learning-induced inhibitory synaptogenesis in murine barrel field." *The Journal of Neuroscience* 30(3): 1176–1184.

Jeffers, M. S., S. Karthikeyan, and D. Corbett (2016). "Proportional stroke recovery in the rat: Evidence for a cross-species biological recovery process that incorporates initial impairment, infarct volume and rehabilitation intensity." Program No. 790.24/N4 2016 Neuroscience Meeting Planner. San Diego, CA: Society for Neuroscience.

Jenq, R. R., and M. R. M. van den Brink (2010). "Allogeneic haematopoietic stem cell transplantation: individualized stem cell and immune therapy of cancer." *Nature Reviews Cancer* 10(3): 213–221.

Jin, K., X. Wang, L. Xie, X. O. Mao, and D. A. Greenberg (2010). "Transgenic ablation of doublecortin-expressing cells suppresses adult neurogenesis and worsens stroke outcome in mice." *Proceedings of the National Academy of Sciences* 107(17): 7993–7998.

Jin, K., X. Wang, L. Xie, X. O. Mao, W. Zhu, Y. Wang, J. Shen, Y. Mao, S. Banwait, and D. A. Greenberg (2006). "Evidence for stroke-induced neurogenesis in the human brain." *Proceedings of the National Academy of Sciences* 103(35): 13198–13202.

Johnson, R., and G. Halder (2014). "The two faces of Hippo: Targeting the Hippo pathway for regenerative medicine and cancer treatment." *Nature Reviews Drug Discovery* 13(1): 63–79.

Johansson, B. B., and A. L. Ohlsson (1996). "Environment, social interaction, and physical activity as determinants of functional outcome after cerebral infarction in the rat." *Experimental Neurology* 139(2):322–327.

Jolkkonen, J., K. Puurunen, S. Rantakömi, J. Sirviö, A. Haapalinna, and J. Sivenius (2000). "Effects-of fluoxetine on sensorimotor and spatial learning deficits following focal cerebral ischemia in rats." *Restorative Neurology and Neuroscience* 17(4): 211–216.

Jørgensen, H. S., H. Nakayama, H. O. Raaschou, and T. S. Olsen (1999). "Stroke: Neurologic and functional recovery the Copenhagen stroke study." *Physical Medicine and Rehabilitation Clinics of North America* 10(4): 887–906.

Jørgensen, H. S., H. Nakayama, H. O. Raaschou, J. Vive-Larsen, M. Støier, and T. S. Olsen (1995). "Outcome and time course of recovery in stroke. Part II: Time course of recovery. The Copenhagen Stroke Study." *Archives of Physical Medicine and Rehabilitation* 76(5): 406–412.

Josselyn, S. A., S. Köhler, and P. W. Frankland (2015). "Finding the engram." *Nature Reviews Neuroscience* 16(9): 521–534.

Joy, M. T., M. Zhou, Y. Cai, A. J. Silva, and S. T. Carmichael ST (2015). "The Role Of C-C Chemokine Receptor 5 In Neural Repair After Stroke." Program No. 403.29/H22 2015 *Neuroscience Meeting Planner.* Chicago, IL: Society for Neuroscience.

Käenmäki, M., A. Tammimäki, T. Myöhänen, K. Pakarinen, C. Amberg, M. Karayiorgou, J. A. Gogos, and P. T. Männistö (2010). "Quantitative role of COMT in dopamine clearance in the prefrontal cortex of freely moving mice." *Journal of Neurochemistry* 114(6):1745–1755.

Kahle, M. P., and G. J. Bix (2013). "Neuronal restoration following ischemic stroke: Influences, barriers, and therapeutic potential." *Neurorehabilitation and Neural Repair* 27(5): 469–478.

Kaito, M., S.-I. Araya, Y. Gondo, M. Fujita, N. Minato, M. Nakanishi, and M. Matsui (2013). "Relevance of distinct monocyte subsets to clinical course of ischemic stroke patients." *PLoS ONE* 8(8): e69409.

Kalladka, D., J. Sinden, K. Pollock, C. Haig, J. McLean, W. Smith, A. McConnachie, C. Santosh, P. M. Bath, L. Dunn, and K. W. Muir (2016). "Human neural stem cells in patients with chronic ischaemic stroke (PISCES): A phase 1, first-in-man study." *Lancet* 388(10046): 787–796.

Kallenberg, L. A. C., and H. J. Hermens (2009). "Motor unit properties of biceps brachii in chronic stroke patients assessed with high-density surface EMG." *Muscle & Nerve* 39(2): 177–185.

Kamper, D. G., H. C. Fischer, E. G. Cruz, and W. Z. Rymer (2006). "Weakness is the primary contributor to finger impairment in chronic stroke." *Archives of Physical Medicine and Rehabilitation* 87(9): 1262–1269.

Kandel, E. R., Schwartz, JH, T. Jessell, S. Siegelbaum, and A. Hudspeth (2012). *Principles of neural science: Part VIII. Development and the emergence of behavior.* 5th ed. New York: McGraw-Hill Medical.

Kang, N., J. J. Summers, and J. H. Cauraugh (2016). "Transcranial direct current stimulation facilitates motor learning post-stroke: A systematic review and meta-analysis." *Journal of Neurology, Neurosurgery & Psychiatry* 87(4): 345–355.

Kasamatsu, T., and J. D. Pettigrew (1976). "Depletion of brain catecholamines: Failure of ocular dominance shift after monocular occlusion in kittens." *Science* 194(4261): 206–209.

Katz, L. C., and C. J. Shatz (1996). "Synaptic activity and the construction of cortical circuits." *Science* 274(5290): 1133–1138.

Kawakami, T., S. Ren, and J. S. Duffield (2013). "Wnt signalling in kidney diseases: Dual roles in renal injury and repair." *The Journal of Pathology* 229(2): 221–231.

Keith, R. A., C. V. Granger, B. B. Hamilton, and F. S. Sherwin (1987). "The functional independence measure." *Advances in Clinical Rehabilitation* 1: 6–18.

Keizer, K., and H. G. J. M. Kuypers (1989). "Distribution of corticospinal neurons with collaterals to the lower brain stem reticular formation in monkey (Macaca fascicularis)." *Experimental Brain Research* 74(2): 311–318.

Kelley, M. S., and O. Steward (1997). "Injury-induced physiological events that may modulate gene expression in neurons and glia." *Reviews in the Neurosciences* 8(3–4): 147–178.

Kelley-Loughnane, N., G. E. Sabla, C. Ley-Ebert, B. J. Aronow, and J. A. Bezerra (2002). "Independent and overlapping transcriptional activation during liver development and regeneration in mice." *Hepatology* 35(3): 525–534.

Kelly, M. P., W. Adamowicz, S. Bove, A. J. Hartman, A. Mariga, G. Pathak, V. Reinhart, A. Romegialli, and R. J. Kleiman (2014). "Select 3′,5′–cyclic nucleotide phosphodiesterases exhibit altered expression in the aged rodent brain." *Cellular Signalling* 26(2): 383–397.

Kelly, M. P., S. F. Logue, J. Brennan, J. P. Day, S. Lakkaraju, L. Jiang, X. Zhong, M. Tam, S. J. S. Rizzo, B. J. Platt, J. M. Dwyer, S. Neal, V. L. Pulito, M. J. Agostino, S. M. Grauer, R. L. Navarra, C. Kelley, T. A. Comery, R. J. Murrills, M. D. Houslay, and N. J. Brandon (2010). "Phosphodiesterase 11A in brain is enriched in ventral hippocampus and deletion causes psychiatric disease-related phenotypes." *Proceedings of the National Academy of Sciences* 107(18): 8457–8462.

Kelly, S., T. M. Bliss, A. K. Shah, G. H. Sun, M. Ma, W. C. Foo, J. Masel, M. A. Yenari, I. L. Weissman, N. Uchida, T. Palmer, and G. K. Steinberg (2004). "Transplanted human fetal neural stem cells survive, migrate, and differentiate in ischemic rat cerebral cortex." *Proceedings of the National Academy of Sciences of the United States of America* 101(32): 11839–11844.

Kernan, W. N., C. M. Viscoli, L. M. Brass, T. M. Gill, P. M. Sarrel, and R. I. Horwitz (2005). "Decline in physical performance among women with a recent transient ischemic attack or ischemic stroke." *Stroke* 36(3): 630–634.

Kernie, S. G., and J. M. Parent (2010). "Forebrain neurogenesis after focal ischemic and traumatic brain injury." *Neurobiology of Disease* 37(2): 267–274.

Kfoury, Y., and David T. Scadden (2015). "Mesenchymal cell contributions to the stem cell niche." *Cell Stem Cell* 16(3): 239–253.

Khabbal, J., E. Kerkelä, B. Mitkari, M. Raki, J. Nystedt, V. Mikkonen, K. Bergström, S. Laitinen, M. Korhonen, and J. Jolkkonen (2015). "Differential clearance of rat and human bone marrow-derived mesenchymal stem cells from the brain after intra-arterial infusion in rats." *Cell Transplantation* 24(5): 819–828.

Khakh, B. S., and M. V. Sofroniew (2015). "Diversity of astrocyte functions and phenotypes in neural circuits." *Nature Neuroscience* 18(7): 942–952.

Kharlamov, E. A., K. L. Downey, P. I. Jukkola, D. R. Grayson, and K. M. Kelly (2008). "Expression of GABAA receptor α1 subunit mRNA and protein in rat neocortex following photothrombotic infarction." *Brain Research* 1210: 29–38.

Khedr, E. M., M. A. Ahmed, N. Fathy, and J. C. Rothwell (2005). "Therapeutic trial of repetitive transcranial magnetic stimulation after acute ischemic stroke." *Neurology* 65(3): 466–468.

Kim, B., and C. Winstein (2016). "Can neurological biomarkers of brain impairment be used to predict poststroke motor recovery? A systematic review." *Neurorehabilitation and Neural Repair.* 31(1):3–24.

Kim, H., M. J. Cooke, and M. S. Shoichet (2012). "Creating permissive microenvironments for stem cell transplantation into the central nervous system." *Trends in Biotechnology* 30(1): 55–63.

Kim, J., J.-T. Kwon, H.-S. Kim, and J.-H. Han (2013). "CREB and neuronal selection for memory trace." *Frontiers in Neural Circuits* 7:44.

Kim, Y.-H., S. H. You, M.-H. Ko, J.-W. Park, K. H. Lee, S. H. Jang, W.-K. Yoo, and M. Hallett (2006). "Repetitive transcranial magnetic stimulation–induced corticomotor excitability and associated motor skill acquisition in chronic stroke." *Stroke* 37(6): 1471–1476.

Kim, Y. K., E. J. Yang, K. Cho, J. Y. Lim, and N.-J. Paik (2014). "Functional recovery after ischemic stroke is associated with reduced GABAergic inhibition in the cerebral cortex: A GABA PET study." *Neurorehabilitation and Neural Repair* 28(6): 576–583.

Kistemaker, D. A., J. D. Wong, and P. L. Gribble (2010). "The central nervous system does not minimize energy cost in arm movements." *Journal of Neurophysiology* 104(6): 2985–2994.

Kitago, T., J. Goldsmith, M. Harran, L. Kane, J. Berard, S. Huang, S. L. Ryan, P. Mazzoni, J. W. Krakauer, and V. S. Huang (2015). "Robotic therapy for chronic stroke: General recovery of impairment or improved task-specific skill?" *Journal of Neurophysiology* 114(3): 1885–1894.

Kitago, T., J. Liang, V. S. Huang, S. Hayes, P. Simon, L. Tenteromano, R. M. Lazar, R. S. Marshall, P. Mazzoni, L. Lennihan, and J. W. Krakauer (2013). "Improvement after constraint-induced movement therapy: Recovery of normal motor control or task-specific compensation?" *Neurorehabilitation and Neural Repair* 27(2): 99–109.

Klamroth-Marganska, V., J. Blanco, K. Campen, A. Curt, V. Dietz, T. Ettlin, M. Felder, B. Fellinghauer, M. Guidali, A. Kollmar, A. Luft, T. Nef, C. Schuster-Amft, W. Stahel, and R. Riener (2014). "Three-dimensional, task-specific robot therapy of the arm after stroke: A multicentre, parallel-group randomised trial." *Lancet Neurology* 13(2): 159–166.

Kleim, J. A., S. Barbay, and R. J. Nudo (1998). "Functional reorganization of the rat motor cortex following motor skill learning." *Journal of Neurophysiology* 80(6): 3321–3325.

Kleim, J. A., T. M. Hogg, P. M. VandenBerg, N. R. Cooper, R. Bruneau, and M. Remple (2004). "Cortical synaptogenesis and motor map reorganization occur during late, but not early, phase of motor skill learning." *Journal of Neuroscience* 24(3): 628–633.

Klein, A., L.-A. R. Sacrey, I. Q. Whishaw, and S. B. Dunnett (2012). "The use of rodent skilled reaching as a translational model for investigating brain damage and disease." *Neuroscience & Biobehavioral Reviews* 36(3): 1030–1042.

Kloppenborg, R. P., P. J. Nederkoorn, A. M. Grool, K. L. Vincken, W. Mali, M. Vermeulen, Y. van der Graaf, M. I. Geerlings, and S. S. Group (2012). "Cerebral small-vessel disease and progression of brain atrophy: The SMART-MR study." *Neurology* 79(20): 2029–2036.

Kofler, B., C. Erhart, P. Erhart, and G. Harrer (1990). "A multidimensional approach in testing nootropic drug effects (Cerebrolysin®)." *Archives of Gerontology and Geriatrics* 10(2): 129–140.

Kollen, B. J., S. Lennon, B. Lyons, L. Wheatley-Smith, M. Scheper, J. H. Buurke, J. Halfens, A. C. H. Geurts, and G. Kwakkel (2009). "The effectiveness of the bobath concept in stroke rehabilitation what is the evidence?" *Stroke* 40(4): e89–e97.

Korhonen, M., and J. Jolkkonen (2013). "Intravascular cell therapy in stroke patients: Where the cells go and what they do." *Regenerative Medicine* 8(2): 93–95.

Koton, S., A. L. C. Schneider, W. D. Rosamond, E. Shahar, Y. Sang, R. F. Gottesman, and J. Coresh (2014). "Stroke incidence and mortality trends in US communities, 1987 to 2011." *JAMA* 312(3): 259–268.

Kozlowski, D. A., D. C. James, and T. Schallert (1996). "Use-dependent exaggeration of neuronal injury after unilateral sensorimotor cortex lesions." *Journal of Neuroscience* 16(15): 4776–4786.

Krakauer, J. W., S. T. Carmichael, D. Corbett, and G. F. Wittenberg (2012). "Getting neurorehabilitation right what can be learned from animal models?" *Neurorehabilitation and Neural Repair* 26(8): 923–931.

Krakauer, J. W., A. A. Ghazanfar, A. Gomez-Marin, M. A. MacIver, and D. Poeppel (2017). "Neuroscience needs behavior: Correcting a reductionist bias." *Neuron* 93(3): 10.

Krakauer, J. W., M.-F. Ghilardi, and C. Ghez (1999). "Independent learning of internal models for kinematic and dynamic control of reaching." *Nature Neuroscience* 2(11): 1026–1031.

Krakauer, J. W., Z. M. Pine, M.-F. Ghilardi, and C. Ghez (2000). "Learning of visuomotor transformations for vectorial planning of reaching trajectories." *Journal of Neuroscience* 20(23): 8916–8924.

Krakauer, J. W., and R. Shadmehr (2006). "Consolidation of motor memory." *Trends in Neurosciences* 29(1): 58–64.

Krebs, H. I., M. Krams, D. K. Agrafiotis, A. DiBernardo, J. C. Chavez, G. S. Littman, E. Yang, G. Byttebier, L. Dipietro, A. Rykman, K. McArthur, K. Hajjar, K. R. Lees, and B. T. Volpe (2014). "Robotic measurement of arm movements after stroke establishes biomarkers of motor recovery." *Stroke: A Journal of Cerebral Circulation* 45(1): 200–204.

Krueger, F., M. Pardini, E. D. Huey, V. Raymont, J. Solomon, R. H. Lipsky, C. A. Hodgkinson, D. Goldman, and J. Grafman (2011). "The role of the Met66 brain-derived neurotrophic factor allele in the recovery of executive functioning after combat-related traumatic brain injury." *Journal of Neuroscience* 31(2): 598–606.

Krupinski, J., J. Kaluza, P. Kumar, S. Kumar, and J. M. Wang (1994). "Role of angiogenesis in patients with cerebral ischemic stroke." *Stroke* 25(9): 1794–1798.

Kurtzer, I., J. A. Pruszynski, T. M. Herter, and S. H. Scott (2006). "Primate upper limb muscles exhibit activity patterns that differ from their anatomical action during a postural task." *Journal of Neurophysiology* 95(1): 493–504.

Kurtzer, I. L., J. A. Pruszynski, and S. H. Scott (2008). "Long-latency reflexes of the human arm reflect an internal model of limb dynamics." *Current Biology* 18(6): 449–453.

Kuypers, H. G. J. M. (1964). "The descending pathways to the spinal cord, their anatomy and function." *Progress in Brain Research* 11: 178–202.

Kwah, L. K., and R. D. Herbert (2016). "Prediction of walking and arm recovery after stroke: A critical review." *Brain Sciences* 6(4): 53.

Kwakkel, G., B. Kollen, and J. Twisk (2006). "Impact of time on improvement of outcome after stroke." *Stroke* 37(9): 2348–2353.

Kwakkel, G., J. M. Veerbeek, E. E. H. van Wegen, and S. L. Wolf (2015). "Constraint-induced movement therapy after stroke." *The Lancet Neurology* 14(2): 224–234.

Kwakkel, G., C. Winters, E. E. H. Van Wegen, R. H. M. Nijland, A. A. A. Van Kuijk, A. Visser-Meily, J. De Groot, E. De Vlugt, J. H. Arendzen, and A. C. H. Geurts (2016). "Effects of unilateral upper limb training in two distinct prognostic groups early after stroke: The EXPLICIT-stroke randomized clinical trial." *Neurorehabilitation and Neural Repair* 30(9): 804–816.

Lackland, D. T., E. J. Roccella, A. F. Deutsch, M. Fornage, M. G. George, G. Howard, B. M. Kissela, S. J. Kittner, J. H. Lichtman, L. D. Lisabeth, L. H. Schwamm, E. E. Smith, and A. Towfighi (2014). "Factors influencing the decline in stroke mortality." *Stroke* 45(1): 315–353.

Lai, S., A. Panarese, C. Spalletti, C. Alia, A. Ghionzoli, M. Caleo, and S. Micera (2015). "Quantitative kinematic characterization of reaching impairments in mice after a stroke." *Neurorehabilitation and Neural Repair* 29(4): 382–392.

Lai, T. W., S. Zhang, and Y. T. Wang (2014). "Excitotoxicity and stroke: Identifying novel targets for neuroprotection." *Progress in Neurobiology* 115: 157–188.

Lake, E. M. R., J. Chaudhuri, L. Thomason, R. Janik, M. Ganguly, M. Brown, J. McLaurin, D. Corbett, G. J. Stanisz, and B. Stefanovic (2015). "The effects of delayed reduction of tonic inhibition on ischemic lesion and sensorimotor function." *Journal of Cerebral Blood Flow & Metabolism* 35(10): 1601–1609.

Lamouille, S., J. Xu, and R. Derynck (2014). "Molecular mechanisms of epithelial–mesenchymal transition." *Nature Reviews: Molecular Cell Biology* 15(3): 178–196.

Lampron, A., A. ElAli, and S. Rivest (2013). "Innate immunity in the CNS: Redefining the relationship between the CNS and its environment." *Neuron* 78(2): 214–232.

Lance, J. W. (1980) "Symposium synopsis." In: R. G. Feldman, R. R. Young, W. P Koella *Spasticity: disordered motor control*. Year Book Medical Publishers, Chicago, USA.

Lander, C., H. Zhang, and S. Hockfield (1998). "Neurons produce a neuronal cell surface-associated chondroitin sulfate proteoglycan." *The Journal of Neuroscience* 18(1): 174–183.

Lang, B. T., J. M. Cregg, M. A. DePaul, A. P. Tran, K. Xu, S. M. Dyck, K. M. Madalena, B. P. Brown, Y.-L. Weng, S. Li, S. Karimi-Abdolrezaee, S. A. Busch, Y. Shen, and J. Silver (2015). "Modulation of the proteoglycan receptor PTPσ promotes recovery after spinal cord injury." *Nature* 518(7539): 404–408.

Lang, C., X. Guo, M. Kerschensteiner, and F. M. Bareyre (2012). "Single collateral reconstructions reveal distinct phases of corticospinal remodeling after spinal cord injury." *PLoS ONE* 7(1): e30461.

Lang, C. E., K. R. Lohse, and R. L. Birkenmeier (2015). "Dose and timing in neurorehabilitation: Prescribing motor therapy after stroke." *Current Opinion in Neurology* 28(6): 549–555.

Lang, C. E., J. R. MacDonald, D. S. Reisman, L. Boyd, T. Jacobson Kimberley, S. M. Schindler-Ivens, T. G. Hornby, S. A. Ross, and P. L. Scheets (2009a). "Observation of amounts of movement practice provided during stroke rehabilitation." *Archives of Physical Medicine and Rehabilitation* 90(10): 1692–1698.

Lang, C. E., S. L. DeJong, and J. A. Beebe (2009b). "Recovery of Thumb and Finger Extension and Its Relation to Grasp Performance After Stroke." *Journal of Neurophysiology* 102(1): 451–459.

Lang, C. E., M. J. Strube, M. D. Bland, K. J. Waddell, K. M. Cherry-Allen, R. J. Nudo, A. W. Dromerick, and R. L. Birkenmeier (2016). "Dose response of task-specific upper limb training in people at least 6 months poststroke: A phase II, single-blind, randomized, controlled trial." *Annals of Neurology* 80(3): 342–354.

Latifi, S., A. M. Estrada-Sánchez, E. Donzis, M. S. Levine, P. Golshani, and S. T. Carmichael (2016). "In vivo calcium imaging reveals distributed cortical network dynamics after stroke." Program No. 520.07 / III2 2016 *Neuroscience Meeting Planner*. San Diego, CA: Society for Neuroscience.

Lawrence, D. G., and H. G. J. M. Kuypers (1968a). "The functional organization of the motor system in the monkey: I. The effects of bilateral pyramidal lesions." *Brain* 91(1): 1–14.

Lawrence, D. G., and H. G. J. M. Kuypers (1968b). "The functional organization of the motor system in the monkey: II. The effects of lesions of the descending brainstem pathways." *Brain* 91(1): 15–36.

Lazar, R. M., M. F. Berman, J. R. Festa, A. E. Geller, T. G. Matejovsky, and R. S. Marshall (2010). "GABAergic but not anti-cholinergic agents re-induce clinical deficits after stroke." *Journal of the Neurological Sciences* 292(1–2): 72–76.

Lazar, R. M., B. Minzer, D. Antoniello, J. R. Festa, J. W. Krakauer, and R. S. Marshall (2010). "Improvement in aphasia scores after stroke is well predicted by initial severity." *Stroke* 41(7): 1485–1488.

Le Belle, J. E., N. M. Orozco, A. A. Paucar, J. P. Saxe, J. Mottahedeh, A. D. Pyle, H. Wu, and H. I. Kornblum (2011). "Proliferative neural stem cells have high endogenous ROS levels that regulate self-renewal and neurogenesis in a PI3K/Akt-dependent manner." *Cell Stem Cell* 8(1): 59–71.

Lee, A. S., C. Tang, M. S. Rao, I. L. Weissman, and J. C. Wu (2013). "Tumorigenicity as a clinical hurdle for pluripotent stem cell therapies." *Nature Medicine* 19(8): 998–1004.

Lee, J.-K., J.-E. Kim, M. Sivula, and S. M. Strittmatter (2004). "Nogo receptor antagonism promotes stroke recovery by enhancing axonal plasticity." *Journal of Neuroscience* 24(27): 6209–6217.

Lee, J. S., J. M. Hong, G. J. Moon, P. H. Lee, Y. H. Ahn, and O. Y. Bang (2010). "A long-term follow-up study of intravenous autologous mesenchymal stem cell transplantation in patients with ischemic stroke." *Stem Cells* 28(6): 1099–1106.

Lee, S.-R., H.-Y. Kim, J. Rogowska, B.-Q. Zhao, P. Bhide, J. M. Parent, and E. H. Lo (2006). "Involvement of matrix metalloproteinase in neuroblast cell migration from the subventricular zone after stroke." *Journal of Neuroscience* 26(13): 3491–3495.

Lemmens, R., T. Jaspers, W. Robberecht, and V. N. Thijs (2013). "Modifying expression of EphA4 and its downstream targets improves functional recovery after stroke." *Human Molecular Genetics.* 22(11): 2214–2220.

Lemmens, R., and G. K. Steinberg (2013). "Stem cell therapy for acute cerebral injury: What do we know and what will the future bring?" *Current Opinion in Neurology* 26(6): 617–625.

Lemon, R., and J. Griffiths (2002). "Is the rat a good model for human neurological disease?" *Nature Reviews Neuroscience* 3(7):580.

Lemon, R. N. (1993). "The GL Brown Prize Lecture." *Experimental Physiology* 78: 263–301.

Lemon, R. N. (2008). "Descending pathways in motor control." *Annual Review of Neuroscience* 31(1): 195–218.

Levelt, C. N., and M. Hübener (2012). "Critical-period plasticity in the visual cortex." *Annual Review of Neuroscience* 35: 309–330.

Levin, M. F., J. A. Kleim, and S. L. Wolf (2009). "What do motor 'recovery' and 'compensation' mean in patients following stroke?" *Neurorehabilitation and Neural Repair* 23(4): 313–319.

Levin, M. F., R. W. Selles, M. H. G. Verheul, and O. G. Meijer (2000). "Deficits in the coordination of agonist and antagonist muscles in stroke patients: Implications for normal motor control." *Brain Research* 853(2): 352–369.

Levy, R., S. Ruland, M. Weinand, D. Lowry, R. Dafer, and R. Bakay (2008). "Cortical stimulation for the rehabilitation of patients with hemiparetic stroke: A multicenter feasibility study of safety and efficacy." *Journal of Neurosurgery* 108(4):707–714.

Levy, R. M., R. L. Harvey, B. M. Kissela, C. J. Winstein, H. L. Lutsep, T. B. Parrish, S. C. Cramer, and L. Venkatesan (2016). "Epidural electrical stimulation for stroke rehabilitation: Results of the prospective, multicenter, randomized, single-blinded everest trial." *Neurorehabilitation and Neural Repair* 30(2): 107–119.

Lewis, G. N., and J. A. Rosie (2012). "Virtual reality games for movement rehabilitation in neurological conditions: How do we meet the needs and expectations of the users?" *Disability and Rehabilitation* 34(22): 1880–1886.

Leyton, A. S. F., and C. S. Sherrington (1917). "Observations on the excitable cortex of the chimpanzee, orangutan, and gorilla." *Quarterly Journal of Experimental Physiology* 11(2): 135–222.

Li, S., and S. T. Carmichael (2006). "Growth-associated gene and protein expression in the region of axonal sprouting in the aged brain after stroke." *Neurobiology of Disease* 23(2): 362–373.

Li, S., B. P. Liu, S. Budel, M. Li, B. Ji, L. Walus, W. Li, A. Jirik, S. Rabacchi, E. Choi, D. Worley, D. W. Y. Sah, B. Pepinsky, D. Lee, J. Relton, and S. M. Strittmatter (2004). "Blockade of Nogo-66, myelin-associated glycoprotein, and oligodendrocyte myelin glycoprotein by soluble Nogo-66 receptor promotes axonal sprouting and recovery after spinal injury." *Journal of Neuroscience* 24(46): 10511–10520.

Li, S., J. Liu, M. Bhadane, P. Zhou, and W. Z. Rymer (2014). "Activation deficit correlates with weakness in chronic stroke: Evidence from evoked and voluntary EMG recordings." *Clinical Neurophysiology* 125(12): 2413–2417.

Li, S., E. H. Nie, Y. Yin, L. I. Benowitz, S. Tung, H. V. Vinters, F. R. Bahjat, M. P. Stenzel-Poore, R. Kawaguchi, G. Coppola, and S. T. Carmichael (2015). "GDF10 is a signal for axonal sprouting and functional recovery after stroke." *Nature Neuroscience* 18(12): 1737–1745.

Li, S., J. J. Overman, D. Katsman, S. V. Kozlov, C. J. Donnelly, J. L. Twiss, R. J. Giger, G. Coppola, D. H. Geschwind, and S. T. Carmichael (2010). "An age-related sprouting transcriptome provides molecular control of axonal sprouting after stroke." *Nature Neuroscience* 13(12): 1496–1504.

Li, W.-L., H.-H. Cai, B. Wang, L. Chen, Q.-G. Zhou, C.-X. Luo, N. Liu, X.-S. Ding, and D.-Y. Zhu (2009). "Chronic fluoxetine treatment improves ischemia-induced spatial cognitive deficits through increasing hippocampal neurogenesis after stroke." *Journal of Neuroscience Research* 87(1): 112–122.

Lidell, J. A. (1873). "*A treatise on apoplexy, cerebral hemorrhage, cerebral embolism, cerebral gout, cerebral rheumatism, and epidemic cerebro-spinal meningitis.*" W. Wood & Company, New York, USA.

Liepert, J., A. Heller, G. Behnisch, and A. Schoenfeld (2013). "Catechol-O-methyltransferase polymorphism influences outcome after ischemic stroke: A prospective double-blind study." *Neurorehabilitation and Neural Repair* 27(6): 491–496.

Lim, D. A., and A. Alvarez-Buylla (2014). "Adult neural stem cells stake their ground." *Trends in Neurosciences* 37(10): 563–571.

Lim, D. H., J. M. LeDue, M. H. Mohajerani, and T. H. Murphy (2014). "Optogenetic mapping after stroke reveals network-wide scaling of functional connections and heterogeneous recovery of the peri-infarct." *Journal of Neuroscience* 34(49): 16455–16466.

Lindau, N. T., B. J. Bänninger, M. Gullo, N. A. Good, L. C. Bachmann, M. L. Starkey, and M. E. Schwab (2014). "Rewiring of the corticospinal tract in the adult rat after unilateral stroke and anti-Nogo-A therapy." *Brain* 137(3): 739–756.

Lindemann, P. G., and C. E. Wright (1998). "Skill acquisition and plans for actions: Learning to write with your other hand." *Methods, Models, and Conceptual Issues: An Invitation to Cognitive Science* 4: 523–584.

Lino, M. M., and A. Merlo (2011). "PI3Kinase signaling in glioblastoma." *Journal of Neuro-Oncology* 103(3): 417–427.

Liu, J., S. C. Cramer, and D. J. Reinkensmeyer (2006). "Learning to perform a new movement with robotic assistance: Comparison of haptic guidance and visual demonstration." *Journal of Neuroengineering and Rehabilitation* 3(1): 20.

Liu, J., Y. Wang, Y. Akamatsu, C. C. Lee, R. A. Stetler, M. T. Lawton, and G.-Y. Yang (2014). "Vascular remodeling after ischemic stroke: Mechanisms and therapeutic potentials." *Progress in Neurobiology* 115: 138–156.

Liu, X. S., Z. G. Zhang, R. L. Zhang, S. R. Gregg, L. Wang, T. Yier, and M. Chopp (2007). "Chemokine ligand 2 (CCL2) induces migration and differentiation of subventricular zone cells after stroke." *Journal of Neuroscience Research* 85(10): 2120–2125.

Liu, Y., and E. M. Rouiller (1999). "Mechanisms of recovery of dexterity following unilateral lesion of the sensorimotor cortex in adult monkeys." *Experimental Brain Research* 128(1–2): 149–159.

Lo, A. C., P. D. Guarino, L. G. Richards, J. K. Haselkorn, G. F. Wittenberg, D. G. Federman, R. J. Ringer, T. H. Wagner, H. I. Krebs, B. T. Volpe, C. T. Bever, Jr., D. M. Bravata, P. W. Duncan, B. H. Corn, A. D. Maffucci, S. E. Nadeau, S. S. Conroy, J. M. Powell, G. D. Huang, and P. Peduzzi (2010). "Robot-assisted therapy for long-term upper-limb impairment after stroke." *New England Journal of Medicine* 362(19): 1772–1783.

Lokk, J., R. S. Roghani, and A. Delbari (2011). "Effect of methylphenidate and/or levodopa coupled with physiotherapy on functional and motor recovery after stroke—a randomized, double-blind, placebo-controlled trial." *Acta Neurologica Scandinavica* 123(4): 266–273.

López-Valdés, H. E., A. N. Clarkson, Y. Ao, A. C. Charles, S. T. Carmichael, M. V. Sofroniew, and K. C. Brennan (2014). "Memantine enhances recovery from stroke." *Stroke* 45(7): 2093–2100.

Lorens, S. A., J. P. Sorensen, and L. M. Yunger (1971). "Behavioral and neurochemical effects of lesions in the raphe system of the rat." *Journal of Comparative and Physiological Psychology* 77(1): 48.

Lovelace Jr, B. (2016). "Pokemon Go now the biggest mobile game in US history." Online. Accessed: 2/14/2017. URL: http://www.cnbc.com/2016/07/13/pokemon-go-now-the-biggest-mobile-game-in-us-history.html.

Lu, P., G. Woodruff, Y. Wang, L. Graham, M. Hunt, D. Wu, E. Boehle, R. Ahmad, G. Poplawski, J. Brock, Lawrence S. B. Goldstein, and Mark H. Tuszynski (2014). "Long-distance axonal growth from human induced pluripotent stem cells after spinal cord injury." *Neuron* 83(4): 789–796.

Lu, Y., S. Belin, and Z. He (2014). "Signaling regulations of neuronal regenerative ability." *Current Opinion in Neurobiology* 27: 135–142.

Lüttgen, J., and H. Heuer (2012). "The influence of haptic guidance on the production of spatio-temporal patterns." *Human Movement Science* 31(3): 519–528.

Lyle, R. C. (1981). "A performance test for assessment of upper limb function in physical rehabilitation treatment and research." *International Journal of Rehabilitation Research* 4(4): 483–492.

Lynch, G., C. D. Cox, and C. M. Gall (2014). "Pharmacological enhancement of memory or cognition in normal subjects." *Frontiers in Systems Neuroscience* 8:90.

Ma, J., W.-M. Tian, S.-P. Hou, Q.-Y. Xu, M. Spector, and F.-Z. Cui (2007). "An experimental test of stroke recovery by implanting a hyaluronic acid hydrogel carrying a Nogo receptor antibody in a rat model." *Biomedical Materials* 2(4): 233.

Mabuchi, T., K. Kitagawa, T. Ohtsuki, K. Kuwabara, Y. Yagita, T. Yanagihara, M. Hori, and M. Matsumoto (2000). "Contribution of microglia/macrophages to expansion of infarction and response of oligodendrocytes after focal cerebral ischemia in rats." *Stroke* 31(7): 1735–1743.

Macas, J., C. Nern, K. H. Plate, and S. Momma (2006). "Increased generation of neuronal progenitors after ischemic injury in the aged adult human forebrain." *Journal of Neuroscience* 26(50): 13114–13119.

MacDonald, E., H. Van der Lee, D. Pocock, C. Cole, N. Thomas, P. M. VandenBerg, R. Bourtchouladze, and J. A. Kleim (2007). "A novel phosphodiesterase type 4 inhibitor, HT-0712, enhances rehabilitation-dependent motor recovery and cortical reorganization after focal cortical ischemia." *Neurorehabilitation and Neural Repair* 21(6): 486–496.

Machado, A. G., J. Cooperrider, H. T. Furmaga, K. B. Baker, H.-J. Park, Z. Chen, and J. T. Gale (2013). "Chronic 30-Hz deep cerebellar stimulation coupled with training enhances post-ischemia motor recovery and peri-infarct synaptophysin expression in rodents." *Neurosurgery* 73(2): 344–353.

Macklis, Jeffrey D. (2012). "Human adult olfactory bulb neurogenesis? Novelty is the best policy." *Neuron* 74(4): 595–596.

MacLeod, M. J., and M. Turner (2015). "Stroke: Stroke outcomes after 90 days—out of sight, out of mind?" *Nature Reviews Neurology* 11(4): 187–188.

Macrez, R., C. Ali, O. Toutirais, B. Le Mauff, G. Defer, U. Dirnagl, and D. Vivien (2011). "Stroke and the immune system: From pathophysiology to new therapeutic strategies." *The Lancet Neurology* 10(5): 471–480.

Maier, M. A., E. Olivier, S. N. Baker, P. A. Kirkwood, T. Morris, and R. N. Lemon (1997). "Direct and indirect corticospinal control of arm and hand motoneurons in the squirrel monkey (Saimiri sciureus)." *Journal of Neurophysiology* 78(2): 721–733.

Mani, S., P. K. Mutha, A. Przybyla, K. Y. Haaland, D. C. Good, and R. L. Sainburg (2013). "Contralesional motor deficits after unilateral stroke reflect hemisphere-specific control mechanisms." *Brain* 136(4):1288–1303.

Manley, H., P. Dayan, and J. Diedrichsen (2014). "When money is not enough: Awareness, success, and variability in motor learning." *PLoS ONE* 9(1): e86580.

Manohar, S. G., T. T. J. Chong, M. A. J. Apps, A. Batla, M. Stamelou, P. R. Jarman, K. P. Bhatia, and M. Husain (2015). "Reward pays the cost of noise reduction in motor and cognitive control." *Current Biology* 25(13): 1707–1716.

Mansur, C. G., F. Fregni, P. S. Boggio, M. Riberto, J. Gallucci-Neto, C. M. Santos, T. Wagner, S. P. Rigonatti, M. A. Marcolin, and A. Pascual-Leone (2005). "A sham stimulation-controlled trial of rTMS of the unaffected hemisphere in stroke patients." *Neurology* 64(10): 1802–1804.

Maroof, Asif M., S. Keros, Jennifer A. Tyson, S.-W. Ying, Yosif M. Ganat, Florian T. Merkle, B. Liu, A. Goulburn, Edouard G. Stanley, Andrew G. Elefanty, Hans R. Widmer, K. Eggan, Peter A. Goldstein, Stewart A. Anderson, and L. Studer (2013). "Directed differentiation and functional maturation of cortical interneurons from human embryonic stem cells." *Cell Stem Cell* 12(5): 559–572.

Martin, L. J., A. A. Zurek, J. F. MacDonald, J. C. Roder, M. F. Jackson, and B. A. Orser (2010). "α5GABAA receptor activity sets the threshold for long-term potentiation and constrains hippocampus-dependent memory." *Journal of Neuroscience* 30(15): 5269–5282.

Martinsson, L., H.-G. Hårdemark, and N. G. Wahlgren (2003). "Amphetamines for improving stroke recovery." *Stroke* 34(11): 2766–2766.

Martinsson, L., and N. G. Wahlgren (2003). "Safety of dexamphetamine in acute ischemic stroke." *Stroke* 34(2): 475–481.

Mascaro, A. L. A., P. Cesare, L. Sacconi, G. Grasselli, G. Mandolesi, B. Maco, G. W. Knott, L. Huang, V. D. Paola, P. Strata, and F. S. Pavone (2013). "In vivo single branch axotomy induces GAP-43–dependent sprouting and synaptic remodeling in cerebellar cortex." *Proceedings of the National Academy of Sciences* 110(26): 10824–10829.

Masiero, S., M. Armani, G. Ferlini, G. Rosati, and A. Rossi (2014). "Randomized trial of a robotic assistive device for the upper extremity during early inpatient stroke rehabilitation." *Neurorehabilitation and Neural Repair* 28(4): 377–386.

Matsusue, E., S. Sugihara, S. Fujii, T. Kinoshita, E. Ohama, and T. Ogawa (2007). "Wallerian degeneration of the corticospinal tracts: Postmortem MR–pathologic correlations." *Acta Radiologica* 48(6): 690–694.

Matsuyama, K., K. Takakusaki, K. Nakajima, and S. Mori (1997). "Multi-segmental innervation of single pontine reticulospinal axons in the cervico-thoracic region of the cat: Anterograde PHA-L tracing study." *Journal of Comparative Neurology* 377(2): 234–250.

Maurer, M. H., W. R. Schabitz, and A. Schneider (2008). "old friends in new constellations—the hematopoietic growth factors G-CSF, GMCSF, and EPO for the treatment of neurological diseases." *Current Medicinal Chemistry* 15(14): 1407–1411.

Mayer, N. H. (1997). "Clinicophysiologic concepts of spasticity and motor dysfunction in adults with an upper motoneuron lesion." *Muscle & Nerve* 20(S6): 1–14.

McCabe, J., M. Monkiewicz, J. Holcomb, S. Pundik, and J. J. Daly (2015). "Comparison of robotics, functional electrical stimulation, and motor learning methods for treatment of persistent upper extremity dysfunction after stroke: A randomized controlled trial." *Archives of Physical Medicine and Rehabilitation* 96(6): 981–990.

McDonnell, M. N., S. Koblar, N. S. Ward, J. C. Rothwell, B. Hordacre, and M. C. Ridding (2015). "An investigation of cortical neuroplasticity following stroke in adults: is there evidence for a critical window for rehabilitation?" *BMC Neurology* 15(1): 109.

McDonnell, M. N., and C. M.Stinear (2017). "TMS measures of motor cortex function after stroke: A meta-analysis. " *Brain Stimulation* pii:S1935-861X(17)30653-30658.

McKinney, R. A., D. Debanne, B. H. Gähwiler, and S. M. Thompson (1997). "Lesion-induced axonal sprouting and hyperexcitability in the hippocampus in vitro: implications for the genesis of posttraumatic epilepsy." *Nature Medicine* 3(9): 990–996.

McMorland, A. J. C., K. D. Runnalls, and W. D. Byblow (2015). "A neuroanatomical framework for upper limb synergies after stroke." *Frontiers in Human Neuroscience* 9:82.

McNeal, D. W., W. G. Darling, J. Ge, K. S. Stilwell-Morecraft, K. M. Solon, S. M. Hynes, M. A. Pizzimenti, D. L. Rotella, T. Vanadurongvan, and R. J. Morecraft (2010). "Selective long-term reorganization of the corticospinal projection from the supplementary motor cortex following recovery from lateral motor cortex injury." *Journal of Comparative Neurology* 518(5): 586–621.

Mechner, F., and M. Latranyi (1963). "Behavioral effects of caffeine, methamphetamine, and methylphenidate in the rat." *Journal of the Experimental Analysis of Behavior* 6:331–342.

Ménard, C., P. Gaudreau, and R. Quirion (2015). "Signaling pathways relevant to cognition-enhancing drug targets." *Handbook of Experimental Pharmacology* 228:59–98.

Méndez, P., A. Pazienti, G. Szabó, and A. Bacci (2012). "Direct alteration of a specific inhibitory circuit of the hippocampus by antidepressants." *Journal of Neuroscience* 32(47): 16616–16628.

Menniti, F. S., J. Ren, T. M. Coskran, J. Liu, D. Morton, D. K. Sietsma, A. Som, D. T. Stephenson, B. A. Tate, and S. P. Finklestein (2009). "Phosphodiesterase 5A inhibitors improve functional recovery after stroke in rats: Optimized dosing regimen with implications for mechanism." *Journal of Pharmacology and Experimental Therapeutics* 331(3): 842–850.

Michelsen, Kimmo A., S. Acosta-Verdugo, M. Benoit-Marand, I. Espuny-Camacho, N. Gaspard, B. Saha, A. Gaillard, and P. Vanderhaeghen (2015). "Area-specific reestablishment of damaged circuits in the adult cerebral cortex by cortical neurons derived from mouse embryonic stem cells." *Neuron* 85(5): 982–997.

Milner, B. (1962). "Les troubles de la memoire accompagnant les lesions hippocampiques bilaterales." In *Physiologie de l'hippocampe*, edited by P. Passouant. Paris: Éditions Recherche Scientifique: 257–272.

Miyamoto, N., L.-D. D. Pham, K. Hayakawa, T. Matsuzaki, J. H. Seo, C. Magnain, C. Ayata, K.-W. Kim, D. Boas, E. H. Lo, and K. Arai (2013). "Age-related decline in oligodendrogenesis retards white matter repair in mice." *Stroke* 44(9): 2573–2578.

Miyata, S., Y. Komatsu, Y. Yoshimura, C. Taya, and H. Kitagawa (2012). "Persistent cortical plasticity by upregulation of chondroitin 6-sulfation." *Nature Neuroscience* 15(3): 414–422.

Modo, M., F. Ambrosio, R. M. Friedlander, S. F. Badylak, and L. R. Wechsler (2013). "Bioengineering solutions for neural repair and recovery in stroke." *Current Opinion in Neurology* 26(6): 626–631.

Mohajerani, M. H., K. Aminoltejari, and T. H. Murphy (2011). "Targeted mini-strokes produce changes in interhemispheric sensory signal processing that are indicative of disinhibition within minutes." *Proceedings of the National Academy of Sciences* 108(22): E183–E191.

Mohapatra, S., R. Harrington, E. Chan, A. W. Dromerick, E. Y. Breceda, and M. Harris-Love (2016). "Role of contralesional hemisphere in paretic arm reaching in patients with severe arm paresis due to stroke: A preliminary report." *Neuroscience Letters* 617: 52–58.

Moisello, C., D. Crupi, E. Tunik, A. Quartarone, M. Bove, G. Tononi, and M. F. Ghilardi (2009). "The serial reaction time task revisited: A study on motor sequence learning with an arm-reaching task." *Experimental Brain Research* 194(1): 143–155.

Molina-Luna, K., B. Hertler, M. M. Buitrago, and A. R. Luft (2008). "Motor learning transiently changes cortical somatotopy." *Neuroimage* 40(4): 1748–1754.

Molina-Luna, K., A. Pekanovic, S. Röhrich, B. Hertler, M. Schubring-Giese, M.-S. Rioult-Pedotti, and A. R. Luft (2009). "Dopamine in motor cortex is necessary for skill learning and synaptic plasticity." *PLoS ONE* 4(9): e7082.

Molofsky, A. V., R. Krenick, E. Ullian, H.-h. Tsai, B. Deneen, W. D. Richardson, B. A. Barres, and D. H. Rowitch (2012). "Astrocytes and disease: A neurodevelopmental perspective." *Genes & Development* 26(9): 891–907.

Moniche, F., I. Escudero, E. Zapata-Arriaza, M. Usero-Ruiz, M. Prieto-León, J. de la Torre, M.-A. Gamero, J. A. Tamayo, J.-J. Ochoa-Sepúlveda, J. Maestre, M. Carmona, P. Piñero, C. Calderón-Cabrera, M.-D. Jimenez, A. Gonzalez, and J. Montaner (2015). "Intra-arterial bone marrow mononuclear cells (BM-MNCs) transplantation in acute ischemic stroke (IBIS trial): Protocol of a phase II, randomized, dose-finding, controlled multicenter trial." *International Journal of Stroke* 10(7): 1149–1152.

Moniche, F., A. Gonzalez, J.-R. Gonzalez-Marcos, M. Carmona, P. Piñero, I. Espigado, D. Garcia-Solis, A. Cayuela, J. Montaner, C. Boada, A. Rosell, M.-D. Jimenez, A. Mayol, and A. Gil-Peralta (2012). "Intra-arterial bone marrow mononuclear cells in ischemic stroke." *Stroke* 43(8): 2242–2244.

Monnier, P. P., A. Sierra, J. M. Schwab, S. Henke-Fahle, and B. K. Mueller (2003). "The Rho/ROCK pathway mediates neurite growth-inhibitory activity associated with the chondroitin sulfate proteoglycans of the CNS glial scar." *Molecular and Cellular Neuroscience* 22(3): 319–330.

Moon, S.-K., M. Alaverdashvili, A. R. Cross, and I. Q. Whishaw (2009). "Both compensation and recovery of skilled reaching following small photothrombotic stroke to motor cortex in the rat." *Experimental Neurology* 218(1): 145–153.

Moore, C. S., S. L. Abdullah, A. Brown, A. Arulpragasam, and S. J. Crocker (2011). "How factors secreted from astrocytes impact myelin repair." *Journal of Neuroscience Research* 89(1): 13–21.

Moreau, D., and A. R. Conway (2014). "The case for an ecological approach to cognitive training." *Trends in Cognitive Science* 18(7): 334–336.

Moreau, D., and A. R. A. Conway (2014). "The case for an ecological approach to cognitive training." *Trends in Cognitive Sciences* 18(7): 334–336.

Mori, F., M. Ribolsi, H. Kusayanagi, F. Monteleone, V. Mantovani, F. Buttari, E. Marasco, G. Bernardi, M. Maccarrone, and D. Centonze (2012). "TRPV1 channels regulate cortical excitability in humans." *Journal of Neuroscience* 32(3): 873–879.

Mori, F., M. Ribolsi, H. Kusayanagi, A. Siracusano, V. Mantovani, E. Marasco, G. Bernardi, and D. Centonze (2011). "Genetic variants of the NMDA receptor influence cortical excitability and plasticity in humans." *Journal of Neurophysiology* 106(4): 1637–1643.

Morrison, H. W., and J. A. Filosa (2013). "A quantitative spatiotemporal analysis of microglia morphology during ischemic stroke and reperfusion." *Journal of Neuroinflammation* 10: 4.

Moshayedi, P., and S. T. Carmichael (2013). "Hyaluronan, neural stem cells and tissue reconstruction after acute ischemic stroke." *Biomatter* 3(1): e23863.

Moshayedi, P., L. R. Nih, I. L. Llorente, A. R. Berg, J. Cinkornpumin, W. E. Lowry, T. Segura, and S. T. Carmichael (2016). "Systematic optimization of an engineered hydrogel allows for selective control of human neural stem cell survival and differentiation after transplantation in the stroke brain." *Biomaterials* 105: 145–155.

Moskowitz, M. A., E. H. Lo, and C. Iadecola (2010). "The science of stroke: Mechanisms in search of treatments." *Neuron* 67(2): 181–198.

Mostany, R., T. G. Chowdhury, D. G. Johnston, S. A. Portonovo, S. T. Carmichael, and C. Portera-Cailliau (2010). "Local hemodynamics dictate long-term dendritic plasticity in peri-infarct cortex." *The Journal of Neuroscience* 30(42): 14116–14126.

Mostert, J. P., M. W. Koch, M. Heerings, D. J. Heersema, and J. De Keyser (2008). "Therapeutic potential of fluoxetine in neurological disorders." *CNS Neuroscience & Therapeutics* 14(2): 153–164.

Mower, A. F., D. S. Liao, E. J. Nestler, R. L. Neve, and A. S. Ramoa (2002). "cAMP/Ca2+response element-binding protein function is essential for ocular dominance plasticity." *Journal of Neuroscience* 22(6): 2237–2245.

Muir, R. B., and R. N. Lemon (1983). "Corticospinal neurons with a special role in precision grip." *Brain Research* 261(2): 312–316.

Mullen, R. J., C. R. Buck, and A. M. Smith (1992). "NeuN, a neuronal specific nuclear protein in vertebrates." *Development* 116(1):201–211.

Mullick, A. A., S. K. Subramanian, and M. F. Levin (2015). "Emerging evidence of the association between cognitive deficits and arm motor recovery after stroke: A meta-analysis." *Restorative Neurology Neuroscience* 33(3): 389–403.

Murase, N., J. Duque, R. Mazzocchio, and L. G. Cohen (2004). "Influence of interhemispheric interactions on motor function in chronic stroke." *Annals of Neurology* 55(3): 400–409.

Murata, Y., N. Higo, T. Hayashi, Y. Nishimura, Y. Sugiyama, T. Oishi, H. Tsukada, T. Isa, and H. Onoe (2015). "Temporal plasticity involved in recovery from manual dexterity deficit after motor cortex lesion in macaque monkeys." *Journal of Neuroscience* 35(1): 84–95.

Murata, Y., N. Higo, T. Oishi, A. Yamashita, K. Matsuda, M. Hayashi, and S. Yamane (2008). "Effects of motor training on the recovery of manual dexterity after primary motor cortex lesion in macaque monkeys." *Journal of Neurophysiology* 99(2): 773–786.

Muresanu, D. F., W.-D. Heiss, V. Hoemberg, O. Bajenaru, C. D. Popescu, J. C. Vester, V. W. Rahlfs, E. Doppler, D. Meier, H. Moessler, and A. Guekht (2016). "Cerebrolysin and recovery after stroke (CARS)." *Stroke* 47(1): 151–159.

Murphy, T. H., and D. Corbett (2009). "Plasticity during stroke recovery: From synapse to behaviour." *Nature Reviews Neuroscience* 10(12): 861–872.

Murry, C. E., and G. Keller (2008). "Differentiation of embryonic stem cells to clinically relevant populations: Lessons from embryonic development." *Cell* 132(4): 661–680.

Mutha, P. K., K. Y. Haaland, and R. L. Sainburg (2012). "The effects of brain lateralization on motor control and adaptation." *Journal of Motor Behavior* 44(6): 455–469.

Nacu, E., and E. M. Tanaka (2011). "Limb regeneration: A new development?" *Annual Review of Cell and Developmental Biology* 27: 409–440.

Naguib, A., and L. C. Trotman (2013). "PTEN plasticity: How the taming of a lethal gene can go too far." *Trends in Cell Biology* 23(8): 374–379.

Nakayama, H., H. S. Jørgensen, H. O. Raaschou, and T. S. Olsen (1994). "Recovery of upper extremity function in stroke patients: The Copenhagen stroke study." *Archives of Physical Medicine and Rehabilitation* 75(4): 394–398.

Nathan, P. W., and M. C. Smith (1955). "Long descending tracts in man." *Brain* 78(2): 248–303.

Neumann-Haefelin, T., G. Hagemann, and O. W. Witte (1995). "Cellular correlates of neuronal hyperexcit-ability in the vicinity of photochemically induced cortical infarcts in rats in vitro." *Neuroscience Letters* 193(2): 101–104.

Newport, R., and T. Schenk (2012). "Prisms and neglect: What have we learned?" *Neuropsychologia* 50(6): 1080–1091.

Nexstim. (2016). "Nexstim Plc provides key update on phase III NICHE stroke therapy trial." http://www
.nexstim.com/news-and-events/press-releases/2014/nexstim-plc-provides-key-update-on-phase-iii-niche
-stroke-therapy-trial/.

Ng, K. L., E. M. Gibson, R. Hubbard, J. Yang, B. Caffo, R. J. O'Brien, J. W. Krakauer, and S. R. Zeiler
(2015). "Fluoxetine maintains a state of heightened responsiveness to motor training early after stroke in a
mouse model." *Stroke* 46(10): 2951–2960.

Ng, S.-C., S. M. de la Monte, G. L. Conboy, L. R. Karns, and M. C. Fishman (1988). "Cloning of human GAP
43: Growth association and ischemic resurgence." *Neuron* 1(2): 133–139.

Ng, S. S. M., and R. B. Shepherd (2013). "Weakness in patients with stroke: implications for strength training
in neurorehabilitation." *Physical Therapy Reviews* 5(4):227–238.

Nijland, R., E. van Wegen, R. Verbunt, R. van Wijk, J. van Kordelaar, and G. Kwakkel (2010). "A comparison
of two validated tests for upper limb function after stroke: The Wolf Motor Function Test and the Action
Research Arm Test." *Journal of Rehabilitation Medicine* 42(7): 694–696.

Nishibe, M., E. T. R. Urban, S. Barbay, and R. J. Nudo (2015). "Rehabilitative training promotes rapid motor
recovery but delayed motor map reorganization in a rat cortical ischemic infarct model." *Neurorehabilitation
and Neural Repair* 29(5): 472–482.

Nishimura, Y., S. I. Perlmutter, R. W. Eaton, and E. E. Fetz (2013). "Spike-timing-dependent plasticity in
primate corticospinal connections induced during free behavior." *Neuron* 80(5): 1301–1309.

Nissen, M. J., and P. Bullemer (1987). "Attentional requirements of learning: Evidence from performance
measures." *Cognitive Psychology* 19(1): 1–32.

Noble, M., J. E. Davies, M. Mayer-Pröschel, C. Pröschel, and S. J. A. Davies (2011). "Precursor cell biology
and the development of astrocyte transplantation therapies: Lessons from spinal cord injury." *Neurotherapeutics* 8(4): 677–693.

Noskin, O., J. W. Krakauer, R. M. Lazar, J. R. Festa, C. Handy, K. A. O'Brien, and R. S. Marshall (2008).
"Ipsilateral motor dysfunction from unilateral stroke: Implications for the functional neuroanatomy of
hemiparesis." *Journal of Neurology, Neurosurgery, and Psychiatry* 79(4): 401–406.

Nudo, R. J., and G. W. Milliken (1996). "Reorganization of movement representations in primary motor
cortex following focal ischemic infarcts in adult squirrel monkeys." *Journal of Neurophysiology* 75(5):
2144–2149.

Nudo, R. J., B. M. Wise, F. SiFuentes, and G. W. Milliken (1996). "Neural substrates for the effects of
rehabilitative training on motor recovery after ischemic infarct." *Science* 272(5269): 1791–1794.

O'Collins, V. E., M. R. Macleod, G. A. Donnan, L. L. Horky, B. H. van der Worp, and D. W. Howells (2006).
"1,026 Experimental treatments in acute stroke." *Annals of Neurology* 59(3): 467–477.

O'Dwyer, N. J., L. Ada, and P. D. Neilson (1996). "Spasticity and muscle contracture following stroke." *Brain*
119(5): 1737–1749.

O'Mahony, P. G., R. G. Thomson, R. Dobson, H. Rodgers, and O. F. W. James (1999). "The prevalence of
stroke and associated disability." *Journal of Public Health* 21(2): 166–171.

OED online (2017). "virtual reality, n." Oxford University Press. May 2017 http://en.oxforddictionaries.com
/definition/virtual_reality

Ogden, R., and S. I. Franz (1917). "On cerebral motor control: The recovery from experimentally produced
hemiplegia." *Psychobiology* 1(1): 33.

Ohab, J. J., and S. T. Carmichael (2008). "Poststroke neurogenesis: Emerging principles of migration and
localization of immature neurons." *The Neuroscientist* 14(4): 369–380.

Ohab, J. J., S. Fleming, A. Blesch, and S. T. Carmichael (2006). "A neurovascular niche for neurogenesis after
stroke." *Journal of Neuroscience* 26(50): 13007–13016.

Oki, K., J. Tatarishvili, J. Wood, P. Koch, S. Wattananit, Y. Mine, E. Monni, D. Tornero, H. Ahlenius, J.
Ladewig, O. Brüstle, O. Lindvall, and Z. Kokaia (2012). "Human-induced pluripotent stem cells form functional
neurons and improve recovery after grafting in stroke-damaged brain." *Stem Cells* 30(6): 1120–1133.

Omura, T., K. Omura, A. Tedeschi, P. Riva, Michio W. Painter, L. Rojas, J. Martin, V. Lisi, Eric A. Huebner,
A. Latremoliere, Y. Yin, L. B. Barrett, B. Singh, S. Lee, Thomas J. Crisman, F. Gao, S. Li, K. Kapur,

Daniel H. Geschwind, Kenneth S. Kosik, G. Coppola, Z. He, S. T. Carmichael, Larry I. Benowitz, M. Costigan, and Clifford J. Woolf (2015). "Robust axonal regeneration occurs in the injured CAST/Ei mouse CNS." *Neuron* 86(5): 1215–1227.

Onodera, S., and T. P. Hicks (2010). "Carbocyanine dye usage in demarcating boundaries of the aged human red nucleus." *PLoS ONE* 5(12): e14430.

Orrell, A.J., F.F. Eves, R.S. Masters, and K.M. MacMahon. (2007). "Implicit sequence learning processes after unilateral stroke." *Neuropsychological Rehabilitation* 17(3):335–54.

Ottenbacher, K. J., Y. Hsu, C. V. Granger, and R. C. Fiedler (1996). "The reliability of the functional independence measure: a quantitative review." *Archives of Physical Medicine and Rehabilitation* 77(12): 1226–1232.

Overman, J. J., and S. T. Carmichael (2014). "Plasticity in the injured brain more than molecules matter." *The Neuroscientist* 20(1): 15–28.

Overman, J. J., A. N. Clarkson, I. B. Wanner, W. T. Overman, I. Eckstein, J. L. Maguire, I. D. Dinov, A. W. Toga, and S. T. Carmichael (2012). "A role for ephrin-A5 in axonal sprouting, recovery, and activity-dependent plasticity after stroke." *Proceedings of the National Academy of Sciences* 109(33): E2230–E2239.

Page, S. J., G. D. Fulk, and P. Boyne (2012). "Clinically important differences for the upper-extremity Fugl-Meyer scale in people with minimal to moderate impairment due to chronic stroke." *Physical Therapy* 92(6): 791–798.

Park, C.-H., N. Kou, and N. S. Ward (2016). "The contribution of lesion location to upper limb deficit after stroke." *Journal of Neurology, Neurosurgery, and Psychiatry* 87:1283–1286.

Park, H.-J., H. Furmaga, J. Cooperrider, J. T. Gale, K. B. Baker, and A. G. Machado (2015). "Modulation of cortical motor evoked potential after stroke during electrical stimulation of the lateral cerebellar nucleus." *Brain Stimulation* 8(6): 1043–1048.

Park, K. I., Y. D. Teng, and E. Y. Snyder (2002). "The injured brain interacts reciprocally with neural stem cells supported by scaffolds to reconstitute lost tissue." *Nature Biotechnology* 20(11): 1111–1117.

Park, K. K., K. Liu, Y. Hu, P. D. Smith, C. Wang, B. Cai, B. Xu, L. Connolly, I. Kramvis, M. Sahin, and Z. He (2008). "Promoting axon regeneration in the adult CNS by modulation of the PTEN/mTOR pathway." *Science* 322(5903): 963–966.

Parker, V. M., D. T. Wade, and R. L. Hewer (1986). "Loss of arm function after stroke: Measurement, frequency, and recovery." *International Rehabilitation Medicine* 8(2): 69–73.

Partin, K. M. (2015). "AMPA receptor potentiators: From drug design to cognitive enhancement." *Current Opinion in Pharmacology* 20: 46–53.

Patton, J. L., M. E. Stoykov, M. Kovic, and F. A. Mussa-Ivaldi (2006). "Evaluation of robotic training forces that either enhance or reduce error in chronic hemiparetic stroke survivors." *Experimental Brain Research* 168(3): 368–383.

Paulus, W., J. Classen, L. G. Cohen, C. H. Large, V. Di Lazzaro, M. Nitsche, A. Pascual-Leone, F. Rosenow, J. C. Rothwell, and U. Ziemann (2008). "State of the art: Pharmacologic effects on cortical excitability measures tested by transcranial magnetic stimulation." *Brain Stimulation* 1(3): 151–163.

Pavo, N., S. Charwat, N. Nyolczas, A. Jakab, Z. Murlasits, J. Bergler-Klein, M. Nikfardjam, I. Benedek, T. Benedek, I. J. Pavo, B. J. Gersh, K. Huber, G. Maurer, and M. Gyöngyösi (2014). "Cell therapy for human ischemic heart diseases: Critical review and summary of the clinical experiences." *Journal of Molecular and Cellular Cardiology* 75: 12–24.

Pearson-Fuhrhop, K. M., E. Burke, and S. C. Cramer (2012). "The influence of genetic factors on brain plasticity and recovery after neural injury." *Current Opinion in Neurology* 25(6): 682–688.

Pendharkar, A. V., J. Y. Chua, R. H. Andres, N. Wang, X. Gaeta, H. Wang, A. De, R. Choi, S. Chen, B. K. Rutt, S. S. Gambhir, and R. Guzman (2010). "Biodistribution of neural stem cells after intravascular therapy for hypoxic–ischemia." *Stroke* 41(9): 2064–2070.

Pereira, V. M., H. Yilmaz, A. Pellaton, L.-A. Slater, T. Krings, and K.-O. Lovblad (2015). "Current status of mechanical thrombectomy for acute stroke treatment." *Journal of Neuroradiology* 42(1): 12–20.

Perini, F., M. Morra, M. Alecci, E. Galloni, M. Marchi, and V. Toso (2001). "Temporal profile of serum anti-inflammatory and pro-inflammatory interleukins in acute ischemic stroke patients." *Neurological Sciences* 22(4): 289–296.

Peters, N., S. Müller-Schunk, T. Freilinger, M. Düring, T. Pfefferkorn, and M. Dichgans (2009). "Ischemic stroke of the cortical 'hand knob' area: stroke mechanisms and prognosis." *Journal of Neurology* 256(7): 1146–1151.

Pierrot-Deseilligny, E., and D. Burke (2012). *The circuitry of the human spinal cord: spinal and corticospinal mechanisms of movement*. Cambridge, UK: Cambridge University Press.

Pietrobon, D., and M. A. Moskowitz (2014). "Chaos and commotion in the wake of cortical spreading depression and spreading depolarizations." *Nature Reviews Neuroscience* 15(6): 379–393.

Pirotte, B., P. Francotte, E. Goffin, and P. d. Tullio (2013). "AMPA receptor positive allosteric modulators: A patent review." *Expert Opinion on Therapeutic Patents* 23(5): 615–628.

Pizzorusso, T., P. Medini, N. Berardi, S. Chierzi, J. W. Fawcett, and L. Maffei (2002). "Reactivation of ocular dominance plasticity in the adult visual cortex." *Science* 298(5596):1248–1251.

Platz, T., C. Pinkowski, F. van Wijck, I.-H. Kim, P. Di Bella, and G. Johnson (2005). "Reliability and validity of arm function assessment with standardized guidelines for the Fugl-Meyer Test, Action Research Arm Test and Box and Block Test: A multicentre study." *Clinical Rehabilitation* 19(4): 404–411.

Plautz, E. J., S. Barbay, S. B. Frost, K. M. Friel, N. Dancause, E. V. Zoubina, A. M. Stowe, B. M. Quaney, and R. J. Nudo (2003). "Post-infarct cortical plasticity and behavioral recovery using concurrent cortical stimulation and rehabilitative training: A feasibility study in primates." *Neurological Research* 25(8): 801–810.

Plautz, E. J., G. W. Milliken, and R. J. Nudo (2000). "Effects of repetitive motor training on movement representations in adult squirrel monkeys: Role of use versus learning." *Neurobiology of Learning and Memory* 74(1): 27–55.

Plow, E. B., D. A. Cunningham, N. Varnerin, and A. Machado (2015). "Rethinking stimulation of the brain in stroke rehabilitation: Why higher motor areas might be better alternatives for patients with greater impairments." *The Neuroscientist* 21(3): 225–240.

Podda, M. V., S. Cocco, A. Mastrodonato, S. Fusco, L. Leone, S. A. Barbati, C. Colussi, C. Ripoli, and C. Grassi (2016). "Anodal transcranial direct current stimulation boosts synaptic plasticity and memory in mice via epigenetic regulation of Bdnf expression." *Scientific Reports* 6:22180.

Polanyi, M. (1966). *The tacit dimension*. Chicago: University of Chicago Press.

Poldrack, R. A., F. W. Sabb, K. Foerde, S. M. Tom, R. F. Asarnow, S. Y. Bookheimer, and B. J. Knowlton (2005). "The neural correlates of motor skill automaticity." *The Journal of Neuroscience: The Official Journal of the Society for Neuroscience* 25(22): 5356–5364.

Pollock, A., S. E. Farmer, M. C. Brady, P. Langhorne, G. E. Mead, J. Mehrholz, and F. van Wijck (2014). "Interventions for improving upper limb function after stroke." *Cochrane Database of Systematic Reviews* (11): CD010820.

Porter, R., and R. Lemon (1993). *Corticospinal function and voluntary movement*. Oxford, UK: Oxford University Press.

Powers, R. K., D. L. Campbell, and W. Z. Rymer (1989). "Stretch reflex dynamics in spastic elbow flexor muscles." *Annals of Neurology* 25(1): 32–42.

Prabhakaran, S., E. Zarahn, C. Riley, A. Speizer, J. Y. Chong, R. M. Lazar, R. S. Marshall, and J. W. Krakauer (2008). "Inter-individual variability in the capacity for motor recovery after ischemic stroke." *Neurorehabilitation and Neural Repair* 22(1): 64–71.

Prasad, K., A. Sharma, A. Garg, S. Mohanty, S. Bhatnagar, S. Johri, K. K. Singh, V. Nair, R. S. Sarkar, S. P. Gorthi, K. M. Hassan, S. Prabhakar, N. Marwaha, N. Khandelwal, U. K. Misra, J. Kalita, and S. Nityanand (2014). "Intravenous autologous bone marrow mononuclear stem cell therapy for ischemic stroke." *Stroke* 45(12): 3618–3624.

Pruszynski, J. A., and S. H. Scott (2012). "Optimal feedback control and the long-latency stretch response." *Experimental Brain Research* 218(3): 341–359.

Qü, M., I. Buchkremer-Ratzmann, K. Schiene, M. Schroeter, O. W. Witte, and K. Zilles (1998). "Bihemispheric reduction of GABAA receptor binding following focal cortical photothrombotic lesions in the rat brain." *Brain Research* 813(2): 374–380.

Rabadi, M. H., and F. M. Rabadi (2006). "Comparison of the Action Research Arm Test and the Fugl-Meyer assessment as measures of upper-extremity motor weakness after stroke." *Archives of Physical Medicine and Rehabilitation* 87(7): 962–966.

Raghavan, P., M. Santello, A. M. Gordon, and J. W. Krakauer (2010). "Compensatory motor control after stroke: An alternative joint strategy for object-dependent shaping of hand posture." *Journal of Neurophysiology* 103(6): 3034–3043.

Raghuraman, S., I. Donkin, S. Versteyhe, R. Barrès, and D. Simar (2016). "The emerging role of epigenetics in inflammation and immunometabolism." *Trends in Endocrinology & Metabolism* 27(11): 782–795.

Rashid, A. J., C. Yan, V. Mercaldo, H.-L. Hsiang, S. Park, C. J. Cole, A. D. Cristofaro, J. Yu, C. Ramakrishnan, S. Y. Lee, K. Deisseroth, P. W. Frankland, and S. A. Josselyn (2016). "Competition between engrams influences fear memory formation and recall." *Science* 353(6297): 383–387.

Redecker, C., W. Wang, J.-M. Fritschy, and O. W. Witte (2002). "Widespread and long-lasting alterations in GABAA-receptor subtypes after focal cortical infarcts in rats: Mediation by NMDA-dependent processes." *Journal of Cerebral Blood Flow & Metabolism* 22(12): 1463–1475.

Redzic, Z. B., T. Rabie, B. A. Sutherland, and A. M. Buchan (2013). "Differential effects of paracrine factors on the survival of cells of the neurovascular unit during oxygen glucose deprivation." *International Journal of Stroke* 10(3): 407–414.

Rennerfeldt, D. A., and K. J. van Vliet (2016). "Concise Review: When Colonies Are Not Clones: Evidence and Implications of Intracolony Heterogeneity in Mesenchymal Stem Cells." *Stem Cells* 34(5):1135–1141.

Reinkensmeyer, D. J., and J. L. Patton (2009). "Can robots help the learning of skilled actions?" *Exercise and Sport Sciences Reviews* 37(1): 43.

Reis, J., H. M. Schambra, L. G. Cohen, E. R. Buch, B. Fritsch, E. Zarahn, P. A. Celnik, and J. W. Krakauer (2009). "Noninvasive cortical stimulation enhances motor skill acquisition over multiple days through an effect on consolidation." *Proceedings of the National Academy of Sciences* 106(5): 1590–1595.

Reisman, D. S., H. McLean, J. Keller, K. A. Danks, and A. J. Bastian (2013). "Repeated split-belt treadmill training improves poststroke step length asymmetry." *Neurorehabilitation and Neural Repair* 27(5): 460–468.

Reisman, D. S., R. Wityk, K. Silver, and A. J. Bastian (2007). "Locomotor adaptation on a split-belt treadmill can improve walking symmetry post-stroke." *Brain* 130(7): 1861–1872.

Ren, J., P. L. Kaplan, M. F. Charette, H. Speller, and S. P. Finklestein (2000). "Time window of intracisternal osteogenic protein-1 in enhancing functional recovery after stroke." *Neuropharmacology* 39(5): 860–865.

Riddle, C. N., and S. N. Baker (2010). "Convergence of pyramidal and medial brain stem descending pathways onto macaque cervical spinal interneurons." *Journal of Neurophysiology* 103(5): 2821–2832.

Riddle, C. N., S. A. Edgley, and S. N. Baker (2009). "Direct and indirect connections with upper limb motoneurons from the primate reticulospinal tract." *The Journal of Neuroscience: The Official Journal of the Society for Neuroscience* 29(15): 4993–4999.

Ringelstein, E. B., V. Thijs, B. Norrving, A. Chamorro, F. Aichner, M. Grond, J. Saver, R. Laage, A. Schneider, F. Rathgeb, G. Vogt, G. Charissé, J. B. Fiebach, S. Schwab, W. R. Schäbitz, R. Kollmar, M. Fisher, M. Brozman, D. Skoloudik, F. Gruber, J. S. Leal, R. Veltkamp, M. Köhrmann, and J. Berrouschot (2013). "Granulocyte colony–stimulating factor in patients with acute ischemic stroke." *Stroke* 44(10): 2681–2687.

Robak, L. A., K. Venkatesh, H. Lee, S. J. Raiker, Y. Duan, J. Lee-Osbourne, T. Hofer, R. G. Mage, C. Rader, and R. J. Giger (2009). "Molecular basis of the interactions of the Nogo-66 receptor and its homolog NgR2 with myelin-associated glycoprotein: Development of NgROMNI-Fc, a novel antagonist of CNS myelin inhibition." *Journal of Neuroscience* 29(18): 5768–5783.

Robertson, E. M. (2007). "The serial reaction time task: Implicit motor skill learning?" *Journal of Neuroscience* 27(38): 10073–10075.

Roby-Brami, A., A. Feydy, M. Combeaud, E. V. Biryukova, B. Bussel, and M. F. Levin (2003). "Motor compensation and recovery for reaching in stroke patients." *Acta Neurologica Scandinavica* 107(5): 369–381.

Rodrigues, F. B., J. B. Neves, D. Caldeira, J. M. Ferro, J. J. Ferreira, and J. Costa (2016). "Endovascular treatment versus medical care alone for ischaemic stroke: Systematic review and meta-analysis." *BMJ* 353: i1754.

Roh, J., W. Z. Rymer, and R. F. Beer (2015). "Evidence for altered upper extremity muscle synergies in chronic stroke survivors with mild and moderate impairment." *Frontiers in Human Neuroscience* 9:6.

Rolls, A., R. Shechter, and M. Schwartz (2009). "The bright side of the glial scar in CNS repair." *Nature Reviews Neuroscience* 10(3): 235–241.

Rosales, R. L., F. Efendy, E. S. A. Teleg, M. M. D. D. Santos, M. C. E. Rosales, M. Ostrea, M. J. Tanglao, and A. R. Ng (2016). "Botulinum toxin as early intervention for spasticity after stroke or non-progressive brain lesion: A meta-analysis." *Journal of the Neurological Sciences* 371: 6–14.

Rosenzweig, S., and S. T. Carmichael (2013). "Age-dependent exacerbation of white matter stroke outcomes." *Stroke* 44(9): 2579–2586.

Rosenzweig, S., and S. T. Carmichael (2015). "The axon–glia unit in white matter stroke: Mechanisms of damage and recovery." *Brain Research* 1623: 123–134.

Rossetti, Y., G. Rode, L. Pisella, A. Farné, L. Li, D. Boisson, and M.-T. Perenin (1998). "Prism adaptation to a rightward optical deviation rehabilitates left hemispatial neglect." *Nature* 395(6698): 166–169.

Rosso, C., and Y. Samson (2014). "The ischemic penumbra: The location rather than the volume of recovery determines outcome." *Current Opinion in Neurology* 27(1): 35–41.

Roth, B. L. (2016). "DREADDs for neuroscientists." *Neuron* 89(4): 683–694.

Rothi, L. J., and J. Horner (1983). "Restitution and substitution: Two theories of recovery with application to neurobehavioral treatment." *Journal of Clinical Neuropsychology* 5(1): 73–81.

Rudolph, U., and H. Möhler (2014). "GABAA receptor subtypes: Therapeutic potential in Down syndrome, affective disorders, schizophrenia, and autism." *Annual review of pharmacology and toxicology* 54: 483–507.

Rymer, W. Z., J. C. Houk, and P. E. Crago (1979). "Mechanisms of the clasp-knife reflex studied in an animal model." *Experimental Brain Research* 37(1): 93–113.

Sacrey, L.-A. R., M. Alaverdashvili, and I. Q. Whishaw (2009). "Similar hand shaping in reaching-for-food (skilled reaching) in rats and humans provides evidence of homology in release, collection, and manipulation movements." *Behavioural Brain Research* 204(1): 153–161.

Saeki, S., H. Ogata, T. Okubo, K. Takahashi, and T. Hoshuyama (1995). "Return to work after stroke a follow-up study." *Stroke* 26(3): 399–401.

Sakai, S. T., A. G. Davidson, and J. A. Buford (2009). "Reticulospinal neurons in the pontomedullary reticular formation of the monkey (Macaca fascicularis)." *Neuroscience* 163(4): 1158–1170.

Sales, V.M., A. C. Ferguson-Smith, and M. E. Patti (2017). "Epigenetic Mechanisms of Transmission of Metabolic Disease across Generations." *Cell Metabolism* 25(3):559–571.

Salmelin, R., N. Forss, J. Knuutila, and R. Hari (1995). "Bilateral activation of the human somatomotor cortex by distal hand movements." *Electroencephalography and Clinical Neurophysiology* 95(6): 444–452.

Sampaio-Baptista, C., N. Filippini, C. J. Stagg, J. Near, J. Scholz, and H. Johansen-Berg (2015). "Changes in functional connectivity and GABA levels with long-term motor learning." *NeuroImage* 106: 15–20.

Sanderson, T. M., and E. Sher (2013). "The role of phosphodiesterases in hippocampal synaptic plasticity." *Neuropharmacology* 74: 86–95.

Sanin, V., C. Heeß, H. A. Kretzschmar, and U. Schüller (2013). "Recruitment of neural precursor cells from circumventricular organs of patients with cerebral ischaemia." *Neuropathology and Applied Neurobiology* 39(5): 510–518.

Sano, Y., Justin L. Shobe, M. Zhou, S. Huang, T. Shuman, Denise J. Cai, P. Golshani, M. Kamata, and Alcino J. Silva (2014). "CREB regulates memory allocation in the insular cortex." *Current Biology* 24(23): 2833–2837.

Saposnik, G., L. G. Cohen, M. Mamdani, S. Pooyania, M. Ploughman, D. Cheung, J. Shaw, J. Hall, P. Nord, and S. Dukelow (2016). "Efficacy and safety of non-immersive virtual reality exercising in stroke rehabilitation (EVREST): A randomised, multicentre, single-blind, controlled trial." *The Lancet Neurology* 15(10): 1019–1027.

Sauer, H., and M. Wartenberg (2005). "Reactive oxygen species as signaling molecules in cardiovascular differentiation of embryonic stem cells and tumor-induced angiogenesis." *Antioxidants & Redox Signaling* 7(11–12): 1423–1434.

Savitz, S. I. (2015). "Developing cellular therapies for stroke." *Stroke: A Journal of Cerebral Circulation* 46(7): 2026–2031.

Savitz, S. I., V. Misra, M. Kasam, H. Juneja, C. S. Cox, S. Alderman, I. Aisiku, S. Kar, A. Gee, and J. C. Grotta (2011). "Intravenous autologous bone marrow mononuclear cells for ischemic stroke." *Annals of Neurology* 70(1): 59–69.

Scadden, D. T. (2006). "The stem-cell niche as an entity of action." *Nature* 441(7097): 1075–1079.

Schaefer, S. Y., C. B. Patterson, and C. E. Lang (2013). "Transfer of training between distinct motor tasks after stroke: Implications for task-specific approaches to upper-extremity neurorehabilitation." *Neurorehabilitation and Neural Repair* 27(7): 602–612.

Scheibe, F., J. Ladhoff, J. Huck, M. Grohmann, K. Blazej, A. Oersal, N. Baeva, M. Seifert, and J. Priller (2012). "Immune effects of mesenchymal stromal cells in experimental stroke." *Journal of Cerebral Blood Flow & Metabolism* 32(8): 1578–1588.

Scheidt, R. A., and C. Ghez (2007). "Separate adaptive mechanisms for controlling trajectory and final position in reaching." *Journal of Neurophysiology* 98(6): 3600–3613.

Scheidt, R. A., C. Ghez, and S. Asnani (2011). "Patterns of hypermetria and terminal cocontraction during point-to-point movements demonstrate independent action of trajectory and postural controllers." *Journal of Neurophysiology* 106(5): 2368–2382.

Scheidtmann, K., W. Fries, F. Müller, and E. Koenig (2001). "Effect of levodopa in combination with physiotherapy on functional motor recovery after stroke: A prospective, randomised, double-blind study." *The Lancet* 358(9284): 787–790.

Schepens, B., and T. Drew (2006). "Descending signals from the pontomedullary reticular formation are bilateral, asymmetric, and gated during reaching movements in the cat." *Journal of Neurophysiology* 96(5): 2229–2252.

Schiene, K., C. Bruehl, K. Zilles, M. Qu, G. Hagemann, M. Kraemer, and O. W. Witte (1996). "Neuronal hyperexcitability and reduction of GABAA-receptor expression in the surround of cerebral photothrombosis." *Journal of Cerebral Blood Flow & Metabolism* 16(5): 906–914.

Schlaug, G., V. Renga, and D. Nair (2008). "Transcranial direct current stimulation in stroke recovery." *Archives of Neurology* 65(12): 1571–1576.

Schmidt, R. A., and T. Lee (1988). *Motor control and learning.* Human Kinetics, Champaign, IL, USA.

Schofield, R. (1977). "The relationship between the spleen colony-forming cell and the haemopoietic stem cell." *Blood Cells* 4(1–2): 7–25.

Schofield, Z. V., T. M. Woodruff, R. Halai, M. C.-L. Wu, and M. A. Cooper (2013). "Neutrophils—a key component of ischemia-reperfusion injury." *Shock* 40(6): 463–470.

Schulz, R., C.-H. Park, M.-H. Boudrias, C. Gerloff, F. C. Hummel, and N. S. Ward (2012). "Assessing the integrity of corticospinal pathways from primary and secondary cortical motor areas after stroke." *Stroke* 43(8): 2248–2251.

Schwab, J. M., E. Postler, T. D. Nguyen, M. Mittelbronn, R. Meyermann, and H. J. Schluesener (2000). "Connective tissue growth factor is expressed by a subset of reactive astrocytes in human cerebral infarction. " *Neuropathology and Applied Neurobiology* 26(5):434–440.

Schwam, E. M., T. Nicholas, R. Chew, C. B. Billing, W. Davidson, D. Ambrose, and L. D. Altstiel (2014). "A multicenter, double-blind, placebo-controlled trial of the pde9a inhibitor, PF-04447943, in Alzheimer's disease." *Current Alzheimer*

Scott, S. H. (2012). "The computational and neural basis of voluntary motor control and planning." *Trends in Cognitive Sciences* 16(11): 541–549.

See, J., L. Dodakian, C. Chou, V. Chan, A. McKenzie, D. J. Reinkensmeyer, and S. C. Cramer (2013). "A standardized approach to the Fugl-Meyer assessment and its implications for clinical trials." *Neurorehabilitation and Neural Repair* 27(8): 732–741.

Sehm, B., M. A. Perez, B. Xu, J. Hidler, and L. G. Cohen (2009). "Functional neuroanatomy of mirroring during a unimanual force generation task." *Cerebral Cortex.*

Semrau, J. A., T. M. Herter, S. H. Scott, and S. P. Dukelow (2015). "Examining differences in patterns of sensory and motor recovery after stroke with robotics." *Stroke: A Journal of Cerebral Circulation* 46(12): 3459–3469.

Shadmehr, R. (2017). "Distinct neural circuits for control of movement vs. holding still." *Journal of Neurophysiology* epub ahead of print.

Shadmehr, R., and J. W. Krakauer (2008). "A computational neuroanatomy for motor control." *Experimental Brain Research* 185(3): 359–381.

Shadmehr, R., and F. A. Mussa-Ivaldi (1994). "Adaptive representation of dynamics during learning of a motor task." *Journal of Neuroscience* 14(5): 3208–3224.

Shadmehr, R., M. A. Smith, and J. W. Krakauer (2010). "Error correction, sensory prediction, and adaptation in motor control." *Annual Review of Neuroscience* 33: 89–108.

Sharma, K., M. E. Selzer, and S. Li (2012). "Scar-mediated inhibition and CSPG receptors in the CNS." *Experimental Neurology* 237(2): 370–378.

Sharp, F. R., A. Lu, Y. Tang, and D. E. Millhorn (2000). "Multiple molecular penumbras after focal cerebral ischemia." *Journal of Cerebral Blood Flow & Metabolism* 20(7): 1011–1032.

Sheean, G. L. (2001). "Botulinum treatment of spasticity: Why is it so difficult to show a functional benefit? [Review]." *Current Opinion in Neurology* 14(6): 771–776.

Shelton, F. d. N. A. P., B. T. Volpe, and M. Reding (2001). "Motor impairment as a predictor of functional recovery and guide to rehabilitation treatment after stroke." *Neurorehabilitation and Neural Repair* 15(3): 229–237.

Sherrington, C. S. (1898). "Decerebrate rigidity, and reflex coordination of movements." *The Journal of Physiology* 22(4): 319–332.

Sherrington, C. S. (1910). Had man had wings, from: Physiology of brain. *Encyclopedia Britannica* 4: 403–413.

Shimada, I. S., A. Borders, A. Aronshtam, and J. L. Spees (2011). "Proliferating reactive astrocytes are regulated by Notch-1 in the peri-infarct area after stroke." *Stroke* 42(11): 3231–3237.

Shmuelof, L., V. S. Huang, A. M. Haith, R. J. Delnicki, P. Mazzoni, and J. W. Krakauer (2012). "Overcoming motor 'forgetting' through reinforcement of learned actions." *Journal of Neuroscience* 32(42): 14617–14621a.

Shmuelof, L., and J. W. Krakauer (2011). "Are we ready for a natural history of motor learning?" *Neuron* 72(3): 469–476.

Shmuelof, L., J. W. Krakauer, and P. Mazzoni (2012). "How is a motor skill learned? Change and invariance at the levels of task success and trajectory control." *Journal of Neurophysiology* 108(2): 578–594.

Shuldiner, A. R., J. R. O'Connell, K. P. Bliden, A. Gandhi, K. Ryan, R. B. Horenstein, C. M. Damcott, R. Pakyz, U. S. Tantry, and Q. Gibson (2009). "Association of cytochrome P450 2C19 genotype with the antiplatelet effect and clinical efficacy of clopidogrel therapy." *JAMA* 302(8): 849–857.

Shyu, W.-C., S.-Z. Lin, M.-F. Chiang, C.-Y. Su, and H. Li (2006). "Intracerebral peripheral blood stem cell (CD34+) implantation induces neuroplasticity by enhancing β1 integrin-mediated angiogenesis in chronic stroke rats." *Journal of Neuroscience* 26(13): 3444–3453.

Sidney, L. E., M. J. Branch, S. E. Dunphy, H. S. Dua, and A. Hopkinson (2014). "Concise review: Evidence for CD34 as a common marker for diverse progenitors." *Stem Cells* 32(6): 1380–1389.

Sigrist, R., G. Rauter, L. Marchal-Crespo, R. Riener, and P. Wolf (2015). "Sonification and haptic feedback in addition to visual feedback enhances complex motor task learning." *Experimental Brain Research* 233(3): 909–925.

Silasi, G., and Timothy H. Murphy (2014). "Stroke and the connectome: How connectivity guides therapeutic intervention." *Neuron* 83(6): 1354–1368.

Simo, L. S., D. Piovesan, L. Botzer, M. Bengtson, C. P. Ghez, R. A. Scheidt (2013). "Arm kinematics during blind and visually guided movements in hemiparetic stroke survivors." Program No. 652.06/VV4 2013 *Neuroscience Meeting Planner.* San Diego, CA: Society for Neuroscience.

Sist, B., K. Fouad, and I. R. Winship (2014). "Plasticity beyond peri-infarct cortex: Spinal up regulation of structural plasticity, neurotrophins, and inflammatory cytokines during recovery from cortical stroke." *Experimental Neurology* 252: 47–56.

Skilbeck, C. E., D. T. Wade, R. L. Hewer, and V. A. Wood (1983). "Recovery after stroke." *Journal of Neurology, Neurosurgery, and Psychiatry* 46(1): 5–8.

Smith, M. A., A. Ghazizadeh, and R. Shadmehr (2006). "Interacting adaptive processes with different timescales underlie short-term motor learning." *PLoS Biology* 4(6): e179.

Sobrino, T., M. Pérez-Mato, D. Brea, M. Rodríguez-Yáñez, M. Blanco, and J. Castillo (2012). "Temporal profile of molecular signatures associated with circulating endothelial progenitor cells in human ischemic stroke." *Journal of Neuroscience Research* 90(9): 1788–1793.

Sofroniew, M. (2012). "Transgenic techniques for cell ablation or molecular deletion to investigate functions of astrocytes and other GFAP-expressing cell types." *Methods in Molecular Biology* 814: 531–544.

Sonde, L., H. Kalimo, S. E. Fernaeus, and M. Viitanen (2000). "Low TENS treatment on post-stroke paretic arm: A three-year follow-up." *Clinical Rehabilitation* 14(1): 14–19.

Sonde, L., and J. Lökk (2007). "Effects of amphetamine and/or l-dopa and physiotherapy after stroke—a blinded randomized study." *Acta Neurologica Scandinavica* 115(1): 55–59.

Soteropoulos, D. S., S. A. Edgley, and S. N. Baker (2011). "Lack of evidence for direct corticospinal contributions to control of the ipsilateral forelimb in monkey." *The Journal of Neuroscience* 31(31): 11208–11219.

Southwell, D. G., R. C. Froemke, A. Alvarez-Buylla, M. P. Stryker, and S. P. Gandhi (2010). "Cortical plasticity induced by inhibitory neuron transplantation." *Science* 327(5969): 1145–1148.

Southwell, D. G., C. R. Nicholas, A. I. Basbaum, M. P. Stryker, A. R. Kriegstein, J. L. Rubenstein, and A. Alvarez-Buylla (2014). "Interneurons from embryonic development to cell-based therapy." *Science* 344(6180): 1240622.

Sozmen, E. G., J. D. Hinman, and S. T. Carmichael (2012). "Models that matter: White matter stroke models." *Neurotherapeutics* 9(2): 349–358.

Sozmen, E. G., A. Kolekar, L. A. Havton, and S. T. Carmichael (2009). "A white matter stroke model in the mouse: Axonal damage, progenitor responses and MRI correlates." *Journal of Neuroscience Methods* 180(2): 261–272.

Sozmen, E. G., S. Rosenzweig, I. L. Llorente, D. J. DiTullio, M. Machnicki, H. V. Vinters, L. A. Havton, R. J. Giger, J. D. Hinman, and S. T. Carmichael (2016). "Nogo receptor blockade overcomes remyelination failure after white matter stroke and stimulates functional recovery in aged mice." *Proceedings of the National Academy of Sciences* 113(52): E8453–E8462.

Span, P. (2016). "F.T.C.'s lumosity penalty doesn't end brain training debate." *The New York Times* 1/19/2016.

Spencer, H. (1855). *The principles of psychology*. Longman, Brown, Green and Longmans, London, England.

Stagg, C. J., V. Bachtiar, and H. Johansen-Berg (2011). "What are we measuring with GABA magnetic resonance spectroscopy?" *Communicative & Integrative Biology* 4(5): 573–575.

Stanley, J., and J. W. Krakauer (2013). "Motor skill depends on knowledge of facts." *Frontiers in Human Neuroscience* 7: 503.

Starkey, M. L., and M. E. Schwab (2014). "How plastic is the brain after a stroke?" *The Neuroscientist* 20(4): 359–371.

Stein, J. (2009). "VECTORS study: Are we going in the right direction? NEJM Journal Watch." http://www.jwatch.org/jn200910270000001/2009/10/27/vectors-study-are-we-going-right-direction

Steinbeck, Julius A., and L. Studer (2015). "Moving stem cells to the clinic: Potential and limitations for brain repair." *Neuron* 86(1): 187–206.

Steinberg, G. K., D. Kondziolka, L. R. Wechsler, L. D. Lunsford, M. L. Coburn, J. B. Billigen, A. S. Kim, J. N. Johnson, D. Bates, B. King, C. Case, M. McGrogan, E. W. Yankee, and N. E. Schwartz (2016). "Clinical outcomes of transplanted modified bone marrow–derived mesenchymal stem cells in stroke: A phase 1/2a study." *Stroke* 48(6):1–8.

Stellwagen, D., and C. J. Shatz (2002). "An instructive role for retinal waves in the development of retinogeniculate connectivity." *Neuron* 33(3): 357–367.

Stephany, C.-É., M. G. Frantz, and A. W. McGee (2016). "Multiple roles for Nogo receptor 1 in visual system plasticity." *The Neuroscientist* 22(6): 653–666.

Stephenson, D. T., T. M. Coskran, M. P. Kelly, R. J. Kleiman, D. Morton, S. M. O'Neill, C. J. Schmidt, R. J. Weinberg, and F. S. Menniti (2012). "The distribution of phosphodiesterase 2A in the rat brain." *Neuroscience* 226: 145–155.

Stinear, C., S. Ackerley, and W. Byblow (2013). "Rehabilitation is initiated early after stroke, but most motor rehabilitation trials are not." *Stroke* 44(7): 2039–2045.

Stinear, C. M., P. A. Barber, P. R. Smale, J. P. Coxon, M. K. Fleming, and W. D. Byblow (2007). "Functional potential in chronic stroke patients depends on corticospinal tract integrity." *Brain* 130(1): 170–180.

Stinear, C. M., M. A. Petoe, S. Anwar, P. A. Barber, and W. D. Byblow (2014). "Bilateral priming accelerates recovery of upper limb function after stroke: A randomized controlled trial." *Stroke* 45(1): 205–210.

Stinear, C. M., and N. S. Ward (2013). "How useful is imaging in predicting outcomes in stroke rehabilitation?" *International Journal of Stroke* 8(1): 33–37.

Strathern, M. (1997). "Improving ratings': audit in the British University system." *European Review* 5(3): 305–321.

Stroemer, P., A. Hope, S. Patel, K. Pollock, and J. Sinden (2007). "Development of a human neural stem cell line for use in recovery from disability after stroke." *Frontiers in Bioscience* 13: 2290–2292.

Sugiyama, Y., N. Higo, K. Yoshino-Saito, Y. Murata, Y. Nishimura, T. Oishi, and T. Isa (2013). "Effects of early versus late rehabilitative training on manual dexterity after corticospinal tract lesion in macaque monkeys." *Journal of Neurophysiology* 109(12): 2853–2865.

Sukal, T. M., M. D. Ellis, and J. P. Dewald (2007). "Shoulder abduction-induced reductions in reaching work area following hemiparetic stroke: Neuroscientific implications." *Experimental Brain Research* 183(2): 215–223.

Summers, J. J., F. A. Kagerer, M. I. Garry, C. Y. Hiraga, A. Loftus, and J. H. Cauraugh (2007). "Bilateral and unilateral movement training on upper limb function in chronic stroke patients: A TMS study." *Journal of the Neurological Sciences* 252(1): 76–82.

Sun, F., K. K. Park, S. Belin, D. Wang, T. Lu, G. Chen, K. Zhang, C. Yeung, G. Feng, B. A. Yankner, and Z. He (2011). "Sustained axon regeneration induced by co-deletion of PTEN and SOCS3." *Nature* 480(7377): 372–375.

Sun, X., Z. Zhou, T. Liu, M. Zhao, S. Zhao, T. Xiao, J. Jolkkonen, and C. Zhao (2016). "Fluoxetine enhances neurogenesis in aged rats with cortical infarcts, but this is not reflected in a behavioral recovery." *Journal of Molecular Neuroscience* 58(2): 233–242.

Sunderland, A., M. P. Bowers, S.-M. Sluman, D. J. Wilcock, and M. E. Ardron (1999). "Impaired dexterity of the ipsilateral hand after stroke and the relationship to cognitive deficit." *Stroke* 30(5): 949–955.

Sundet, K., A. Finset, and I. Reinvang (1988). "Neuropsychological predictors in stroke rehabilitation." *Journal of Clinical and Experimental Neuropsychology* 10(4): 363–379.

Suzuki, S., K. Tanaka, and N. Suzuki (2009). "Ambivalent aspects of interleukin-6 in cerebral ischemia: Inflammatory versus neurotrophic aspects." *Journal of Cerebral Blood Flow & Metabolism* 29(3): 464–479.

Swayne, O. B. C., J. C. Rothwell, N. S. Ward, and R. J. Greenwood (2008). "Stages of motor output reorganization after hemispheric stroke suggested by longitudinal studies of cortical physiology." *Cerebral Cortex* 18(8): 1909–1922.

Tagawa, Y., P. O. Kanold, M. Majdan, and C. J. Shatz (2005). "Multiple periods of functional ocular dominance plasticity in mouse visual cortex." *Nature Neuroscience* 8(3): 380–388.

Taguchi, A., C. Sakai, T. Soma, Y. Kasahara, D. M. Stern, K. Kajimoto, M. Ihara, T. Daimon, K. Yamahara, K. Doi, N. Kohara, H. Nishimura, T. Matsuyama, H. Naritomi, N. Sakai, and K. Nagatsuka (2015). "Intravenous autologous bone marrow mononuclear cell transplantation for stroke: Phase1/2a clinical trial in a homogeneous group of stroke patients." *Stem Cells and Development* 24(19): 2207–2218.

Taguchi, A., T. Soma, H. Tanaka, T. Kanda, H. Nishimura, H. Yoshikawa, Y. Tsukamoto, H. Iso, Y. Fujimori, D. M. Stern, H. Naritomi, and T. Matsuyama (2004). "Administration of CD34+ cells after stroke enhances neurogenesis via angiogenesisin a mouse model." *The Journal of Clinical Investigation* 114(3): 330–338.

Takahashi, K., and S. Yamanaka (2006). "Induction of pluripotent stem cells from mouse embryonic and adult fibroblast cultures by defined factors." *Cell* 126(4): 663–676.

Takekazu, K., H. Katsuhiko, Y. Atsushi, and Y. Toshihide (2007). "Rho-ROCK inhibitors as emerging strategies to promote nerve regeneration." *Current Pharmaceutical Design* 13(24): 2493–2499.

Takesian, A. E., and T. K. Hensch (2013). "Balancing plasticity/stability across brain development." *Progress in Brain Resesrach* 207: 3–34.

Takeuchi, N., T. Chuma, Y. Matsuo, I. Watanabe, and K. Ikoma (2005). "Repetitive transcranial magnetic stimulation of contralesional primary motor cortex improves hand function after stroke." *Stroke* 36(12): 2681–2686.

Tang, A., and W. Z. Rymer (1981). "Abnormal force–EMG relations in paretic limbs of hemiparetic human subjects." *Journal of Neurology, Neurosurgery & Psychiatry* 44(8): 690–698.

Tanna, T., and V. Sachan (2014). "Mesenchymal stem cells: potential in treatment of neurodegenerative diseases." *Current Stem Cell Research and Therapy* 9(6):513–521.

Taub, E. (1980). "Somatosensory deafferentation research with monkeys: implications for rehabilitation medicine." In: *Behavioral psychology in rehabilitation medicine: clinical applications.* Ince L.P., editor. Williams & Wilkins, New York, USA.

Taub, E., J. E. Crago, L. D. Burgio, T. E. Groomes, E. W. Cook, S. C. DeLuca, and N. E. Miller (1994). "An operant approach to rehabilitation medicine: overcoming learned nonuse by shaping." *Journal of the Experimental Analysis of Behavior* 61(2): 281–293.

Taylor, J. A., J. W. Krakauer, and R. B. Ivry (2014). "Explicit and implicit contributions to learning in a sensorimotor adaptation task." *Journal of Neuroscience* 34(8): 3023–3032.

Tennant, K. A., D. L. Adkins, M. D. Scalco, N. A. Donlan, A. L. Asay, N. Thomas, J. A. Kleim, and T. A. Jones (2012). "Skill learning induced plasticity of motor cortical representations is time and age-dependent." *Neurobiology of Learning and Memory* 98(3): 291–302.

Tennant, K. A., and C. E. Brown (2013). "Diabetes augments in vivo microvascular blood flow dynamics after stroke." *The Journal of Neuroscience* 33(49): 19194–19204.

Terasaki, Y., Y. Liu, K. Hayakawa, L. D. Pham, E. H. Lo, X. Ji, and K. Arai (2014). "Mechanisms of neurovascular dysfunction in acute ischemic brain." *Current Medicinal Chemistry* 21(18): 2035–2042.

Teskey, G. C., C. Flynn, C. D. Goertzen, M. H. Monfils, and N. A. Young (2003). "Cortical stimulation improves skilled forelimb use following a focal ischemic infarct in the rat." *Neurological Research* 25(8): 794–800.

Therrien, A. S., D. M. Wolpert, and A. J. Bastian (2016). "Effective reinforcement learning following cerebellar damage requires a balance between exploration and motor noise." *Brain* 139(1): 101–114.

Thompson, L. H., and A. Björklund (2015). "Reconstruction of brain circuitry by neural transplants generated from pluripotent stem cells." *Neurobiology of Disease* 79: 28–40.

Thoroughman, K. A., and R. Shadmehr (2000). "Learning of action through adaptive combination of motor primitives." *Nature* 407(6805): 742–747.

Thorsén, A.-M., L. Widén Holmqvist, and L. von Koch (2006). "Early supported discharge and continued rehabilitation at home after stroke: 5-year follow-up of resource use." *Journal of Stroke and Cerebrovascular Diseases* 15(4): 139–143.

Tonchev, A. B., T. Yamashima, J. Guo, G. N. Chaldakov, and N. Takakura (2007). "Expression of angiogenic and neurotrophic factors in the progenitor cell niche of adult monkey subventricular zone." *Neuroscience* 144(4): 1425–1435.

Tornero, D., S. Wattananit, M. G. Madsen, P. Koch, J. Wood, J. Tatarishvili, Y. Mine, R. Ge, E. Monni, K. Devaraju, R. F. Hevner, O. Brüstle, O. Lindvall, and Z. Kokaia (2013). "Human induced pluripotent stem cell-derived cortical neurons integrate in stroke-injured cortex and improve functional recovery." *Brain* 136(12): 3561–3577.

Tower, S. S. (1940). "Pyramidal lesion in the monkey." *Brain* 63(1): 36–90.

Trachtenberg, J. T., B. E. Chen, G. W. Knott, G. Feng, J. R. Sanes, E. Welker, and K. Svoboda (2002). "Long-term in vivo imaging of experience-dependent synaptic plasticity in adult cortex." *Nature* 420(6917): 788–794.

Turton, A., S. Wroe, N. Trepte, C. Fraser, and R. N. Lemon (1996). "Contralateral and ipsilateral EMG responses to transcranial magnetic stimulation during recovery of arm and hand function after stroke." *Electroencephalography and Clinical Neurophysiology/Electromyography and Motor Control* 101(4): 316–328.

Tuszynski, M. H., Y. Wang, L. Graham, M. Gao, D. Wu, J. Brock, A. Blesch, E. S. Rosenzweig, L. A. Havton, B. Zheng, J. M. Conner, M. Marsala, and P. Lu (2014). "Neural stem cell dissemination after grafting to CNS injury sites." *Cell* 156(3): 388–389.

Twitchell, T. E. (1951). "The restoration of motor function following hemiplegia in man." *Brain* 74(4): 443–480.

Vallence, A. M., and M. C. Ridding (2014). "Non-invasive induction of plasticity in the human cortex: uses and limitations." *Cortex* 58:261–271.

van der Lee, J. H., H. Beckerman, G. J. Lankhorst, and L. M. Bouter (2001). "The responsiveness of the Action Research Arm test and the Fugl-Meyer Assessment scale in chronic stroke patients." *Journal of Rehabilitation Medicine* 33(3): 110–113.

van Kordelaar, J., E. van Wegen, and G. Kwakkel (2014). "Impact of time on quality of motor control of the paretic upper limb after stroke." *Archives of Physical Medicine and Rehabilitation* 95(2): 338–344.

van Kordelaar, J., E. E. H. van Wegen, R. H. M. Nijland, A. Daffertshofer, and G. Kwakkel (2013). "Understanding adaptive motor control of the paretic upper limb early poststroke: the EXPLICIT-stroke program." *Neurorehabilitation and Neural Repair* 27(9): 854–863.

van Kordelaar, J., E. E. H. van Wegen, R. H. M. Nijland, J. H. de Groot, C. G. M. Meskers, J. Harlaar, and G. Kwakkel (2012). "Assessing longitudinal change in coordination of the paretic upper limb using on-site 3-dimensional kinematic measurements." *Physical Therapy* 92(1): 142.

van Kuijk, A. A., J. W. Pasman, H. T. Hendricks, M. J. Zwarts, and A. C. H. Geurts (2009). "Predicting hand motor recovery in severe stroke: The role of motor evoked potentials in relation to early clinical assessment." *Neurorehabilitation and Neural Repair* 23(1): 45–51.

Veerbeek, J. M., A. C. Langbroek-Amersfoort, E. E. H. van Wegen, C. G. M. Meskers, and G. Kwakkel (2016). "Effects of robot-assisted therapy for the upper limb after stroke: A systematic review and meta-analysis." *Neurorehabilitation and Neural Repair.* 31(2):107–121.

Veerbeek, J. M., E. v. Wegen, R. v. Peppen, P. J. v. d. Wees, E. Hendriks, M. Rietberg, and G. Kwakkel (2014). "What is the evidence for physical therapy poststroke? A systematic review and meta-analysis." *PLoS ONE* 9(2): e87987.

Verstynen, T., and P. N. Sabes (2011). "How each movement changes the next: An experimental and theoretical study of fast adaptive priors in reaching." *Journal of Neuroscience* 31(27): 10050–10059.

Vetencourt, J. F. M., A. Sale, A. Viegi, L. Baroncelli, R. De Pasquale, O. F. O'Leary, E. Castrén, and L. Maffei (2008). "The antidepressant fluoxetine restores plasticity in the adult visual cortex." *Science* 320(5874): 385–388.

Via, A. G., A. Frizziero, and F. Oliva (2012). "Biological properties of mesenchymal Stem Cells from different sources." *Muscles, Ligaments and Tendons Journal* 2(3): 154–162.

Vieira, H. L. A., P. M. Alves, and A. Vercelli (2011). "Modulation of neuronal stem cell differentiation by hypoxia and reactive oxygen species." *Progress in Neurobiology* 93(3): 444–455.

Virley, D., S. J. Hadingham, J. C. Roberts, B. Farnfield, H. Elliott, G. Whelan, J. Golder, C. David, A. A. Parsons, and A. J. Hunter (2004). "A new primate model of focal stroke: endothelin-1—induced middle cerebral artery occlusion and reperfusion in the common marmoset." *Journal of Cerebral Blood Flow & Metabolism* 24(1): 24–41.

von Bornstädt, D., T. Houben, J. L. Seidel, Y. Zheng, E. Dilekoz, T. Qin, N. Sandow, S. Kura, K. Eikermann-Haerter, M. Endres, David A. Boas, Michael A. Moskowitz, Eng H. Lo, Jens P. Dreier, S. Woitzik, S. Sakadžić, and C. Ayata (2015). "Supply-demand mismatch transients in susceptible peri-infarct hot zones explain the origins of spreading injury depolarizations." *Neuron* 85(5): 1117–1131.

von Monakow, C. (1914). *Die Lokalisation im Grosshirn und der Abbau der Funktion durch kortikale Herde.* JF Bergmann. Wiesbaden, Germany.

Vu, Q., K. Xie, M. Eckert, W. Zhao, and S. C. Cramer (2014). "Meta-analysis of preclinical studies of mesenchymal stromal cells for ischemic stroke." *Neurology* 82(14): 1277–1286.

Waddell, K. J., R. L. Birkenmeier, J. L. Moore, T. G. Hornby, and C. E. Lang (2014). "Feasibility of high-repetition, task-specific training for individuals with upper-extremity paresis." *American Journal of Occupational Therapy* 68(4): 444–453.

Wagner, J. M., C. E. Lang, S. A. Sahrmann, Q. Hu, A. J. Bastian, D. F. Edwards, and A. W. Dromerick (2006). "Relationships between sensorimotor impairments and reaching deficits in acute hemiparesis." *Neurorehabilitation and Neural Repair* 20(3): 406–416.

Wahl, A. S., W. Omlor, J. C. Rubio, J. L. Chen, H. Zheng, A. Schröter, M. Gullo, O. Weinmann, K. Kobayashi, and F. Helmchen (2014). "Asynchronous therapy restores motor control by rewiring of the rat corticospinal tract after stroke." *Science* 344(6189): 1250–1255.

Wahl, A. S., W. Omlor, J. C. Rubio, J. L. Chen, H. Zheng, A. Schröter, M. Gullo, O. Weinmann, K. Kobayashi, F. Helmchen, B. Ommer, and M. E. Schwab (2014). "Asynchronous therapy restores motor control by rewiring of the rat corticospinal tract after stroke." *Science* 344(6189): 1250–1255.

Walshe, F. M. R. (1919). "On the genesis and physiological significance of spasticity and other disorders of motor innervation: With a consideration of the functional relationships of the pyramidal system." *Brain* 42(1): 1–28.

Walshe, F. M. R. (1947). "On the role of the pyramidal system in willed movements." *Brain* 70(3): 329–354.

Wan, C. Y., X. Zheng, S. Marchina, A. Norton, and G. Schlaug (2014). "Intensive therapy induces contralateral white matter changes in chronic stroke patients with Broca's aphasia." *Brain and Language* 136: 1–7.

Wan, L., K. Pantel, and Y. Kang (2013). "Tumor metastasis: Moving new biological insights into the clinic." *Nature Medicine* 19(11): 1450–1464.

Wang, D., and J. Fawcett (2012). "The perineuronal net and the control of CNS plasticity." *Cell and Tissue Research* 349(1): 147–160.

Wang, J., and I. Conboy (2010). "Embryonic vs. adult myogenesis: Challenging the 'regeneration recapitulates development' paradigm." *Journal of Molecular Cell Biology* 2(1): 1–4.

Wang, S., J. Bates, X. Li, S. Schanz, D. Chandler-Militello, C. Levine, N. Maherali, L. Studer, K. Hochedlinger, M. Windrem, and Steven A. Goldman (2013). "Human iPSC-derived oligodendrocyte progenitor cells can myelinate and rescue a mouse model of congenital hypomyelination." *Cell Stem Cell* 12(2): 252–264.

Wanner, I. B., M. A. Anderson, B. Song, J. Levine, A. Fernandez, Z. Gray-Thompson, Y. Ao, and M. V. Sofroniew (2013). "Glial scar borders are formed by newly proliferated, elongated astrocytes that interact to corral inflammatory and fibrotic cells via STAT3-dependent mechanisms after spinal cord injury." *The Journal of Neuroscience* 33(31): 12870–12886.

Ward, N. S., and L. G. Cohen (2004). "Mechanisms underlying recovery of motor function after stroke." *Archives of Neurology* 61(12): 1844–1848.

Waters-Metenier, S., M. Husain, T. Wiestler, and J. Diedrichsen (2014). "Bihemispheric transcranial direct current stimulation enhances effector-independent representations of motor synergy and sequence learning." *Journal of Neuroscience* 34(3): 1037–1050.

Wei, L., D.-J. Ying, L. Cui, J. Langsdorf, and S. Ping Yu (2004). "Necrosis, apoptosis and hybrid death in the cortex and thalamus after barrel cortex ischemia in rats." *Brain Research* 1022(1–2): 54–61.

Weiss, P. H., S. D. Ubben, S. Kaesberg, E. Kalbe, J. Kessler, T. Liebig, and G. R. Fink (2016). "Where language meets meaningful action: A combined behavior and lesion analysis of aphasia and apraxia." *Brain Structure and Function* 221(1): 563–576.

Westover, A. N., and E. A. Halm (2012). "Do prescription stimulants increase the risk of adverse cardiovascular events? A systematic review." *BMC Cardiovascular Disorders* 12(1): 41.

Whishaw, I. Q., S. M. Pellis, and B. P. Gorny (1992). "Skilled reaching in rats and humans: evidence for parallel development or homology." *Behavioural Brain Research* 47(1): 59–70.

Whitehead, J., K. Bolland, E. Valdès-Márquez, A. Lihic, M. Ali, and K. Lees (2009). "Using historical lesion volume data in the design of a new phase II clinical trial in acute stroke." *Stroke* 40(4): 1347–1352.

Wilems, T. S., and S. E. Sakiyama-Elbert (2015). "Sustained dual drug delivery of anti-inhibitory molecules for treatment of spinal cord injury." *Journal of Controlled Release* 213:103–111.

Willi, R., and M. E. Schwab (2013). "Nogo and Nogo receptor: Relevance to schizophrenia?" *Neurobiology of Disease* 54: 150–157.

Windle, V., and D. Corbett (2005). "Fluoxetine and recovery of motor function after focal ischemia in rats." *Brain Research* 1044(1): 25–32.

Winstein, C. J., and P. S. Pohl (1995). "Effects of unilateral brain damage on the control of goal-directed hand movements." *Experimental Brain Research* 105(1): 163–174.

Winstein, C. J., P. S. Pohl, and R. Lewthwaite (1994). "Effects of physical guidance and knowledge of results on motor learning: Support for the guidance hypothesis." *Research Quarterly for Exercise and Sport* 65(4): 316–323.

Winstein, C. J., S. L. Wolf, A. W. Dromerick, C. J. Lane, M. A. Nelsen, R. Lewthwaite, S. Y. Cen, and S. P. Azen (2016). "Effect of a task-oriented rehabilitation program on upper extremity recovery following motor stroke: The ICARE randomized clinical trial." *JAMA* 315(6): 571–581.

Winstein, C. J., S. L. Wolf, and N. Schweighofer (2014). "Task-oriented training to promote upper extremity recovery." *Stroke Recovery and Rehabilitation* 2: 320–343.

Winters, C., E. E. H. v. Wegen, A. Daffertshofer, and G. Kwakkel (2015). "Generalizability of the proportional recovery model for the upper extremity after an ischemic stroke." *Neurorehabilitation and Neural Repair* 29(7): 614–622.

Withers, G. S., D. Higgins, M. Charette, and G. Banker (2000). "Bone morphogenetic protein-7 enhances dendritic growth and receptivity to innervation in cultured hippocampal neurons." *European Journal of Neuroscience* 12(1): 106–116.

Witte, O. W., and G. Stoll (1997). "Delayed and remote effects of focal cortical infarctions: Secondary damage and reactive plasticity." *Advances in Neurology* 73: 207–227.

Wolf, S. L., P. A. Catlin, M. Ellis, A. L. Archer, B. Morgan, and A. Piacentino (2001). "Assessing Wolf motor function test as outcome measure for research in patients after stroke." *Stroke* 32(7): 1635–1639.

Wolf, S. L., P. A. Thompson, C. J. Winstein, J. P. Miller, S. R. Blanton, D. S. Nichols-Larsen, D. M. Morris, G. Uswatte, E. Taub, K. E. Light, and L. Sawaki (2010). "The EXCITE stroke trial: Comparing early and delayed constraint-induced movement therapy." *Stroke* 41(10): 2309–2315.

Wonders, C. P., and S. A. Anderson (2006). "The origin and specification of cortical interneurons." *Nature Reviews Neuroscience* 7(9): 687–696.

Wong, A. L., M. A. Lindquist, A. M. Haith, and J. W. Krakauer (2015). "Explicit knowledge enhances motor vigor and performance: Motivation versus practice in sequence tasks." *Journal of Neurophysiology* 114(1): 219–232.

World Health Organization (2002). "Towards a common language for functioning, disability and health: ICF." Geneva, Switzerland.

Wright, Z. A., W. Z. Rymer, and M. W. Slutzky (2014). "Reducing abnormal muscle coactivation after stroke using a myoelectric-computer interface: A pilot study." *Neurorehabilitation and Neural Repair* 28(5): 443–451.

Wu, H., Y. Zhou, and Z.-Q. Xiong (2007). "Transducer of regulated CREB and late phase long-term synaptic potentiation." *FEBS Journal* 274(13): 3218–3223.

Wyller, T. B., U. Sveen, K. M. Sødring, A. M. Pettersen, and E. Bautz-Holter (1997). "Subjective well-being one year after stroke." *Clinical Rehabilitation* 11(2): 139–145.

Xie, Y., J. Zhang, Y. Lin, X. Gaeta, X. Meng, Dona R. R. Wisidagama, J. Cinkornpumin, C. M. Koehler, C. S. Malone, M. A. Teitell, and W. E. Lowry (2014). "Defining the role of oxygen tension in human neural progenitor fate." *Stem Cell Reports* 3(5): 743–757.

Xu, B., D. Park, Y. Ohtake, H. Li, U. Hayat, J. Liu, M. E. Selzer, F. M. Longo, and S. Li (2015). "Role of CSPG receptor LAR phosphatase in restricting axon regeneration after CNS injury." *Neurobiology of Disease* 73: 36–48.

Xu, J., M. Branscheidt, H. Schambra, G. Liuzzi, L. Steiner, N. Kim, T. Kitago, A. Luft, J. W. Krakauer, and P. A. Celnik (2016). "Abnormal interhemispheric interactions are present in the chronic but not in the acute or subacute post-stroke periods." Program No. 436.14/UU11 2016 *Neuroscience Meeting Planner*. San Diego, CA: Society for Neuroscience.

Xu, J., N. Ejaz, B. Hertler, M. Branscheidt, M. Widmer, A. V. Faria, M. Harran, J. C. Cortes, N. Kim, P. A. Celnik, T. Kitago, A. Luft, J. W. Krakauer, and J. Diedrichsen (2017). "Separable systems for finger strength and control." *Journal of Neurophysiology*. In press.

Xu, Y., H.-T. Zhang, and J. M. O'Donnell (2011). "Phosphodiesterases in the central nervous system: Implications in mood and cognitive disorders." In *Phosphodiesterases as drug targets*, edited by S. H. Francis, M. Conti, and M. D. Houslay. Berlin: Springer: 447–485.

Xu-Wilson, M., D. S. Zee, and R. Shadmehr (2009). "The intrinsic value of visual information affects saccade velocities." *Experimental Brain Research* 196(4): 475–481.

Yadav, V., and R. L. Sainburg (2014). "Handedness can be explained by a serial hybrid control scheme." *Neuroscience* 278: 385–396.

Yagi, M., H. Yasunaga, H. Matsui, K. Morita, K. Fushimi, M. Fujimoto, T. Koyama, and J. Fujitani (2017). "Impact of rehabilitation on outcomes in patients with ischemic stroke." *Stroke*. 48:740–746.

Yan, Y.-P., K. A. Sailor, B. T. Lang, S.-W. Park, R. Vemuganti, and R. J. Dempsey (2007). "Monocyte chemoattractant protein-1 plays a critical role in neuroblast migration after focal cerebral ischemia." *Journal of Cerebral Blood Flow & Metabolism* 27(6): 1213–1224.

Yang, B., E. Migliati, K. Parsha, K. Schaar, X. Xi, J. Aronowski, and S. I. Savitz (2013). "Intra-arterial delivery is not superior to intravenous delivery of autologous bone marrow mononuclear cells in acute ischemic stroke." *Stroke* 44(12): 3463–3472.

Yang, X.-D., W. F. Bischof, and P. Boulanger (2008). "Validating the performance of haptic motor skill training." *IEEE: Symposium on Haptic Interfaces for Virtual Environment and Teleoperator Systems.* Reno,USA, pp: 129–135.

York, G. K., and D. A. Steinberg (2011). "Hughlings Jackson's neurological ideas." *Brain* 134(10): 3106–3113.

Young, C. C., O. Al-Dalahmah, N. J. Lewis, K. J. Brooks, M. M. Jenkins, F. Poirier, A. M. Buchan, and F. G. Szele (2014). "Blocked angiogenesis in Galectin-3 null mice does not alter cellular and behavioral recovery after middle cerebral artery occlusion stroke." *Neurobiology of Disease* 63: 155–164.

Yozbatiran, N., L. Der-Yeghiaian, and S. C. Cramer (2008). "A standardized approach to performing the action research arm test." *Neurorehabilitation and Neural Repair* 22(1): 78–90.

Zaaimi, B., S. A. Edgley, D. S. Soteropoulos, and S. N. Baker (2012). "Changes in descending motor pathway connectivity after corticospinal tract lesion in macaque monkey." *Brain* 135(7): 2277–2289.

Zackowski, K. M., A. W. Dromerick, S. A. Sahrmann, W. T. Thach, and A. J. Bastian (2004). "How do strength, sensation, spasticity and joint individuation relate to the reaching deficits of people with chronic hemiparesis?" *Brain* 127(5): 1035–1046.

Zai, L., C. Ferrari, C. Dice, S. Subbaiah, L. A. Havton, G. Coppola, D. Geschwind, N. Irwin, E. Huebner, and S. M. Strittmatter (2011). "Inosine augments the effects of a Nogo receptor blocker and of environmental enrichment to restore skilled forelimb use after stroke." *The Journal of Neuroscience* 31(16): 5977–5988.

Zai, L., C. Ferrari, S. Subbaiah, L. A. Havton, G. Coppola, S. Strittmatter, N. Irwin, D. Geschwind, and L. I. Benowitz (2009). "Inosine alters gene expression and axonal projections in neurons contralateral to a cortical infarct and improves skilled use of the impaired limb." *Journal of Neuroscience* 29(25): 8187–8197.

Zaiou, M., and H. El Amri (2017). "Cardiovascular pharmacogenetics: A promise for genomically-guided therapy and personalized medicine." *Clinical Genetics.* 91(3):355–370

Zamanian, J. L., L. Xu, L. C. Foo, N. Nouri, L. Zhou, R. G. Giffard, and B. A. Barres (2012). "Genomic analysis of reactive astrogliosis." *Journal of Neuroscience* 32(18): 6391–6410.

Zarahn, E., L. Alon, S. L. Ryan, R. M. Lazar, M.-S. Vry, C. Weiller, R. S. Marshall, and J. W. Krakauer (2011). "Prediction of motor recovery using initial impairment and fMRI 48 h poststroke." *Cerebral Cortex* 21(12): 2712–2721.

Zee, D. S., and D. A. Robinson (1979). "A hypothetical explanation of saccadic oscillations." *Annals of Neurology* 5(5): 405–414.

Zeiler, S. R., E. M. Gibson, R. E. Hoesch, M. Y. Li, P. F. Worley, R. J. O'Brien, and J. W. Krakauer (2013). "Medial premotor cortex shows a reduction in inhibitory markers and mediates recovery in a mouse model of focal stroke." *Stroke* 44(2): 483–489.

Zeiler, S. R., R. Hubbard, E. M. Gibson, T. Zheng, K. Ng, R. O'Brien, and J. W. Krakauer (2016). "Paradoxical motor recovery from a first stroke after induction of a second stroke reopening a postischemic sensitive period." *Neurorehabilitation and Neural Repair* 30(8): 794–800.

Zeiler, S. R., and J. W. Krakauer (2013). "The interaction between training and plasticity in the poststroke brain." *Current Opinion in Neurology* 26(6): 609–616.

Zhang, J., and M. Chopp (2013). "Cell-based therapy for ischemic stroke." *Expert Opinion on Biological Therapy* 13(9): 1229–1240.

Zhang, J., T. Tokatlian, J. Zhong, Q. K. T. Ng, M. Patterson, W. E. Lowry, S. T. Carmichael, and T. Segura (2011). "Physically associated synthetic hydrogels with long-term covalent stabilization for cell culture and stem cell transplantation." *Advanced Materials* 23(43): 5098–5103.

Zhang, J., Y. Zhang, S. Xing, Z. Liang, and J. Zeng (2012). "Secondary neurodegeneration in remote regions after focal cerebral infarction." *Stroke* 43(6): 1700–1705.

Zhang, L., M. Chopp, D. H. Meier, S. Winter, L. Wang, A. Szalad, M. Lu, M. Wei, Y. Cui, and Z. G. Zhang (2013). "Sonic hedgehog signaling pathway mediates cerebrolysin-improved neurological function after stroke." *Stroke* 44(7): 1965–1972.

Zhao, Y., and D. A. Rempe (2010). "Targeting astrocytes for stroke therapy." *Neurotherapeutics* 7(4): 439–451.

Zheng, X., and G. Schlaug (2015). "Structural white matter changes in descending motor tracts correlate with improvements in motor impairment after undergoing a treatment course of tDCS and physical therapy." *Frontiers in Human Neuroscience* 9:229.

Zhong, J., A. Chan, L. Morad, H. I. Kornblum, F. Guoping, and S. T. Carmichael (2010). "Hydrogel matrix to support stem cell survival after brain transplantation in stroke." *Neurorehabilitation and Neural Repair* 24(7): 636–644.

Zhou, M., S. Greenhill, S. Huang, T. K. Silva, Y. Sano, S. Wu, Y. Cai, Y. Nagaoka, M. Sehgal, and D. J. Cai (2016). "CCR5 is a suppressor for cortical plasticity and hippocampal learning and memory." *eLife* 5: e20985.

Zhou, Y., J. Won, M. G. Karlsson, M. Zhou, T. Rogerson, J. Balaji, R. Neve, P. Poirazi, and A. J. Silva (2009). "CREB regulates excitability and the allocation of memory to subsets of neurons in the amygdala." *Nature Neuroscience* 12(11): 1438–1443.

Ziganshina, L. E., T. Abakumova, and L. Vernay (2016). "Cerebrolysin for acute ischaemic stroke." *Cochrane Database of Systematic Reviews.* 4: CD007026.

Zijdewind, I., and D. Kernell (2001). "Bilateral interactions during contractions of intrinsic hand muscles." *Journal of Neurophysiology* 85(5): 1907–1913.

Zwinkels, A., C. Geusgens, P. van de Sande, and C. van Heugten (2004). "Assessment of apraxia: inter-rater reliability of a new apraxia test, association between apraxia and other cognitive deficits and prevalence of apraxia in a rehabilitation setting." *Clinical Rehabilitation* 18(7): 819–827.

Index

Printed in the United States
by Baker & Taylor Publisher Services